MARINE PLANKTON ECOLOGY

MARINE PLANKTON ECOLOGY

by
PAUL BOUGIS
*Professor at the University of Paris VI
and at the Oceanographical Institute,
Director of the Zoological Station, Villefranche/Mer, France*

1976

**NORTH-HOLLAND PUBLISHING COMPANY, AMSTERDAM – OXFORD
AMERICAN ELSEVIER PUBLISHING COMPANY, INC., NEW YORK**

© 1976 North-Holland Publishing Company.
All rights reserved. No part of this publication may be reproduced,
stored in a retrieval system, or transmitted, in any form or by any means,
electronic, mechanical, photocopying, recording or otherwise,
without the prior permission of the copyright owner.

This book was published originally in French by
Masson & Cie., Paris
under the title:

Ecologie du Plancton Marin
by P. Bougis

The translation was made by Myriam Achituv

ISBN North-Holland: 0-7204-0391 x
ISBN American Elsevier: 0-444-11033 x

PUBLISHERS:
North-Holland Publishing Company, Amsterdam – Oxford

SOLE DISTRIBUTORS FOR THE U.S.A. AND CANADA:
American Elsevier Publishing Company, Inc.
52 Vanderbilt Avenue, New York, N.Y. 10017

Library of Congress Cataloging in Publication Data
Bougis, Paul.
  Marine plankton ecology.

  Translation of Écologie du plancton marin.
  Includes index.
  1. Marine plankton. 2. Marine ecology. I. Title.
QH91.8.P5B6413   574.92   76-9831
ISBN 0-444-11033-X

Printed in The Netherlands

PREFACE

The word plankton (from Greek $\pi\lambda\alpha\nu\kappa\tau\sigma\varsigma$ = drifting) was introduced by Hensen in 1887 to distinguish all the organisms passively floating in the water. As a result of much progress in oceanography, the study of the plankton has been considerably developed. Systematics has been followed by ecological research; the available data are numerous and important but far from final. Nevertheless, there is a need for them to be collected together for both the plant and animal kingdoms, and their relations to the physical and chemical factors considered in order to allow a logical interpretation of the facts. This work attempts to fulfill this aim and is the result of lectures given to students specializing in biological oceanography. It pre-supposes a general knowledge of marine organisms. It tries to tabulate our present ecological data, and to describe mechanisms and relations rather than to give exhaustive descriptions of particular situations. It is, therefore, less complete as compilation of the phenomena described, and more an attempt to understand planktonic ecology, relying on the most important work. Its aim is to assist the reader to be able to study the ever more numerous specialized investigations, which explore the different aspects of the planktonic ecology and to help him to place each of these in more perspective.

CONTENTS

Preface v

Introduction 1

    0-1. —Definition and divisions of the plankton 1
    0-2. —General characteristics of the plankton 3
    0-3. —Adaptation to pelagic life 3

Chapter 1.—The phytoplankton: general characteristics and systematic outline . . . 7

    1-1. —General 7
    1-2. —The diatoms 8
    1-3. —The Peridinia 11
    1-4. —Other flagellates 12
    1-5. —The Cyanophyceae 14

Chapter 2.—The factors affecting photosynthesis 15

    2-1. —Carbon dioxide 15
    2-2. —Light 17
    2-3. —Utilization of light radiation by pigments 23
    2-4. —Utilization of the light by the phytoplankton 27

Chapter 3.—The nitrogen cycle 35

    3-1. —Necessity for nitrogen 35
    3-2. —Absorption of nitrogen 36
    3-3. —Remineralization of nitrogen 40
    3-4. —Nitrification 42
    3-5. —The nitrogen cycle 45
    3-6. —Dissolved organic matter 46
    3-7. —Estimation of nitrogen 48
    3-8. —Quantitative variation in inorganic nitrogen 48
    3-9. —The nitrogen balance 51
    3-10.—Modes and simulations of the nitrogen cycle 54

Chapter 4.—Phosphorus and its cycle 57

    4-1. —The need for phosphorus 57
    4-2. —Absorption of phosphorus 58
    4-3. —Remineralization of phosphorus 60
    4-4. —The phosphorus cycle 63
    4-5. —Estimation of phosphorus 66
    4-6. —Distribution of phosphorus in the sea 66
    4-7. —Long term variations in amount of phosphorus 70
    4-8. —Regeneration of nitrate and phosphate 71

viii

Chapter 5.—Silicon, oligoelements, and growth factors 75

    5-1. —Silicon and silicate 75
    5-2. —Iron 77
    5-3. —Manganese 80
    5-4. —Copper 81
    5-5. —Vitamin B_{12} 81
    5-6. —Vitamin B_1 83
    5-7. —Other factors 84

Chapter 6.—Quantitative study of the phytoplankton 89

    6-1. —Enumeration 89
    6-2. —Measurement of pigments 90
    6-3. —Other methods 93
    6-4. —Vertical distribution 93
    6-5. —Seasonal fluctuations 95
    6-6. —Ecological analysis of the phytoplanktonic bloom 97
    6-7. —Causes of the phytoplanktonic blooming: Sverdrup theory 100
    6-8. —Succession of the phytoplanktonic populations 104
    6-9. —Spatial variations 111
    6-10.—Regional variations: geographical variations 118
    6-11.—The coloured water 119

Chapter 7.—Primary production 123

    7-1. —Remarks on primary production: methods of measurement 123
    7-2. —The oxygen method 124
    7-3. —Carbon-14 method 126
    7-4. —Particle counting method 129
    7-5. —The phosphate method 132
    7-6. —CO_2 changes method 132
    7-7. —Changes in oxygen method 133
    7-8. —The chlorophyll method 134
    7-9. —Mathematical models 136
    7-10.—Results of measurements 139
    7-11.—Limits of primary production 148
    7-12.—Relation between primary production and biomass 149
    7-13.—Energy yield and energy flow 151
    7-14.—The chemical composition of phytoplankton 153

Chapter 8.—The zooplankton: systematic composition 157

    8-1. —Protozoa 157
    8-2. —Cnidaria 160
    8-3. —Ctenophora 162
    8-4. —The worms–Annelida and Chaetognatha 165
    8-5. —Crustacea 166
    8-6. —Mollusca 169
    8-7. —Tunicata 172
    8-8. —Planktonic larvae 174

Chapter 9.—Quantitative studies and the distribution of zooplankton 179

    9-1. —Collection methods 179

9-2. —Treatment of samples	179
9-3. —Validity of the results	182
9-4. —Sampling and microdistribution	185
9-5. —Quantitative variation with time	195
9-6. —Distribution in space	204

Chapter 10.—Vertical distribution and diurnal migration of zooplankton	215
10-1. —The vertical distribution	215
10-2. —The characteristics of diurnal migration	217
10-3. —Causes of diurnal migration	225
10-4. —Ecological incidence of diurnal migration	231
10-5. —Vertical migration and scattering layers	235

Chapter 11.—Nutrition, metabolism and energetic budget of the zooplankton	237
11-1. —Mode of nutrition and feeding	237
11-2. —Quantitative study of nutrition and grazing	243
11-3. —Digestion and rate of assimilation	251
11-4. —Utilization of assimilated metabolites and energetic budget	254
11-5. —Respiration	256
11-6. —Excretion	264
11-7. —Growth	266

Chapter 12.—The secondary production of plankton	273
12-1. —Levels of production	273
12-2. —Cohort (age-class) methods: Allen curves	274
12-3. —Cumulative growth method	278
12-4. —Turnover time method	283
12-5. —Physiological method	285
12-6. —Use of models	286
12-7. —Results obtained by these methods	290
12-8. —Total comparison of the production	290
12-9. —Chemical composition of the zooplankton	292

Chapter 13.—The plankton in the marine ecosystem	299
13-1. —The plankton within the seston	299
13-2. —The plankton in the food web	303
13-3. —The plankton and man	305

Chapter 14.—Appendix	309
14-1. —Systematic place of principal genera of the marine phytoplankton	309
14-2. —Systematic place of the principal elements of the zooplankton	310
14-3. —Some ideas concerning growth	315
14-4. —Derivation of the Sverdrup equation	317
14-5. —Construction of plankton nets	319
14-6. —Equivalents and correspondencies	325

Bibliography	329
Index	349

INTRODUCTION

0-1—DEFINITION AND DIVISIONS OF THE PLANKTON

The plankton is defined as the whole of those pelagic organisms which float and drift under the action of water movement. It differs from the nekton, also pelagic, in the absence of self-motility, and stands in contrast to the benthos which consists of the bottom-living organisms or those living near the bottom. As in any classification, such distinctions, between plankton on the one hand and nekton and the benthos on the other, are not always so clearly marked. We shall see later that some organisms, considered as planktonic are able to make rapid vertical migrations which are not dependent on water movement. Indeed, they are very close to the animals which form the micronekton i.e., the small-sized nekton. Other organisms which live near the bottom, or on the bottom itself, have a pelagic phase in their life cycle, so that here distinction between the planktonic and the benthic animals is blurred. Even so, most of the planktonic organisms show relatively little movement in the environment and are usually collected with plankton nets; these consist of large cones of silk or nylon gauze filtering the organisms through their meshes, retaining organisms larger than the pores-size, and concentrating them at the end of the cone in a collector. Exceptions from this method are the collection of elements smaller than the finest gauze by special methods and, on the other hand, the collection of larger organisms by hand on the surface, by diving or by using a pelagic trawl.

The above definition of plankton is very general and includes a wide spectrum of animals. For detailed studies we must use more limited categories which may be defined according to various criteria. The first criterion is that of size. Classical terminology considers those planktonic organisms whose size is greater than several millimeters as macroplankton. The microplankton includes forms larger than 50 μm which is the mesh size of a fine plankton net; and the nanoplankton consists of planktonic organisms passing such nets. Some authors have tried to give more precise definitions; Dussart (1965, 1966), for example, suggested the following classification for fresh water plankton.

Ultramicroplankton: <2 μm; nanoplankton: 2–20 μm.
Microplankton: 20–200 μm; macroplankton: 200–2000 μm.
Megaplankton: <2000 μm = <2 mm.

There is, however, still no general agreement on the categories to be established by size: a collective effort to standardize the method of collection brought together an international team of planktologists, (Anonymous, 1968) who, on an essentially pragmentic basis, distinguished the following categories of the zooplankton;

Microzooplankton: <200 μm (grouping the nanoplankton and the microplankton).
Small mesozooplankton: mean size >200 μm but <10 mm and collected on a gauze with 200 μm mesh net.
Large mesozooplankton: organisms collected by 1 mm mesh net.
Macrozooplankton: a range of 2–10 cm and collected as the nekton.

This terminology is, itself, still not precise and it is better to define as closely as possible, the size of the meshes of the nets by which the plankton is collected. We shall indicate these dimensions in parentheses and with an oblique line, thus: (/505 μm) corresponds to plankton collected with a net with mesh width of 505 μm, and (/8-12 μm) to plankton collected by a net which has a pore size ranging between 8 and 12 μm. Later we shall discuss the relation between the mesh size of the gauze and the dimensions of the organisms collected by it (Chapter 9-3). We may draw attention to the works of Sheldon and Sutcliffe (1969) and of Sheldon (1972), concerning the retaining capacity of filters.

The vertical distribution forms another basis for the separation of different groups. Epiplankton is that plankton living in the epipelagic zone (photic zone) including, in shallow water, animals found at a depth of 20–120 m depending on the region and the conditions; mesoplankton (different from the mesoplankton defined by its size) is that plankton living in the mesopelagic zone, i.e., 100 to 300 m depth, infraplankton that in the infrapelagic zone, 500–600 m, and bathyplankton that in the bathypelagic zone deeper than 600 m.

The shallow water plankton well studied by Zaitsev has a distinctive classification (David, 1967). The pleuston consists of those animals living on the surface of the water, partly in air and partly in water and drifted mainly by the wind (*Velella Physalia, Janthina*), while the neuston consists of the remainder of the shallow fauna and includes the epineuston living above the interface and the hyponeuston living below the interface.

On the basis of its nutrition we may divide the plankton into the plant plankton, or phytoplankton, capable of synthesizing some of its own material by photosynthesis, and the animal plankton or zooplankton feeding on existing material.

Another important distinction is based on biological criteria and divides the plankton into holoplankton and meroplankton. The holoplankton consists of organisms which are planktonic throughout all their life cycle. Most of the Chaetognatha are holoplanktonic since the eggs, larvae, juveniles, and adults are pelagic. *Salpa* has an alternation of generation and since both the sexual and the asexual generations are planktonic *Salpa* is, therefore, included within the holoplankton. On the contrary, most of the hydromedusae are meroplanktonic since they are a part of the life cycle of the Hydrozoa, the larvae of which after hatching from the eggs settle on the sea bottom (or other substrata) and give rise to a benthic polyps which by budding give the medusae. An important fraction of the meroplankton is larval, many of them belonging to benthic or nektonic organisms and are only a short stage in the life cycle as for example the pleuteus larvae of sea urchins and the eggs and larvae of most of the bony fish.

Finally, we must distinguish between the plankton and other floating particles, termed seston, which is made up of the living particles–the plankton–and the non-living particles, the tripton, which consists of dead organisms, organic detritus, and particulate minerals (Reid, 1961).

0-2.—GENERAL CHARACTERISTICS OF THE PLANKTON

Despite the extreme diversity of the plankton it is possible to find some general characteristics which gives it its particular characters. These are essentially its colour and the dimensions. Planktonic animals have a tendency to be transparent and are little coloured, pigmentation being restricted to some organs. This is particularly clear in the hydromedusae, the siphonophora, and *Salpa*. Two exceptions should be mentioned; surface planktonic animals often have a deep blue colour (*Velella*) while those living in deep water are often red or brown.

In general planktonic organisms are rather small, although there are some exceptions such as some medusae which reach a diameter of 1 m, or *Pyrosoma* which reaches a length of several meters. Yet the majority of the organisms have dimensions of the order of centimeters or millimeters, in the case of planktonic animals and 100 or 10 μm in the case of plant plankton.

0-3.—ADAPTATION TO PELAGIC LIFE

The planktonic organisms are necessarily adapted to pelagic life and have to remain in the water and not sink to the bottom; in other words, their sinking velocity must be zero for a perfect adaptation.

According to Stoke's law, the sinking velocity of a spherical particle with a radius, r, is as follows:

$$v = \frac{F}{6\pi\eta r}$$

Where F is the difference between the weight of the organism (density of the body multiplied by its volume) and Archimedes upthrust (density of the medium multiplied by volume of the body) acting on the organisms, and η is the viscosity of the sea water. Stoke's law is strictly appreciable only to spheres below a limited diameter (for quartz spheres it is 60 μm); this limit is higher when the difference of density from the milieu is less (Sverdrup, Johnson and Fleming, 1946). This law may serve as a guide when dealing with the physical relations between the plankton and the surrounding sea water. A decrease in F will decrease the sinking velocity and any reduction in the density of the organism relative to sea water will be an advantage to planktonic existence. This may be realised in different ways. In a given group skeletal components may be less resistant and lighter in planktonic species than in the benthic species; there are numerous examples of this in the diatoms, carapaces of crustacean, shells of gastropods. In the Heteropoda we can find a series passing through *Atlantia* and *Carinaria* with shells and ending in the complete disappearance of the shell in *Pterotrachea*. The increase in body tissue water by the development of gelatinous substances is another and frequent method of pelagic adaptation. A comparison of the amount of water in the same group shows this clearly.

Cnidaria:	planktonic form (*Cyanea*)	96.5%
	benthic form (*Anemonia*)	87.2%
Crustacea:	planktonic form (*Calanus*)	85.7%
	benthic form (*Crangon*)	74.5%

Another way of decreasing the weight is to maintain an osmotic equilibrium with the sea water by lighter ions. The monovalent chloride ion (35.5) replaced the divalent sulphate ion ($96/2 = 48$). Frequently floats are found in planktonic organisms, gas floats in the Siphonophora or oil "floats" in fish larvae and some copepods.

A reduction of the organisms dimensions also decreases its sinking velocity. According to Stoke's law F is proportional to the difference between the density of the body and that of the water and the body volume; so $F = kr^3$, where k is a constant. The previous expression becomes:

$$v = \frac{kr^3}{6\pi\eta r} = \frac{kr^2}{6\pi\eta}$$

As the radius become smaller the sinking velocity is lowered. The small size which, as we have noted, is one of the general characteristics of planktonic organisms, contributes to their improved adaptation to a planktonic existence.

The adaptation to pelagic life is improved by any morphological characteristics which increase the resistance to sinking. More or less flattened forms, which reduces the sinking velocity are found in the plankton, for instance *Sapphirinia* (Copepoda) or phyllosoma larvae (lobsters), while the morphology of medusae is very similar to that of a parachute. Another efficient adaptation is the existence of very long and even exaggerated appendages such as those of *Chaetoceros* (a diatom) or some zoea (larval crustaceans).

When the adaptation is imperfect the organism must have a "complementary adaptation". This involves a waste of energy in motor activities such as undulation of the flagellum in peridinians, the beating of the cilia of the ciliary bands of many larvae or the swimming activity of crustaceans.

We shall return now to Stoke's law. The viscosity of sea water, η, may be markedly modified, particularly by temperature: at 0°C it is 18 millipoise, at 10°C 13 millipoise, and at 20°C 10 millipoise. As a consequence, an organism with given dimensions and density, which is in equilibrium with the water at a given temperature will sink when the temperature rises and eventually have to spend a certain amount of energy to complete its adaptation.

For a spherical organism in equilibrium with sea water at 0° the radius must be decreased by a factor of 0.74, i.e., reduced by a quarter in order to maintain its equilibrium at a temperature of 20°C because of the change in the viscosity of sea water. On the other hand, it is well known that the solution of certain substances such as agar–agar considerably increases its viscosity in water and this may be attained with a small quantity of material. It does not seem that the possibility of this kind of reaction has been considered until recently as affecting plankton, despite the fact that many organisms are known to be able to excrete organic substances. (This is probably an interesting field of research).

Concerning the physical meaning of the viscosity this has to be considered in the light of Margalef's ideas (1957) which emphasise the importance of electrostatic charges. For a detailed study on the flotation of the phytoplankton see the review of Smayda (1970).

CHAPTER 1

THE PHYTOPLANKTON: GENERAL CHARACTERISTICS AND SYSTEMATIC OUTLINE

1-1.—GENERAL

The phytoplankton is defined as the plant plankton, i.e., plankton which is able to photosynthesize materials from water, carbon dioxide, using light energy.

The phytoplankton as well as the benthic algae are the source of the primary production of organic matter in the sea. Light penetration, however, rapidly decreases with depth, and in consequence the benthic algae are restricted to the littoral fringe, while the phytoplankton is distributed over the entire superficial zone of the seas, apart from those regions covered by ice. Even these latter regions are not without plant production. An abundant unicellular algal flora is developed on sea ice (Bunt & Lee, 1970), but according to the definition of plankton these algae do not, strictly, belong to the phytoplankton. With one exception, the phytoplankton consists of microscopic algae, single cells, or cells forming chains, their size ranging from several μm to several hundred μm. The ratio of the external surface to the volume is higher in single or loosely attached cells than in grouped cells. As we have already seen the increase of this area/volume ratio improves the buoyancy but this is not the only way used by planktonic organisms to attain this end. A high surface to volume ratio also increases the surface available for the absorption of nutrients needed for photosynthesis, of importance because the concentrations of nutrients in sea water are very low. However, the aggregation of cells in the multicellular benthic algae does not seem to be a handicap in this respect. Finally, in the present state of our knowledge we can only state the fact of the microscopic nature of the phytoplankton without any possibility of showing that this corresponds to the only possible adaptation of plants to a pelagic life.

There is one exception, namely, *Sargassum* a brown alga from the order Fucales which attains a size of several decimeters; it has small spherical floats. *Sargassum* is abundant in the water of the central Atlantic, between 20° and 30°N an area known as Sargasso Sea. The floating *Sargassum* does not produce reproductive organs but continues to grow as the older parts decay. These plants are not, therefore, algae torn from the coast by wave action and accumulating in this zone but are true planktonic algae. Another exception has been described by Womersley & Norris (1959), namely, a floating red algae which is present on the coasts of Australia. This algae, of the genus *Antithamnion*, forms spherical floats about 1 cm in diameter and is cast up on the coast in large quantities.

Among the unicellular planktonic algae we shall first consider the diatoms, with numerous species, and often very abundant. Another important group is the order Peridinales, which are flagellated unicellular algae. A certain number of other groups of flagellated algae, mainly coccolithophorids and silicoflagellates, are also found in the phytoplankton population. The Cyanophyceae (may mainly in tropical waters) contribute to the phytoplankton. These different groups will be reviewed using the classification of Parke & Dixon (1968), and summarized in Chapter 14-1.

1-2.—THE DIATOMS

The Bacillariophyceae, more commonly known as diatoms, is a class of algae with numerous species both benthic and planktonic; the latter are the most important group in the economy of the sea. The diatoms are unicellular algae whose hard cell envelope, termed a frustule, is made up of pectic substances associated with silica; cellulose is absent. The frustule is made of two halves, or valves, which form a pill box-like structure similar to the two halves of Petri dish. The valves are laterally prolonged by connective bands (pleura) forming a girdle, and these can move one over the other. There is a large vacuole in the cytoplasm and this often occupies a large part of the cell. The nucleus is of the usual type. The chromatophores (Fig. 1-1, a, g, i, k), yellow or gold-brown, contain chlorophyll (chlorophyll *a* and *c*) and carotenoids, (xanthophylls among which the most abundant is fucoxanthin and carotins); they are generally small and numerous in planktonic diatoms. The products of synthesis are small drops of lipids, and granules of volutin. This latter substance, enigmatic for a long time, is a phosphatic accumulation, in the form of polyphosphates. Diatoms do not produce starch.

Multiplication is by binary fission. The valves separate from each other, mitosis gives rise to two nuclei, the protoplasm divides and the wall which is formed is split into two halves. Each new valve forms a new pill box with the old valve which accompanies it, the older one becoming the outer half of the pill box. The size of the cells must tend to decrease at each division. This phenomenon is not, however, as general as was at one time thought, and in cultures stable dimensions have sometimes been observed; the frustule can probably grow a little. The diminution in size is compensated by means of a special mode of multiplication, namely, auxosporulation, the characteristics of which often presents some sexual phenomenon; the process differs according to the group, but always gives rise to a voluminous cell called an auxospore which secretes a larger frustule (Fig. 1-1, c, i).

There are other methods of reproduction. In microsporulation small biflagellated cells without a silicous membrane, the microspores, are formed; they swim and two fuse to produce a zygote which gives rise to a normal cell. In the formation of endospores protoplasm is concentrated within

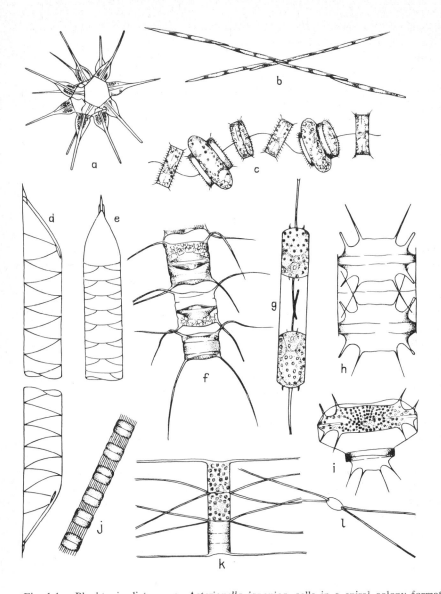

Fig. 1-1.—Planktonic diatoms. a, *Asterionella japonica*, cells in a spiral colony formation showing chromatophores (length of one cell, 50–90 μm); b, *Nitzschia seriata*, cells in chain formation (length of a cell, 90–100 μm); c, *Thalassiosira gravida*, cells in chain formation with two auxospores (diameter, 20–60 μm); d, *Rhizosolenia styliformis*, side view of two extremities of a cell (diameter, 40–100 μm; length, up to 1.5 mm); e, *Rhizosolenia styliformis*, dorsal view of an extremity; f, *Chaetoceros diadema*, with two endospores (diameter of valve, 12–46 μm); g, *Ditylum brightwellii*, cell after division with chromatophores (diameter: 28–46 μm); h, *Biddulphia mobiliensis*, cells beginning to divide (length with spines 60–200 μm); i, *Biddulphia mobiliensis* after forming auxospores showing chromatophores; j, *Skeletonema costatum*, cells in chain formation (diameter, 8–15 μm); k, *Chaetoceros danicum*, cells in chain formation showing chromatophores (diameter of valve, 16–20 μm); l, *Chaetoceros danicum*, cell in valve view (after Hendey, 1967).

the cell walls and becomes surrounded by a hard shell, often much ornamentated and characteristic. These endospores are resting spores and, when conditions become adverse, they sink to the bottom and later produce normal vegetative cells (Fig. 1-1, f).

The diatoms are divided into two groups, the pennate and centric, arranged in eight sub orders (Hendey, 1964) which are distinguished mainly by their morphology. In the pennates the valves are symmetrical relative to two planes perpendicular to each other and to the plane of the valve. Very often there is a secondary asymmetry caused by deformation. Most of the pennates are benthic but some species are typically planktonic and these frequently form chains. From the sub-order Fragilariineae we may mention the genera: *Thalassiothrix*, *Thalassionema*, *Asterionella* (Fig. 1-1, a) and *Nitzschia* from the sub-order Naviculineae.

Nitzschia closterium f. *minutissima* has been cultured since 1907 (by Allen) and the original stock has been distributed to many laboratories and is widely used for feeding larval cultures. Later this culture was replaced by a different organism, *Phaeodactylum tricornutum* whose systematic position is obscure; it exists in three forms, namely, fusiform, triradiate, and oval, and only the oval form has a rudimentary and unique value.

In the Centricae the valves have a typical radial symmetry although it is sometimes much masked. Three suborders are known, Coscinodiscineae, Rhizosoleniineae and Biddulphiineae. The simplest forms of the Coscinodiscineae have the shape of a Petri dish, as in the genus *Coscinodiscus*. In *Melosira* the cells form a chain and are attached by mucus; in *Skeletonema* (Fig. 1-1, j) the cells are attached by a fine siliceous filament and in *Thalassiosira* (Fig. 1-1, c) by a gelatinous filament.

The Rhizosoleniineae have an elongated cylindrical frustule. Besides *Dactyliosolen* the large genus *Rhizosolenia* is well known and is pointed at both extremities; the most common species are *R. styliformis* (Fig. 1-1, d, e), *R. shrubsolei*, *R. stolterfothii*, *R. castracanei*, the last reaching several 100 µm in size and being visible to the naked eye. The Biddulphiineae have a short frustule which has prominent angles, excrescences, horns, and bristles. The common species is *Biddulphia*, with two horns and two spines on each valve. *B. mobiliensis* (Fig. 1-1, h, i), was studied in detail in 1900 in the basin of Arcachon (France). It is abundant on the bottom at the beginning of the autumn and becomes planktonic in October–November (meroplanktonic); it remains abundant in the plankton for six months i.e., until the spring. A closely similar species, *B. sinensis*, appeared in the mouth of the Elbe in 1893, and since then has been rapidly dispersed in the North Sea and adjacent areas, where it is now a common form. The principal genus of the suborder and of the entire planktonic diatoms is *Chaetoceros*, characterized by a valve with an elliptical section, each valve carrying large horns which, in the chain, connect the cells to one another. The species are numerous and form an important part of the phytoplankton (*C. didymum*, *C. danicum*, (Fig. 1-1, f)). We may also mention among the Biddulphiineae the species *Ditylum brightwellii* (Fig. 1-1, g) and the genus *Bacteriastrum*.

1-3.—THE PERIDINIA

The Peridinia, also called the Dinoflagellata or Dinophyceae, are a class of algae, with two flagella, one longitudinal and the other perpendicular to the first and present in two more or less well-developed perpendicular grooves (Fig. 1-2, j). The nucleus is of a peculiar type being always granulated (dinocaryon) (Fig. 1-2, e). The yellow-green chloroplasts or chromatophores contain chlorophyll *a* and *c* associated with carotin and xanthophylls of which the most abundant of them is peridin (Fig. 1-2, e). The reserve materials are lipids and starch. Often a cellulose theca is present. Multiplication is by simple division in the free Peridinia, (numerous parasitic forms exist), and sexual phenomena were rarely observed until relatively recently (Smith, 1955).

Generally, a peridinian has chlorophyll and is autotrophic but heterotrophic species without chlorophyll are known; some of these are predators, able to ingest prey. Intermediate forms termed mixotrophes which can use both kinds of nutrition are known. The Peridinia belong both to the plant and animal kingdoms and, therefore, to the phytoplankton and to the zooplankton. In one order, the Prorocentrales (or Adinidea), the cell is

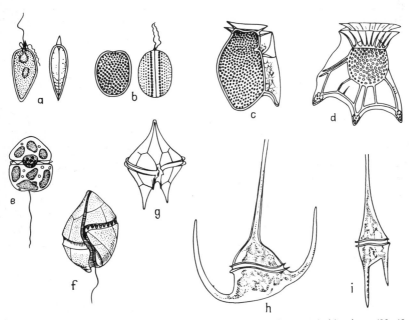

Fig. 1-2.—Planktonic Peridinia. a, *Prorocentrum micans* front and side views (30–40 μm); b, *Exuviella marina* front and side views (20–30 μm); c, *Dinophysis schroederi* (70 μm); d, *Ornithocercus magnificus* (90 μm); e, *Gymnodinium venificum* (10 μm) dorsal view with granulated nucleus in the centre, six chromatophores and two flagella; f, *Gonyaulax* (50 μm), ventral view showing two grooves and two flagella; g, *Peridinium oblongum* (100 μm); h, *Ceratium tripos* (300 μm); i, *Ceratium furca* (150 μm). (After: Lebour, 1925, (a,b left); Schutt in Massuti & Margalef, 1950, (b right); Massuti & Margalef, 1950, (c,d); Ballantine, 1956, (e); Margalef, 1967 (f,g); Hardy, 1958 (h,i)).

oval or lance-shaped and the flagellae are in an apical position with little differentiation. Two common species belong to this group, namely, *Exuviella marina* (Fig. 1-2, b) sometimes found in high densities, and *Prorocentrum micans* (Fig. 1-2, a) often in cultures.

In the other Dinophyceae the two flagellae are inserted laterally, corresponding to well-marked grooves. In the order Dinophysiales a bivalved theca which develops a collar or wings along the groove is present and includes the typical *Dinophysis* (Fig. 1-2, c); *Phalacroma* has an oval form, while on the contrary *Amphisolenia* is very elongated and *Ornithocercus* has a beautiful ornamentation (Fig. 1-2, d). The simplest genera of the Peridiniales are naked, like *Gymnodinium*, with its many species. We may particularly note among these, *Gymnodinium veneficum* described by Ballantine (1956); it was isolated from a sample of sea water from the English Channel and after being cultured at Plymouth was found to be very toxic to fish and molluscs (Fig. 1-2, e). *Noctiluca* is now placed in the Peridiniales, in spite of its particular morphology; its nucleus is of the dinocaryon type, and its spores are of peridinian type. In this order there are a certain number of forms characterized by more or less thick cellulosic theca which are made up of plates. We may mention *Peridinium* with a great diversity of species (Fig. 1-2, g), *Gonyaulax* often grown in culture and luminescent (Fig. 1-2, f) and the large genus *Ceratium* (Fig. 1-2, h and i). *Ceratium* sp. usually have three horns, one forward, and two backward. After dividing, the cells sometimes remain attached in relatively short chains. The numerous species are classified in several sections. Among the most common are *C. furca, C. fusus, C. pentagonum, C. tripos, C. caudenabrum, C. trichoceros.*

1-4.—OTHER FLAGELLATES

Because of their numerous species which are sometimes very abundant the Coccolithophores are of great importance. These flagellates, essentially characterized by a cell covered by small calcerous scales (*Coccolithus*, Fig. 1-3, f) which are of very different shapes and, according to the species, ovoid, bowl, club shaped etc., loosely arranged, or tight forming a protecting shell. They have two flagellae and two large parietal chromatophores; the reserves are lipids and leucosin–a substance very similar to starch. Reproduction is by simple longitudinal division, the shell being divided into two daughter cells which are afterwards completed. There are sporulation phenomena, still not well known. Among the principal genera, we may mention *Pontosphaera, Coccolithus, Syracosphaera* (Fig. 1-3, d, e, f, g).

Recently our knowledge of the coccolithophores has been greatly increased (Paasche, 1968). In *Pontosphaera roscoffensis*, for example, we now know that as well as flagellae there are filaments used for attachment to the substratum. These filaments, called haptonema, have also been found in certain numbers of coccolithophores and in some organisms formerly

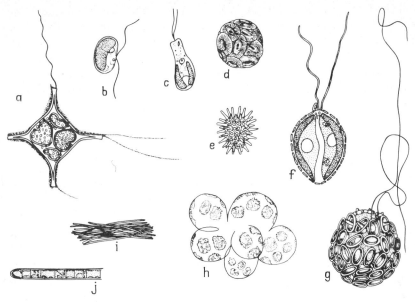

Fig. 1-3.—a, a silicoflagellate *Dictyocha fibula* (50–70 μm) Erysophyceae, showing a flagellum (up), pseudopods (to the right), silicon skeleton and numerous small chromatophores; b, *Hemiselmis rufescens* (4 μm) Cryptophyceae, showing a large chromatophore; c, *Isochrysis galbana* (6 μm) Haptophyceae, showing two chromatophores; d, a coccolithophore *Coccolithus pelagicus* (20 μm) Haptophyceae; e, a coccolithophore *Rabdosphaera subopaca* (20 μm) Haptophyceae; f, a section of a coccolithophore *Pontosphaera roscoffensis* (15–20 μm) showing the peripheral coccoliths and two large parietal chromatophores; g, a coccolithophore *Syracosphaera mediterranea* (20–30 μm); h, a colony of *Phaeocystis* (0.5–1 m) Haptophyceae; i, *Oscillatoria*, Cyanophyceae, a group of filaments (3–4 mm); j, *Oscillatoria*, Cyanophyceae, details of a filament (50 μm). (After: Marshall in Grasse, 1952 (a); Parke, 1952 (b,c); Murray & Blackman in Grasse, 1952 (d); Bernard in Grasse, 1952 (e); Chadefaud & Feldmann in Grasse, 1952 (f); Lecal-Schlander in Grasse, 1952 (g); Gran in Raymont, 1963 (h); Massuti & Margalef, 1950 (i); Dangeard in Massuti & Margalef, 1950 (j)).

considered as Chrysophyceae and for this reason the class Haptophyceae was established. On the other hand, the coccolithophores show an alternation of phases, a flagellated phase and a non-motile stage without a flagellum; both may be planktonic. Parke & Adams (1960) showed that the unflagellated "species" *Coccolithus pelagicus* is a phase of the flagellated form known as *Crystallolithus hyalinus*.

Grouped with the coccolithophores in the Haptophyceae is *Prymnesium* which blooms mainly in brackish water and secretes a toxin; *Phaeocystis poucheti* (Fig. 1-3, h) abundant in colder seas, has cells without flagellae grouped in large gelatinous colonies visible to the naked eye; *Isochrysis* is commonly cultured (Fig. 1-3, c).

Among the Chrysophyceae are found the silicoflagellates which have a hollow internal silicous skeleton, only a single flagellum, and very numerous yellow-brown chromatophores. The skeleton is a relatively simple geometrical shape diamond or crown-like, with more or less much variations even in the same species. These very common forms belong to the only genus *Dictyocha* (Fig. 1-3, a).

Classed now in the Prasinophyceae although formerly in the Xauthophyceae and common in all seas is *Halosphaera viridis*, which takes the form of a large pale green sphere reaching diameter of half a millimeter.

Dunaliella (Order: Volvocales, Class: Chlorophyceae) is known mainly by the species *D. salina*, very euryhaline and capable of multiplying in salt marshes, colouring them red; another species *D. marina*, is well known in culture. In the Prasinophyceae and Chlorophyceae, chlorophyll *b* replaces chlorophyll *c*.

Hemiselmis (Class: Cryptophyceae) represented by *H. rufescens*, has been cultured in laboratory (Fig. 1-3, b); in the same class *Cryptomonas* may be mentioned since it is sometimes very abundant in the phytoplankton (1.6×10^6 cells/l at Roscoff; Grall & Jacques, 1964).

Euglene sp. (Class: Euglenophyceae) with numerous chromatophores, has a body which can change its shape; although generally living near the bottom it may be found in littoral pools; it may also swarm in the phytoplankton (coasts of Norway).

1-5.—THE CYANOPHYCEAE

The Cyanophyceae are characterized by the absence of a "true" nucleus; they have a green-blue pigment diffused throughout their cytoplasm, from which they get their name – blue-green algae or Cyanophyceae. The pigmentation is due to a mixture of biliproteins, similar to phycocyanin, together with chlorophyll *a*, carotene and various xanthophylls. The carbohydrate is stored as glycogen. In a certain number of forms a direct absorption of dissolved nitrogen has been observed.

The Cyanophyceae are very important in freshwater plankton and frequently present in massive amounts: in marine plankton they appear only in warmer waters. The genus *Oscillatoria* (= *Trichodesmium*) is sometimes so abundant that it may visually colour the sea; to this is attributed the colouration of the Red-Sea, due to great development of red pigments. The species is present in the form of bundles of simple filaments, reaching a length of 5 to 6 mm (Sournia, 1968b) (Fig. 1-3, i, j).

According to Bernard & Lecal (1960) *Nostoc* characterized by cells arranged in strings may exist at densities of several thousands/cm^3 in the nanoplankton of the Mediterranean and Indian Oceans.

Close to *Nostoc* is *Richelia intracellularis* which is found inside the cells of the diatom *Rhizosolenia styliformis* and is also intracellular in *Chaetoceros*. It is still not clear whether this association is a form of parasitism or symbiosis.

Fournier (1970) has drawn attention to some enigmatic, small pigmented cells living at depth, but it is not possible yet to include them in any of the known groups.

CHAPTER 2

THE FACTORS AFFECTING PHOTOSYNTHESIS

The general equation for photosynthesis is as follows:

$CO_2 + 2H_2A + light \rightarrow (CH_2O) + H_2O + 2A$

Where H_2A represents any oxidizable compound (or a hydrogen donor) and A is the product of oxidation or of dehydrogenation (which is the same). In some bacteria H_2A may be H_2S, but in plants, including algae, H_2A represent water, so we have

$CO_2 + 2H_2O + light \rightarrow (CH_2O) + H_2O + O_2$

or more simply

$CO_2 + H_2O + light \rightarrow (CH_2O) + O_2$

Photosynthesis requires water, carbon dioxide, and light. Water is always available but this is not always true as regards either carbon dioxide, the source of the carbon or light, the source of energy.

2-1.—CARBON DIOXIDE

Carbon dioxide, CO_2, from the air or from the respiration of marine organisms is dissolved in the water. In the air the concentration of CO_2 is remarkably constant $0.03\% \pm 0.003\%$, and so the partial pressure of CO_2 in the air, pCO_2, is $0.03\% \times 760$ mm Hg $= 0.22$ mm Hg, or 0.03% atm ($= 0.03\%$ atm $= 0.03 \times 10^{-2}$ atm or 30×10^{-5} atm). Sea water in contact with air will dissolve CO_2 until the partial pressure of the gas in the water is equal to its partial pressure in air and so sea water in equilibrium with air will have a pCO_2 of 0.03% atm under any conditions.

Carbon dioxide in solution reacts with the water to give, by hydration, carbonic acid,

$CO_2 + H_2O \rightleftarrows H_2CO_3$ \hfill (1)

At equilibrium the concentration of H_2CO_3 is a thousand times lower than that of the CO_2 and so the quantity of carbonic acid is always very small. On the other hand, the carbonic acid is ionized in water to give bicarbonate and hydrogen ions,

$H_2CO_3 \rightleftarrows HCO_3^- + H^+$ \hfill (2)

According to the law of mass action

$$\frac{[H^+][HCO_3^-]}{[H_2CO_3]} = K_1 \tag{3}$$

Since the concentration of H_2CO_3 is proportional to that of molecular CO_2 this may be written:

$$\frac{[H^+][HCO_3^-]}{[CO_2]} = K_1' \tag{4}$$

According to Henry's law the concentration of CO_2 is proportional to the partial pressure of CO_2 ($CO_2 = \alpha p\, CO_2$) and, so

$$\frac{[H^+][HCO_3^-]}{\alpha p CO_2} = K_1' \tag{5}$$

Since $[H^+]$ is related to the pH (pH = $-\log(H^+)$), this equation gives the relation between the partial pressure of CO_2, pH, and the concentration of bicarbonate. None of these variables can be changed without effecting at least one of the two others. Since the bicarbonate ions are in equilibrium with the carbonate ions we have the following equation:

$$CO_2 + H_2O \rightleftarrows H_2CO_3 \rightleftarrows HCO_3^- + H^+ \rightleftarrows CO_3^{2-} + 2H^+$$

Buch (1951) gave detailed tables for sea water which enables one to calculate the partial pressure of molecular CO_2 in solution, the amount of bicarbonate and carbonate, and the total amount of CO_2 as a function of pH, temperature, and salinity. Table 2-1 gives the amount of CO_2 and its partial pressure in sea water, at 16°C, at salinity of 35‰ and for different pH values. The partial pressure of molecular, or free, CO_2 decreases very rapidly as an inverse function of the pH, which is clearly evident from equation (5). The total CO_2 decreases much more slowly (Fig. 2-1). For a given pH the total amount of CO_2 increases almost proportionally with salinity between 27‰ and 38‰. It decreases about 1% for a temperature rise of 1°C. Harvey (1955) has given the necessary calculations for the following example, which will give a more concrete idea of the mechanisms involved. Water of salinity of 35‰, at 15°C, in equilibrium with the atmosphere will reach a pH of about 8.16 and will contain 45.4 cm³ of CO_2/l in the form of bicarbonate and carbonate ions, 0.26 cm³ of non-dissociated, molecular CO_2 in solution, and a very low concentration of non-dissociated H_2CO_3. The partial pressure of the molecular CO_2 is equal to that of CO_2 in air i.e., 0.033% or 33×10^{-5} atm. Supposing that in one litre of sea water algae are grown and absorb 2.24 cm³ of CO_2 (i.e., 0.1 mmole, since 1 mole = 22.4 l = 22.400 cm³, containing 1 mmole of carbon which is 1.2 mg of C) the following changes will take place:

(1) The pH changes from 8.16 to 8.31;
(2) The molecular CO_2 in solution decreases from 0.26 cm³ to 0.17 cm³ (i.e., less 0.09 cm³), the partial pressure decreasing from 33 to 21×10^{-5} atm;

TABLE 2-1

Variations in the amount of CO_2 as a function of the pH of sea water of salinity 35‰ at 16°C. Total moles CO_2 per litre $\times 10^{-5}$; partial pressure in atm $\times 10^{-5}$.

pH	7.4	7.8	8.0	8.1	8.2	8.3	8.4	8.5
Total CO_2	238	235	216	211	206	199	191	183
CO_2 partial pressure	230	85	50	39	29	22	15	12

(3) Some bicarbonate ions are transformed into carbonate ions, liberating 2.14 cm³ of CO_2 which together with the 0.1 cm³ mentioned above gives 2.24 cm³ of absorbed CO_2, according to the reaction:

$$2HCO_3^- \rightarrow CO_3^{2-} + H_2O + CO_2$$

The quantity of free CO_2 in the water is low, but the preceding mechanisms go hand in hand with the re-supply by the contact with the air and animal respiration which enables the phytoplankton to use the required CO_2. For this problem and the physical-chemistry of it see the work of Ivanoff (1972).

2-2.—LIGHT

Besides water and carbon dioxide light is an essential factor for photosynthesis. We have to see at what level it is available for photosynthesis.

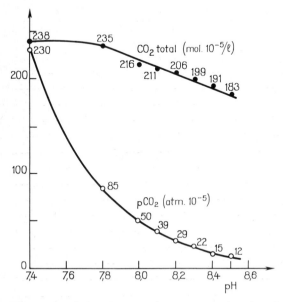

Fig. 2-1.—Variation of pCO_2 and total CO_2 as a function of the pH for sea water of salinity 35‰ at 16°C.

Strickland (1958) has given an excellent review of this problem and a large part of this chapter is based on his work.

a. *Estimation of the light energy reaching the sea*

The radiant energy of the sun measured beyond the earth's atmosphere is equivalent to an irradiance of 1.92 g-cal/cm^2/min. We may remember that the irradiance at a point is defined as the ratio of the energy flow (radiated quantity per unit time) received by an infinitely small element of surface around the considered point to the area of this element (Ivanoff & Morel, 1970). It is symbolized by I and expressed in watts/cm^2, or in g-cal/cm^2/min, with the following equivalents: 1 g-cal/cm^2/min = 0.07 W/cm^2 (since 1 g-cal = 4.185 joules and 1 watt = 1 joule/sec). The gram-calorie is often written cal, and the larger unit kilocal. Some authors use the unit Langley which is g-cal/cm^2 and the irradiance is expressed in Langley/min. It is preferable to use watt/cm^2 as the units of irradiance since it is used in physics.

For a plane parallel to the surface of the earth and turning with it the estimated irradiance is an average of 1100 g-cal/cm^2/day at the poles in the summer and between 800 to 900 cal/cm^2/day at the equator in the equinoctal period. While passing through the atmosphere a more or less important fraction of the radiated energy is absorbed or diffused by the cloud cover, dust, water vapour, and carbon dioxide so that at sea level at 48°N the irradiance is rarely greater than 1.5 cal/cm^2/min and very often less. At the same latitude, the daily total energy varies according to the season from 805 cal/cm^2/day to 160 cal/cm^2/day.

The following relation has been established between the irradiance (coming directly from the sun or diffused by the sky) and reaching the surface of the sea in the absence of clouds (I_0) and the irradiance reaching it under the same conditions (I) except for a cloudy cover equal to c (varying from 0 to 1),

$I = I_0 (1-0.71c)$

The difference between I and I_0 may be important. At 48°N near the island of Vancouver, the average value of I varies from 414 cal/cm^2/day (21st May) to 71 cal/cm^2/day (21st December) (Kimball, 1928). This author estimated the daily radiant energy reaching the surface of the sea for certain number of zones of the globe. The results are given, for different months, as the irradiance from the sun and from the sky, taking into account cloud cover. For the same latitude there are slight differences due to the meteorological conditions. This table enables one readily to estimate average values for a given region but for precise experimental work it cannot replace direct measurements.

In the preceding discussion we have considered the whole energy radiated by the sun and this corresponds to radiation between 300 and 5000 nm. As we shall see later, it is generally accepted that the radiation used in photosynthesis lies between 380 and 720 nm, which is about 50% of the total radiation. So the energy available, for photosynthesis at sea level does not exceed 0.7 to 0.8 cal/cm^2/min (49000 to 56000 μW/cm^2) under the best conditions. Fig. 2-2 illustrates the daily distribution of this energy for two extreme cases during the year in an average latitude. The incident energy utilisable for photosynthesis was 400 cal/cm^2/day on the 17th of July (4 · 10^6 cal/m^2/day) and 10 to 20 cal/cm^2/day for the 14th of December. According to Kimball's table the mean value for a region very near to this (42°N and 66.70°W) is 239 cal/cm^2/day (50% of 477) in June and 60 cal/cm^2/day (50% of 120) in December.

b. *Loss of light energy passing the surface*

A certain fraction of the light reaching the sea is reflected by it and does not penetrate the water. In a quiet sea the reflected fraction is 3 to 4% when the sun is at an altitude of at least 40° above the horizon. When the sun is lower, reflection rapidly increases, being 40% when the sun is at an angle of 5° (Sverdrup et al., 1946). When the sea surface is disturbed the loss of light by reflection is from 5 to 17% with a slight wind and above 30% when the wind is moderate or strong (Strickland, 1958).

c. *Penetration of light energy in the depth**

After the light radiation has passed through the surface it is absorbed by the water and, moreover, diffused by the molecules of water and by particles in suspension. The relative importance of these two phenomena varies with the wavelength. As a consequence the irradiance decreases with increasing depth. When the logarithm of the irradiance as abscissa and the depth as ordinate is plotted we obtain a curve (Fig. 2-3); the slope of which at any

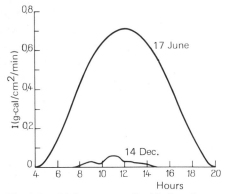

Fig. 2-2.—Light energy utilizable for photosynthesis at the sea surface at Newport, Rhode Island (71°20′W; 41°25′N) on the 17th June and the 14th December, 1954, the longest and shortest days of the year respectively (after Ryther, 1956).

---

*The author thanks A. Ivanoff and A. Morel for their help in the presentation of this work.

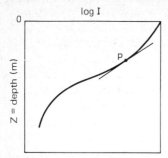

Fig. 2-3.—Extinction coefficient of a point P on the curve relating the logarithm of irradiance and depth.

point P is termed the extinction coefficient at that point. The extinction coefficient, which depends on the diffusing and absorbing properties of the water according to complicated laws, is symbolized by K and expressed as follows:

$$K = -\frac{d \log I}{dz} \text{ or } \left(\text{as } d \log I = \frac{dI}{I}\right) K = \frac{-dI}{I\,dz}$$

and so,

$$\frac{dI}{I} = -K\,dz$$

The extinction coefficient, K, is often constant for a given depth interval i.e., the line in the preceding graph is made of segments of a straight line (Fig. 2-4). It is possible to express the coefficient between two levels according to the equation:

$$K_{1\text{-}2} = -\frac{\ln I_2 - \ln I_1}{Z_2 - Z_1}$$

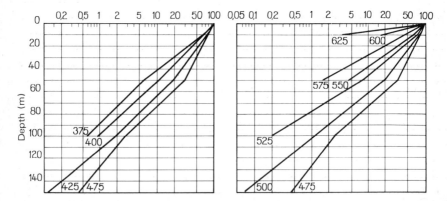

Fig. 2-4.—Penetration of daylight at a station off Bermuda for different wavelengths (in nm): abscissa, percentage as a function of irradiance on the surface (logarithmic scale) (after Jerlov, 1951).

when I_1 and I_2 are the irradiances at the levels Z_1 and Z_2 (the depths Z_1 and Z_2 are expressed positively towards the bottom).

Transferring the logarithmic values to natural values we obtain:

$I_2 = I_1 e^{K_{1\text{-}2}(Z_2-Z_1)}$

If the conditions are such that the coefficient of extinction may be considered as constant, the equation may be simplified as follows;

$I = I_0 e^{-KZ}$

when I is the irradiance at the depth Z and I_0 the irradiance at the surface. This may be written:

$\ln I/I_0 = -KZ$

Fig. 2-5 illustrates the variation of K at different wave lengths for clear and

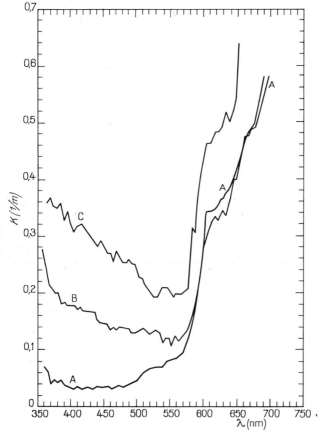

Fig. 2-5.—Variation of the extinction coefficient, K, according to wavelength: A, Gulf Stream, 10 miles west of Bimini, depth 900 m, mean values for the 3rd July 1967: B, Gulf of California, east of Saint-Esprit Island, depth 150 m, values for 9 m level, on 8th May, 1968: C, Gulf of California, same station, values for 17 m level (after Tyler & Smith, 1970).

blue (oceanic) water (A) and for green and productive (coastal) water (B and C). For the latter the extinction coefficient is higher, and, furthermore, the shape of the curve is notably different. In blue waters K varies only little in the violet and blue part of the spectrum (from 400 to 500 nm) but then it increases rapidly. In green waters the coefficient decreases between 400 and 500 nm and then increases, giving a well-defined minimum in the green. The transmission, on the other hand, will be maximal in the blue for blue water and in the green for green water. The difference is due to the abundance of particles and notably of phytoplankton and of detritus rich in derivates of chlorophyll in green waters which by diffusion and by absorption increase the coefficient K and also to the existence in these waters of the yellow substance of Kalle which absorbs the short wave lengths (and more in the U.V.) and causes the observed minimum in the green. In clear and blue waters the blue and violet radiation, 400 to 500 nm, will penetrate most deeply (Fig. 2-6); in productive and green waters the maximum deviation is towards the longer wave lengths of about 500 nm (Fig. 2-7) and often beyond and up to 560 nm. In these last two figures the irradiance is expressed as a percentage of the maximum measured at each level; Figs 2-8 and 2-9 give the true ratio of the

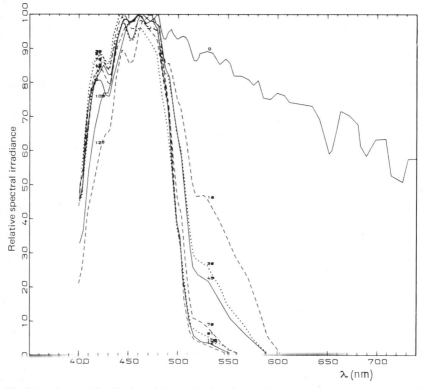

Fig. 2-6.—Spectral distribution of the relative irradiance of a station of clear and blue water in the Sargasso Sea (Station 22, 25°30′ N; 69°45′ W): for every depth the irradiance is related to the maximal irradiance at this depth, considered to equal to 100: the O curve corresponds to the irradiance under the surface (after Morel & Caloumenos, 1972).

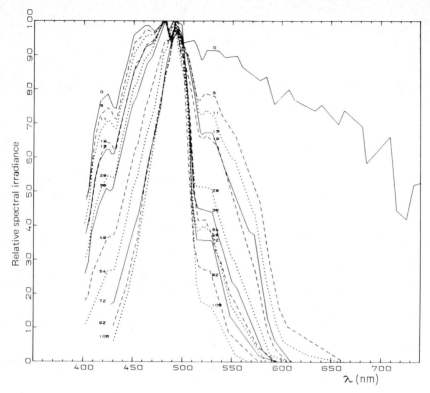

Fig. 2-7.—Spectral distribution of the relative irradiance for a station of green and productive water in the Peru Current (Station 12, 05°37' S; 87°45' W): for every depth the irradiance is related to the maximum irradiance (considered as 100): the O curve corresponds to the irradiance below the surface (after Morel & Caloumenos, 1972).

absolute values of the irradiance according to the wave length at different depths; the contrast between the two stations is clearly seen by a comparison of the irradiance above 100 m, in either absolute values or in a spectral composition.

2-3.—UTILIZATION OF LIGHT RADIATION BY PIGMENTS

We shall now discuss to what extent the elements of the phytoplankton are capable of absorbing and using the light radiation in the marine environment.

Chlorophyll *a*, the most abundant pigment of the phytoplankton (Parsons, 1961) has the absorption spectrum, given in Fig. 2-10, and characterized by a minimum at 450 to 640 nm. If we compare this with the preceding Figs 2-6 to 2-9 it is seen that much of the radiation which is most diffused in sea water is poorly absorbed by chlorophyll and will be useless for photosynthesis. The bulk of the organisms of the phytoplankton, the diatoms and flagellates have, apart from chlorophyll, brown carotenoid pigments, particularly fucoxan-

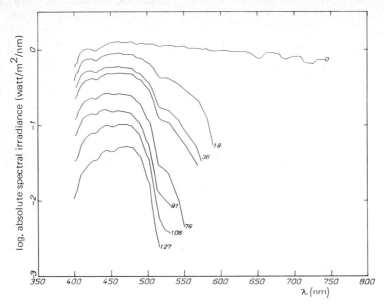

Fig. 2-8.—Variations at different depths of the absolute values of spectral irradiance for a station of clear and blue water (Station 22, 25°30' N; 69°45' W): the O curve corresponds to the irradiance below the surface (after Morel & Caloumenos, 1972).

thin. The absorption spectrum is markedly modified as shown in Fig. 2-11, where the lower peak of absorption is wider and passes over 500 nm. The light energy absorbed by the accessory pigments is transmitted to the chlorophyll which is the only active pigment in photosynthesis. The transfer of energy is at an efficiency of the order of 70% (Rabinovitch, 1956).

The absorption of the pigments as measured on extracts does not always reflect the true absorption of the pigments as they are found in the natural state. Only the absorption of pigments in vivo should be considered, as has been stressed by several authors. Haxo & Blinks (1950) illuminated a narrow thallus of the brown algae *Coilodesma* with light of different wave lengths but of the same energy content and measured the velocity of oxygen production as a measure of photosynthesis. These velocities are given in a graph (Fig. 2-12) where the wave length is the abcissa and give the action spectrum. The curve is very close to the absorption spectrum of the thallus, which is notably different from that of extracts.

For unicellular algae we have only the action spectrum for luminescence established by McLeod (1958), notably for *Phaeodactylum tricornutum* and *Coscinodiscus* sp. (Fig. 2-13). But the action spectra of luminescence emitted by the chlorophyll and the action spectrum of photosynthesis are very close and it is possible to deduce that, for these unicellular algae the photosynthetic activity is increased below 550 nm, and this accompanies the absorption of light energy at such wavelengths by the brown carotenoid pigments.

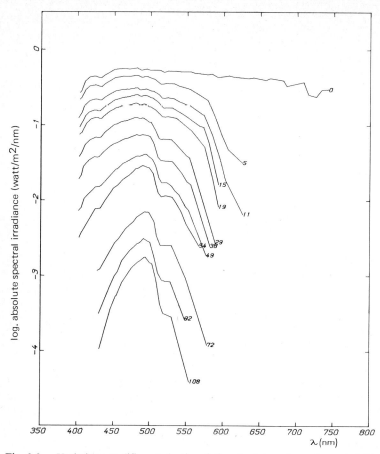

Fig. 2-9.—Variations at different depths of the absolute values of spectral irradiance for a station of green and productive water (Station 12, 05°37' S; 87°45' W): the O curve corresponds to the irradiance below the surface (after Morel & Caloumenos, 1972).

From the data of Jerlov (1951) on the penetration of light in oceanic water, Yentsch (1962) has estimated the action spectra of different depths; these are given in Fig. 2-14. We see that the absorption maximum of chlorophyll at longer wavelengths (in the red) is absent below 10 m. At 50 m, the absorption of chlorophyll in the blue, less than 450 nm, is very reduced. At depths greater than 50 m, the carotenoids (fucoxanthin) are the principal pigments absorbing energy.

Girand (1959) confirmed this for the red algae, *Rhodosorus marinus*, in culture, the additional pigments, phycoerythrin and phycocyanin (biliprotein and phycobiliprotein) using the light energy efficiently.

Stanbury (1931) had previously shown the existence of chromatic adaptation in cultures of *Phaeodactylum* when the culture is placed under selective filters; the cells develop a complementary colour to the light when they multiply. They become dark brown under blue and green screens and green

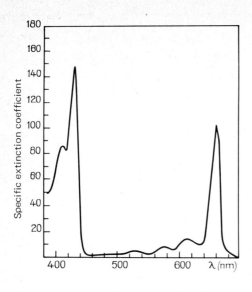

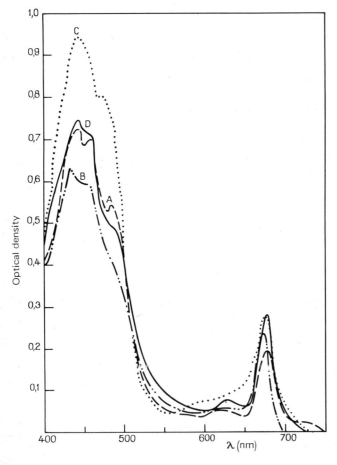

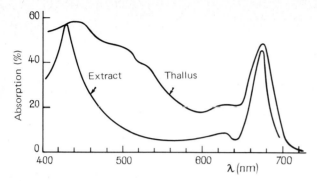

Fig. 2-12.—Light absorption by a thallus and ether extract of the algae *Coilodesma* (after Haxo & Blinks, 1950).

under red and yellow screens. This experiment has not been repeated. It would be interesting to know whether such chromatic adaptation occurs under natural conditions, particularly when there is little turbulence and the cells develop under relatively uniform spectral conditions.

The existence of additional pigments capable of transferring energy to chlorophyll, leads to the possibility (Strickland, 1958) that all the light energy between 380 nm and 720 nm is able to contribute to photosynthesis. This fraction corresponds approximately to the visible part of the spectrum. On the surface it represents, as we have already seen, about half of the total energy received. Because of the small penetration in depth of the long and short waves (infra-red and ultra violet) all the light energy at several meters depth is, in practice, active in photosynthesis (Fig. 2-15).

A consideration of the luminescence action spectrum given by McLeod (Fig. 2-13), shows that below 380 nm this action spectrum cannot be neglected and a possible contribution to photosynthetic action of the ultra violet fraction of the spectrum must be taken into account. It is very quickly absorbed in the green waters, but this is not the case in the blue waters.

2-4.—UTILIZATION OF THE LIGHT BY THE PHYTOPLANKTON

We have now seen how the light energy used in photosynthesis penetrates the sea and how it is absorbed by the phytoplankton. We shall now consider how it is used effectively by it. In the classical work of Jenkin (1937) bottles

---

Fig. 2-10.—Absorption spectrum of chlorophyll *a* (on the ordinate the specific extinction coefficient) (after Zcheile & Comar in Rabinowitch, 1951–1956).

Fig. 2-11.—Absorption spectra of planktonic algae *in vivo*: A, diatom, *Cyclotella* sp.: B, dinoflagellate, *Amphidinium* sp.; C, green flagellate *Chlamydomonas* sp.; D, natural phytoplankton from the water at Woods Hole: a supplementary selective of absorption due to chlorophyll *b* is seen clearly at about 630–645 nm in curve C.

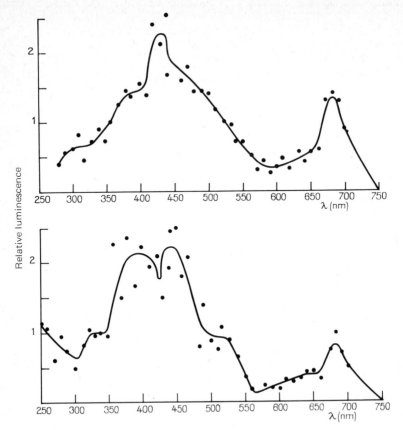

Fig. 2-13.—Luminescent action spectra of *Phaeodactylum tricornutum* (upper) and of *Coscinodiscus* sp. (lower); ordinate, relative luminescence received per quantum (after McLeod, 1958).

containing cultures of the diatom *Coscinodiscus excentricus* (= *concinnus*) were suspended in the sea at different depths. The production of oxygen was measured by Winkler method and corresponds to net photosynthesis. A bottle wrapped with black cloth enables one to estimate the oxygen consumption by respiration. This when added to the net photosynthesis gives the gross photosynthesis. For 22.4 cm of oxygen produced 12 mg of carbon are fixed by photosynthesis. At the same time, the light energy at the different depths was estimated, thus enabling both the production of oxygen (proportional to fixation of carbon in the synthesized organic matter) and the available light energy, as a function of the depth to be examined.

The graph reproduced in Fig. 2-16 shows the results of such an experiment.

It emphasizes the following facts:

(1) At a low light energies the production is proportional to the energy.
(2) When light energy increases production reaches a plateau (saturation).

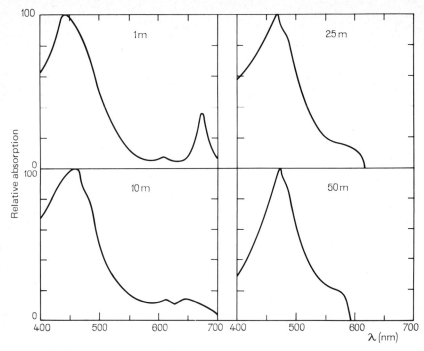

Fig. 2-14.—Relative absorption by phytoplankton (the maximum absorption is equal to 100) of the available light energy at different depths in oceanic water (after Yentsch, 1962).

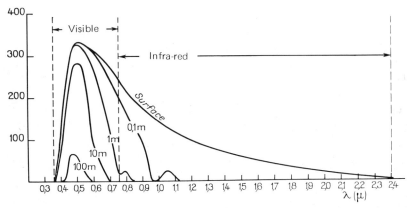

Fig. 2-15.—Schematic representation of the energy radiated by the sun and the sky at different levels in the sea: energy in arbitrary units (after Sverdrup et al., 1946).

(3) With a further increase of light energy, the production decreases (inhibition).

In this experiment the compensation depth, i.e., the level at which the production of oxygen by photosynthesis is balanced by oxygen consumption in respiration (so that the net photosynthesis is zero), is found at the depth of about 45 m with a light energy (compensation irradiance) of

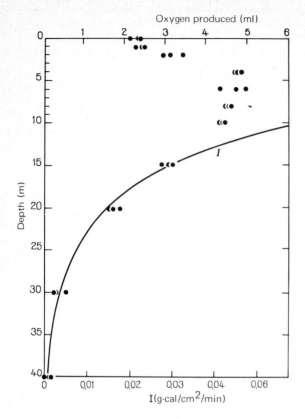

Fig. 2-16.—Oxygen production by gross photosynthesis (ml/10^6 cells) of *Coscinodiscus concinnus* at different levels (solid circles) and irradiance variation as a function of depth (curve) in front of Plymouth on the 25th July, 1933, from 09:45 h to 16:05 h. In this experiment the oxygen uptake was of the order of 0.1 cm^3/10^6 cells and the curve virtually corresponds, therefore, to net photosynthesis (after Jenkin, 1937).

0.002 cal/cm^2/min (or 14×10^{-5} W/cm^2). In general, it is agreed that the compensation depth receives about 1% of the maximum available energy under the surface, but under certain conditions, there is net photosynthesis at lower light levels.

Apart from this it is necessary to establish a distinction between the compensation point determined experimentally under continuous exposure to light over several hours and the mean compensation depth on a 24-hour basis. In the latter the night hours are involved when respiration is not balanced by photosynthesis; this has important ecological implications. The mean compensation depth is naturally less deeper than the compensation depth. The definition of the term "compensation depth" gives rise to confusion and it is recommended that it is not used.

The photic zone is defined as that zone between the surface and the depth where the light is still sufficient for effective photosynthetic processes (Pérès & Deveze, 1963). Strictly speaking, the bottom of the photic zone may be deeper than the level corresponding to 1% of the maximum light or the compensation depth; however, this value is often used and the appropriate level accepted as the lower limit. It is possible to establish a curve of photosynthetic rate as a function of the available light energy. The photosynthesis may be estimated by various methods, such as the production of oxygen or the incorporation of radioactive carbon into the cells; it is convenient, however, in order to facilitate comparison of data to express it as a percentage of the maximum photosynthesis (P_{max}) obtained for saturation irradiance (I_s). We have, therefore, a curve of P/P_{max} against $= f(I)$, i.e., relative photosynthesis as a function of irradiance.

Ryther (1956), using the carbon-14 technique for several diatoms (*Skeletonema costatum*, *Nitzschia closterium*, *Navicula* sp.) obtained results comparable to those of Jenkin (Fig. 2-17), the saturation irradiance and, therefore, maximum photosynthesis corresponded to 0.07 cal/cm²/min.

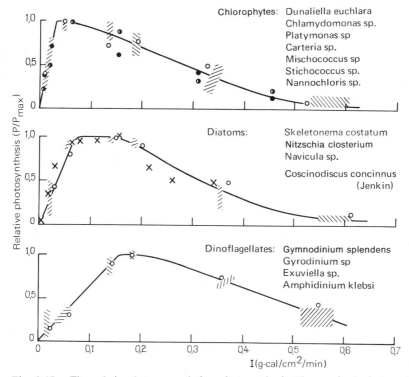

Fig. 2-17.—The relation between relative photosynthesis (the maximal photosynthesis is considered as the unit) and the irradiance for different groups of phytoplankton (under the name chlorophytes are included the Chlorophyceae and the Prasinophyceae): shaded rectangles represent the dispersion of points obtained experimentally using neutral filters in cultures grown with an irradiance of 0.062 cal/cm²/min: open circles correspond to cultures in natural light: solid and half solid circles are cultures at 0.022 and 0.093 cal/cm²/min and measured in the harbour at Woods Hole: crosses correspond to the data of Jenkin (1937): original data are in foot-candles and have been transformed to cal/cm²/min according to Strickland (1958) p. 458 (after Ryther, 1956).

Beyond a plateau, from 0.16 cal/cm²/min photosynthesis decreased. The general shape of the curve is the same but the values of the saturation irradiance may vary notably according to the species. In the chlorophytes (Chlorophyceae and Prasinophyceae: *Dunaliella, Platymonas Nannochloris* etc.) maximum photosynthesis was obtained at 0.03 cal/cm²/min. For the dinoflagellates (*Gymnodinium, Exuviella* etc.,) the maximum is reached at an irradiance of 0.16 cal/cm²/min and the inhibition is slower than for the two preceding groups.

McAllister, Shah & Strickland (1967) did not find any inhibition above the saturation intensity. The curve reached a plateau (Fig. 2-18). The experiment was carried out in an incubator with artificial light and the difference may be due to the almost complete elimination in the incubation conditions, of ultraviolet radiation which might be the source of the inhibition phenomenon (Strickland, 1965).

It is clear that the curve $P/P_{max} = f(I)$ will depend upon temperature. Curl & McLeod (1961) gave the curve shown in Fig. 2-19 for *Skeletonema costatum* above and below 18°C; the saturation irradiance decreased to 0.1 cal/cm³/min at 20°C and should give a marked inhibition while at 18°C the same irradiance is a somewhat under saturation. Our knowledge of the effect of temperature on photosynthesis is far from being adequate, Curl & McLeod's data suggests that *Skeletonema costatum* may be related to terrestrial plants with a low photosynthetic capability (Black, 1971) and characterized by the existence of photorespiration, i.e., respiration catalyzed by light.

The light conditions of the cells before the measurements are made may not be neglected. Cells acclimated to low light reach saturation intensity at lower values than is needed to saturate cells from surface waters (Fig. 2-20).

Finally, in spite of the large amount of work on photosynthesis and in spite of the extensive use of photosynthesis to estimate the production of organic matter in the sea (as will be discussed later), our knowledge of this subject is still far from complete and much research is still needed.

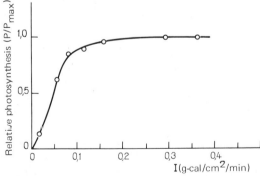

Fig. 2-18.—Relative photosynthesis (maximum photosynthesis is equal to 1.0) of *Skeletonema costatum* as a function of irradiance when incubated at 20°C, measured by oxygen production with high nitrate and phosphate concentrations (after McAllister, Shah & Strickland, 1967).

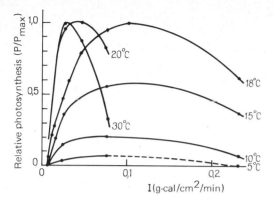

Fig. 2-19.—Relative photosynthesis of *Skeletonema costatum* as a function of irradiance at different temperatures: original data in lux have been transformed to cal/cm^2/min taking 1 cal/cm^2/min = 155 × 10^3 lux (after Curl & McLeod, 1961).

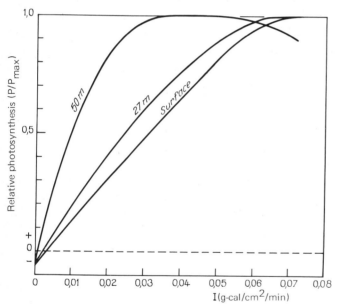

Fig. 2-20.—Relative photosynthesis under different conditions of irradiance for arctic phytoplankton in summer at different depths: original data in lux transformed to cal/cm^2/min taking 1 cal/cm^2/min = 155 × 10^3 lux (after Steemann-Nielsen & Hansen, 1959).

Nevertheless, at the present we may consider that the maximum growth of phytoplankton will be at an irradiance of about 0.15 cal/cm^2/min (0.01 W/cm^2) and that there will be a significant inhibition when this irradiance reaches or exceeds 0.5 cal/cm^2/min (0.035 W/cm^2) (Strickland, 1958). From the data given in Fig. 2-17 Ryther established an "average" curve: knowing the light energy reaching unit surface in 24 h and the extinction coefficient K of sea water, it is possible to calculate the relative

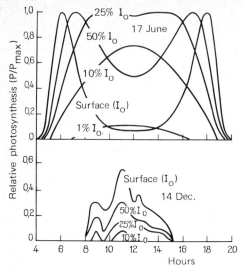

Fig. 2-21.—Relative photosynthesis during the day at different irradiance levels on 17th June and 14th December, 1954, at a latitude and meteorological conditions, comparable to that of Newport, Rhode Island, U.S.A. (after Ryther, 1956).

daily photosynthetic activity at any depth in a column of water where the distribution of the phytoplankton is homogenous. Ryther calculated values at Newport, Rhode Island, for a summer day and a winter day (Figs 2-20 and 2-21) for the surface and depths receiving 50%, 25%, 10% and 1% of the irradiance at the surface. (I_0). At the surface, the maximum photosynthesis is during the morning and evening in the summer while in winter it is far from being reached even in the middle of the day.

CHAPTER 3

THE NITROGEN CYCLE

3-1.—NECESSITY FOR NITROGEN

Analysis of phytoplankton gives a ratio of 10–20 atoms of nitrogen to 100 atoms of carbon. The phytoplankton has to find a source of nitrogen in the marine environment and with a lack of nitrogen growth is limited. Kain & Fogg (1960) demonstrated this experimentally for the dinoflagellate *Prorocentrum micans* maintained in bacteria-free culture. The final number of cells obtained was proportional to the initial amount of nitrogen (Fig. 3-1). The quantity of nitrogen required per cell is relatively constant when the available nitrogen is adequate. In this particular case it was 19 pg-at. (19×10 pg) per cell. Some values calculated by Kain & Fogg, expressed in μg-at. N/cell vol. of 1 mm^3 may be given.

Asterionella japonica	0.27
Peridinium sp.	0.47
Monodus subterraneus (Xanthophyceae)	0.19
Prorocentrum micans	1.1
Isochrysis galbana	1.6

When the cells do not have enough nitrogen in the medium, their nitrogen content decreases markedly. This has been clearly shown for the peridinian *Cachonina niei* cultured in a medium poor in nitrogen (see Table 4-1): for 100 μg-at. of carbon the amount of nitrogen falls from 13 to 7. In *Chlorella* it was noted that the amount falls to a quarter or one-fifth of the normal concentration and the products of photosynthesis are at the same time modified. Carbohydrates and then lipids replace proteins (Syrett, 1962).

The decrease in cellular nitrogen cannot fall below a given threshold, below which cell division is no longer possible. Caperon (1968) showed for *Isochrysis galbana* that this level is reached at 0.031 pg-at./cell.

It is, of course, very interesting to be able to estimate under natural conditions the deficiency caused by a lack of nitrogen. Apart from direct measurements such as the decrease in intracellular nitrogen or the increase in the C/N ratio, indirect measurements have been proposed and the most promising seem to be the decrease of the assimilation ratio (photosynthesis at saturation/amount of chlorophyll) (Curl & Small, 1965), the increase of the ratio between ^{14}C absorption in the dark and at light saturation (Steemann-Nielsen & Alkhaly, 1956), or the increase of absorption of ammonia in the dark (Fitzgerald, 1968). From a study of the assimilation quotient, Thomas (1970) has shown that the phytoplankton of poor superficial waters of the oriental tropical Pacific was near to deficiency (an

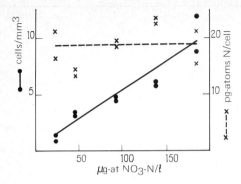

Fig. 3-1.—Final number of cells/1 mm^3 and quantity of nitrogen/cell, in a culture of *Prorocentrum micans*, as a function of the quantity of nitrate (added as potassium nitrate) in the medium (after Kain & Fogg, 1960).

assimilation quotient of 3.15, compared with a value of 4.95 for rich water, a significant different at $p = 1\%$); poor waters lack nitrate but a small amount of ammonia (an average of 0.5 µg-at./l) prevents the nitrogen deficiency becoming extreme.

3-2.—ABSORPTION OF NITROGEN

Nitrogen is found in the sea in the following forms, namely, ammonia (NH_4^+), nitrite (NO_2^-) and nitrate (NO_3^-), with ammonia and nitrates dominant under natural conditions. Nitrogen may be absorbed by the phytoplankton in these three forms*. In general, when ammonia and nitrate are available to the cells the ammonia will be absorbed first as is shown by the experiment reported in Fig. 3-2, which concerns the small dinoflagellate *Cachonina niei*; the experiment was carried out in a large tank (diameter 3 m; depth 10 m) with a concentration of nitrogen of the order of that found in the sea. The absorption of nitrate-nitrogen only began when the amount of ammoniacal nitrogen had fallen below 1 µg-at./l. This was also found in experiments with another peridinian *Gonyaulax polyedra*, and the diatom *Ditylum brightwellii*.

Prochazkova, Blazka & Kralova (1970) confirmed these facts for natural water of Czechoslovakian lakes. The proportion of nitrate nitrogen in the absorbed inorganic nitrogen decreases rapidly with the increasing amount of ammonium nitrogen in the environment and it remains less than 20% for 70 µg/l of NH_4–N.

---

*In the following discussion the amounts will be expressed in nitrogen, the chemical form being on occasions specified when necessary; thus 1 mg (NO_3–N (or N–NO_3) = 1 mg of nitrogen in the nitrate form.

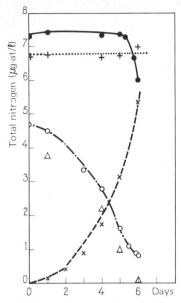

Fig. 3-2.—Changes in the dissolved and particulate nitrogen (cells) with time in a culture of *Cachonina niei* in a deep tank: dissolved nitrogen in three forms, namely, ammonia, nitrate, and organic nitrogen: solid circles, NO_3^-; straight crosses, organic N; oblique crosses, observed particulated N; open circles, NH_4^+ and amino acid (hypochlorite estimation); open triangles, NH_4^+ only (bispyrazolone estimation); broken line (near oblique crosses), particulated nitrogen predicted according to nitrogen uptake (after Strickland, Holm-Hensen, Eppley & Linn, 1969).

The utilization of the nitrate ion requires reduction which consumes energy so that its energetic cost to the cell is higher. The reduction is effected by nitrate reductase (Fig. 3-3), which only appears in quantity when the ammonium ion is almost used up; Fig. 3-4 shows this phenomenon. Some flagellates (*Euglena gracilis* var. *bacillaris*, *Chlamydomonus reinhardti*), probably due to a deficiency in the production of nitrate-reductase (Syrett, 1962), cannot use nitrate; MacIsaac & Dugdale (1969) using the isotope ^{15}N showed more precisely how the absorption of nitrogen by a natural population of phytoplankton is effected. The absorption rate depends upon the nitrogen concentration in the environment, at first increasing rapidly and then reaching a plateau with increasing amounts. This was found (Fig. 3-5)

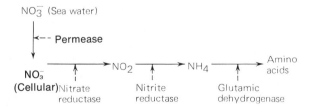

Fig. 3-3.—Scheme of the biochemical processes (successive enzymatic stages) in the incorporation of nitrate ion into the amino acids of the phytoplankton (after Packard et al., 1971).

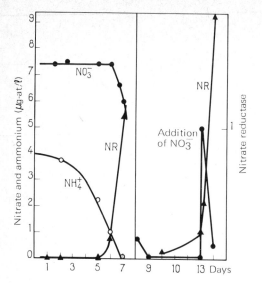

Fig. 3-4.—Variation in the amount of ammonia, nitrate, and nitrate reductase as a function of time in a culture of *Cachonina niei* in a deep tank: on the seventh day the mixing was stopped and the cells migrated to the surface and on the 8th day the tank was almost completely emptied, then filled so that the amount of nitrogen was almost zero: NR, nitrate reductase (after Eppley, Coastworth & Solorzano, 1969).

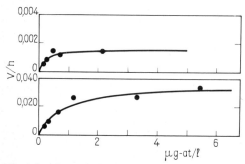

Fig. 3-5.—Rate of nitrate absorption as a function of concentration: upper, Station Thomson 26-7 (oligotrophic); initial amount 0.02 μg-at. NO_3–N/l; amount of particulate nitrogen 0.87 μg at./l; water taken from 25% I_0 level and incubated at 50% I_0 for 24 h: lower, Station Thomson 26-38 (eutrophic); initial amount of nitrogen 0.12 μg-at. NO_3–N/l; amount for particulate nitrogen 4.32 μg-at./l; water was taken from 100% I_0 and incubated at 100% I_0 for 6 h: V/h in units of absorbed nitrogen/unit particulate nitrogen (essentially phytoplankton)/h (after MacIsaac & Dugdale, 1969).

both for poor water (oligotrophic conditions) and for rich water (eutrophic conditions).

It is convenient to use the Michaelis-Menten equation governing enzyme activity to represent these ^{15}N experiments the curve is drawn by eye

$$V = V_{max} \frac{S}{K_s + S}$$

where, V = the absorption velocity of nitrogen
 V_{max} = the maximal velocity
 S = concentration of nitrogen
 K_s = a constant equal to concentration at half saturation for which $V = V_{max}/2$ or constant of half saturation.

For the curve of Station 26-7, V_{max} is equal to 0.0016 and $K_s\,NO_3 = 0.21$; for Station 26-38, $V_{max} = 0.0361$ and $K_s\,NO_3 = 0.98$.

MacIsaac & Dugdale suggested that high values of K_s correspond to eutrophic and low values to oligotrophic water, so that the phytoplankton population of this poor water is adapted to their environment.

We may note that in the absorption of nitrogen other less well known factors may interfere. According to Gerloff & Skoog (1954) cultures of *Microcystis aeruginosa* (Cyanophyceae) cease to grow when a considerable concentration of nitrogen still remains. During the rapid phase of growth, the percentage of nitrogen relative to the dry weight is twice as high as after the declining phase (these cultures, in fresh water, were balanced in phosphate and iron but not in minor elements and vitamins).

More generally, there is the question as to how the absorption of nitrogen and growth are related, i.e., with respect to the multiplication of cells. This relation is not fixed, since we noted in the preceding paragraph important variations in the amount of cellular nitrogen. Eppley & Thomas considered this question and concluded that most often there is a hyperbolic relation between the relative rate of growth and amount of cellular nitrogen, and the rate increases rapidly as a function of amount and then becomes stabilized.

Apart from ammonia, nitrite, and nitrate, gaseous nitrogen is dissolved in the water and might be also a source of nitrogen. It is, indeed, fixed by Cyanophyceae of the genus *Oscillatoria* (= Trichodesmium) which is common in tropical waters. This was proved using the isotope ^{15}N (Dugdale, Goering & Ryther, 1964) in the Sargasso and in the Arabian Seas. This fixation, which does not exclude the absorption of ammonium or nitrate nitrogen, may reach some importance. Goering, Dugdale & Menzel (1966) found the fixation of 0.32 µg N/l/h during a massive multiplication of *Oscillatoria* (its biomass represented 234 µg organic nitrogen/l).

A new method for the study of the fixation of nitrogen by the Cyanophyceae has been proposed by Rusness & Burris (1970): it was shown experimentally that the enzymatic complex concerned in the fixation of nitrogen can reduce acetylene and so the measurement of this reduction gives an estimation of capacity of the phytoplankton to fix nitrogen. By this method these authors were able to show that in eutrophic lakes (under eutrophic conditions) there is a daily cycle of nitrogen fixation with a maximum at noon.

Finally, the question arises as to whether nitrogen can be absorbed by phytoplankton from an organic, as distinct from a mineral, source. Goering, Dungal & Menzel used glycine labelled with ^{15}N, and found an absorption of this amino acid, attributed to *Oscillatoria*. Hellebust (in Riley, 1971) showed that some species of phytoplankton could absorb amino acids dissolved in sea water, even at very low concentrations, from 10^{-2} to 10^{-8} M (1.4–0.14 µg N/l). The possibility that the phytoplankton is able to use organic nitrogen should not be forgotten and needs further investigation. Until then we must accept that essentially it is the mineral form of nitrogen which is absorbed*.

3-3.—REMINERALIZATION OF NITROGEN

The nitrogen present in organic matter can be used by the phytoplankton only after being remineralized. The regeneration of nitrogen can take place in two ways, either directly or through bacteria, and this will now be discussed.

a. *Direct regeneration*

There is a direct return to the marine environment of inorganic nitrogen present in the living matter, and this is essentially concerned with the excretion of waste products of nitrogen metabolism in the form of ammonia (ammonotelic excretion) by zooplankton and other marine organisms. Harris (1959) was the first to study the excretion of zooplankton. He showed that important quantities of ammonium ions were produced and estimated that at least 50% of the regeneration of nitrogen in Long Island Sound was due to the zooplankton. Indeed, an important part of the nitrogenous excretion of the zooplankton consists of ammonia; e.g., the daily excretion of *Euphausia pacifica* at 10°C is 2 µg NH_3–N/mg dry weight (i.e., about 16 µg of ammoniacal nitrogen per mg nitrogen of the body weight per day) (Jawed, 1969).

Fish also excrete a significant amount of ammoniacal nitrogen in their urine: the angler (*Lophius piscatorius*) has 13% of its nitrogenous excretion in this form. Whitledge & Packard (1971) found an excretion of 40.6 µg-at. NH_3–N/g body nitrogen/h and at 15°C (about 13 µg NH_3–N/mg of N of body weight per day) in the pelagic anchovy (*Engraulis ringeus*) from Peru.

A good correlation should exist between areas rich in zooplankton and fish and those rich in ammonia; so far we have not enough information on the relation of fish concentrations and ammonia, but Redfield & Keys (1938) published the maps given in Fig. 3-6, showing there is a satisfactory correlation between high volumes of zooplankton and maxima of NH_4 concentration.

In darkness, the phytoplankton behave like heterotrophic organisms of the zooplankton and are able, therefore, to excrete nitrogen; so far we have no data on this subject. On the contrary, an absorption of ammoniacal nitrogen in the darkness has even been observed and this has been attributed to a deficiency of nitrogen (Fitzgerald, 1968). Prochazkova et al. (1970) also

---

*McCarthy (1972) has recently shown that the absorption of urea by phytoplankton supplies an average of 28% of the nitrogen taken from the coastal waters in California he studied.

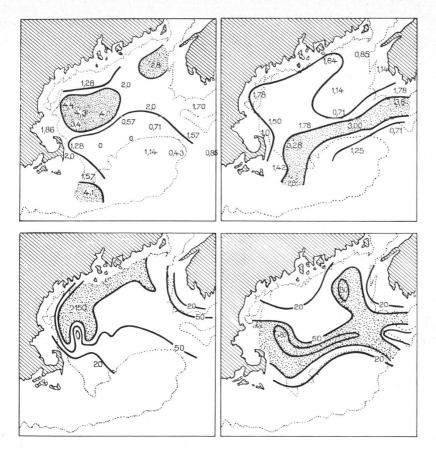

Fig. 3-6.—Gulf of Maine, Atlantic coast of U.S.A.: upper, maximum concentration of NH_4^+ (in µg-at. N/l) in the first 80 m in September 1934 (left) and May 1934 (right): lower, zooplankton volume (in ml of dry plankton/m^2) in September 1934 (left) and May 1934 (right) (after Redfield & Keys, 1938).

found such an absorption in the dark for particulate organic matter (interpreted as representing essentially phytoplankton) when this is poor in protein, but they also noted excretion of ammonia when the amount of protein was high (Fig. 3-7).

b. *Bacterial regeneration*

The bacteria take part in this nitrogen return, regenerating nitrogen from dissolved organic matter or from organic detritus.

There are different sources of dissolved organic matter. It is known that a certain fraction of the organic matter of the phytoplankton passes to the environment, but the proportion of nitrogenous substances present is still not known. Apart from ammonia excretion, the zooplankton excrete urea and mainly amino acids. *Euphausia pacifica* excretes, at 10°C, 0.3 µg

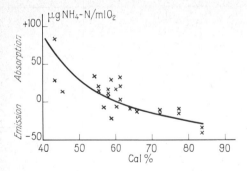

Fig. 3-7.—Changes in uptake and release of ammonia in dark bottle/ml of oxygen produced in the corresponding light bottle (same sample of water) related to caloric percentage of protein in particulate matter in the bottle (essentially phytoplankton): the correlation coefficient, $r = 0.785$, highly significant and calculated for the log ordinate (uptake, release) + 50 and the abscissa (cal % of protein) (after Prochazkova, Blazka & Karlova, 1970).

amino-N/mg dry weight/day. Pelagic fish excrete nitrogen in a variety of different organic forms, namely, urea, creatine, amino acids, and trimethylamine oxide. On the other hand when they die a rapid loss can occur. Krause (1964) observed that in freshwater zooplankton 30% of the nitrogen passed to the environment within 15 to 30 min after death, and without bacterial action.

All these organic nitrogenous substances are able to be decomposed by bacteria with the ultimate formation of ammonia. Budd (1969) has shown that the marine bacterium *Pseudomonas* NCMB 1154 can catabolize the trimethylamine.

The origin of much of the organic detritus is dead animals but in addition faecal pellets, particularly of the very abundant copepods, are common. The nitrogen contained in this detritus is remineralized by the action of marine bacteria of which many are proteolytic (Vaccaro, 1965). Johannes (1968) disagreed, however, on the importance of the last way of remineralization of organic matter. It is, of course, not enough to identify bacteria which are able to transform given materials but also necessary to measure the effective importance of the reactions involved under the natural conditions; this has been little studied.

Finally, it should be mentioned that whatever is the relative importance of the different ways of regeneration the final product is ammonia.

3-4.—NITRIFICATION

We have seen that the ammonium ion can be rapidly absorbed by phytoplankton. If, however, it is not used it is then oxidized to nitrite (NO_2^-) and then to nitrate (NO_3^-). The phenomenon may be illustrated by the classical experiment of von Brand, Rakestraw & Renn (Fig. 3-8).

Phytoplankton was added to sea water contained in large vessels placed

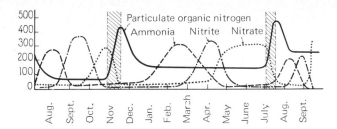

Fig. 3-8.—Decomposition of phytoplankton in sea water and variation of the different forms of nitrogen: ordinate, µg N/l: shaded, light periods (after von Brand, Rakestraw & Renn, 1939).

in the dark and the various forms of nitrogen determined at intervals. The organic nitrogen (organic detritus originating from dead plankton) decreased rapidly in favour of dissolved inorganic nitrogen. The ammonium ion appeared first, reaching a maximum and then was replaced by nitrite and eventually nitrate. The vessels were then exposed to light which induced a multiplication of the phytoplankton with a consequent fall in nitrate. The light was removed and the cycle began again.

Von Brand and his colleagues showed the existence of enzymatic processes in the oxidation of the ammonium ion to nitrate ion but could not prove the existence of nitrifying bacteria in the milieu. Such bacteria, belonging to the same genera as the terrestrial nitrifying bacteria, *Nitrosomonas* and *Nitrobacter*, are known from sea water but only from marine ooze and from coastal waters. A new nitrifying bacteria from the open sea, *Nitrosocystis oceanus*, transforming ammonia to nitrite, has been described and isolated (Murray & Watson, 1963). Our knowledge of the details and the importance of the processes involved in the bacterial reaction is far from satisfactory.

The oxidation of NH_4^+ and NO_2^- takes place only in the zone of active photosynthesis, and below a given depth all the inorganic nitrogen is present in the form of nitrate. Fig. 3-9 illustrates this phenomenon; below 100 m the amount of NH_4^+ and NO_2^- is virtually zero. This conclusion is only relative; below 100 m there are still animals (zooplankton, necton) which excrete nitrogen, and ammonium nitrogen should be present down to the greatest depth but in smaller quantities than in the shallow water.

In moderately deep water nitrification occurs on bottom itself during the winter.

The nitrification may be reversed by action of other bacteria, for instance *Pseudomonas perfectomarinus*, which reduces nitrates and nitrites (denitrifying bacteria); this reaction can be continued until it produces dissolved molecular nitrogen; such denitrification occurs particularly when there is a large decrease in the oxygen concentration. In water depleted of oxygen off Peru, Carlucci & Schubert (1969) identified bacteria which were capable of transforming nitrate to nitrite under anaerobic conditions. An excellent example of denitrification was given recently by Richards, Anderson & Cline (1971) in Golfo Dulce, a gulf which penetrates about 30 miles into the

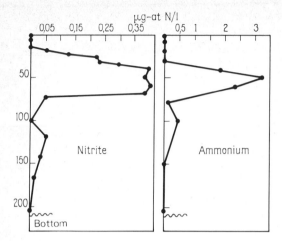

Fig. 3-9.—Vertical distribution of nitrites and ammonia in the Gulf of Maine, Atlantic coast of the U.S.A., in May: ordinate in m (after Redfield & Keys, 1938).

Pacific coast of Costa Rica and has a sill at 60 m separating from the open sea a basin of 200 m depth. As shown in Fig. 3-10, a decrease of oxygen goes hand by hand with a fall in the nitrate nitrogen and is also accompanied by the appearance of nitrites. At the bottom of the basin there is neither oxygen nor nitrates.

Denitrification does not, however, seem to be solely a bacterial activity; Vaccaro & Ryther (1960) have shown that under certain conditions algae of the phytoplankton can produce and excrete nitrites. This occurs particularly at low light intensities when nitrates are available, and may be considered a result of the more rapid reduction of nitrates than the normal, with a further reduction of nitrite to ammonium, and so the excess nitrite passes to the

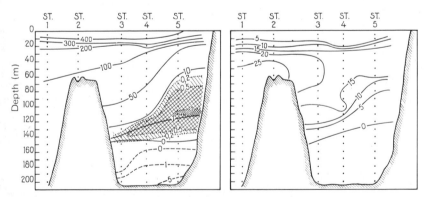

Fig. 3-10.—Transect along the axis of Golfo Dulce, Pacific coast of Costa Rica made on the 10th March, 1969: left, solid isopleths, oxygen (O_2, μg-at./l); dashed isopleths, hydrogen sulphide (μmole/l); dotted isopleths, nitrates (NO_2–N μg-at./l); densest shading, maximum of nitrates: right, isopleths of nitrates (NO_3^--N, μg-at./l) (after Richards, Anderson & Cline).

environment. The first maximum of nitrite found at the level of the thermocline could be so explained. The second maximum is found at several hundred meters depth and, in contrast, is due to anaerobic bacterial action as already described (Carlucci & Schubert, 1969).

3-5.—THE NITROGEN CYCLE

The nitrogen cycle is outlined in Fig. 3-11. In order to complete the cycle, bacteria and benthic algae have been added to the phytoplankton, and zooplankton to the animals from the necton and the benthos. In the following paragraph we shall explain this scheme. The phytoplankton takes up nitrogen essentially as ammonia, nitrite, and nitrate but in addition (indicated by broken lines) there are two other sources of nitrogen whose importance is underestimated, namely, dissolved gaseous nitrogen and dissolved organic nitrogen. As opposed to the uptake of nitrogen by phytoplankton there is also the production of dissolved organic matter by the phytoplankton, which is also indicated by broken lines. In the transformation of ammonia to nitrite and nitrate and in denitrification, bacterial action is involved and this is indicated by the letter B. We have also noted the possibility of dissolved nitrogen being integrated into the cycle by bacterial action since nitrogen fixing bacteria have been identified in sea water (*Azotobacter, Clostridium*); since this fixation is endothermic and requires a quantity of carbohydrates its importance is in doubt (Vaccaro, 1965). Bacteria are involved in the production of ammonia from organic nitrogen or

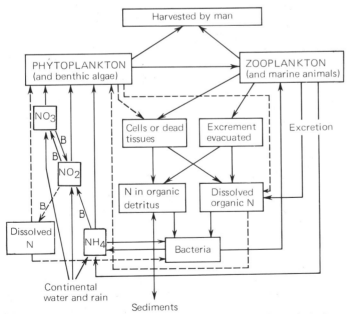

Fig. 3-11.—The nitrogen cycle in sea water (B = bacterial action): for explanation see text.

by the transformation of the excess of nitrogen consumed, after the proliferation had stopped, by rapid decomposition of the cells; but they may also take this ion from the environment in order to build up their own organic substances when they are short of nitrogen; in this case they compete with the algae of the phytoplankton. This is the meaning of the two opposite arrows in the scheme. These bacteria may be ingested by some animals, particularly cilliates; Johannes (1968) has emphasized that it is only indirectly by bacteria but directly by the excretion of such organisms that remineralization may be efficiently effected. We have marked by broken line the contribution by dead cells since it is accepted that the cells of the phytoplankton are always eaten, more or less quickly, by the zooplankton, the faeces of which may sometimes contain numerous fragments of the phytoplankton, more or less broken up but undigested (superfluous feeding).

The nitrogen cycle is not completely closed, since the introduction of nitrogen in rain water cannot, in general, be neglected (Vaccaro, 1965). The ammoniacal nitrogen in rain may reach several mg/l and the average amounts observed in the cold or in temperate zones of the northern hemisphere are of the order of 100 to 400 μg/l. The amount of nitrate-nitrogen varies between the same limits. The rôle of the precipitated nitrogen is not well known, but should be of some importance in shallow and relatively isolated waters. In Colorado semidrainage mountain lakes of several meters depth and without a regular outlet, are mainly supplied by precipitation; Pennak (1969) found an average nitrate-nitrogen content of over 1 mg/l with a N/P ratio of 500.

The continental water of the Mississippi also brings a certain quantity of nitrogen to the sea (Riley, 1937); the annual mean is 20 μg/l of NH_3–N 5 μl/l of NO_2–N, 200 μg/l of NO_3–N and 350 μg/l of organic nitrogen.

Detritus sinks to the bottom and represents, therefore, a loss of nitrogen; this loss is sometimes only temporary since the detritus may be re-suspended by turbulence at least in shallow water. The nitrogen in the detritus is, however, often in the form of organic compounds resistant to bacterial attack.

Finally, fishing by men (or collecting of marine algae) also constitute a certain loss of nitrogen. We shall try to estimate this loss for a region of active fisheries, the North Sea, whose area is about 635 000 km^2 and has an average depth of 100 m. The production, calculated for the years 1950 and 1951, is 1640×10^3 tons of fish/yr (Blanke, 1956). Accepting a value of 3% of the wet weight as nitrogen in the living matter (mean value for herring which is about 80% of the catch; Vinogradov, 1953) we obtain an annual loss of 77.5 mg of nitrogen/m^2, which corresponds to 212 μg/m^2/day or 2.1 μg/m^3/day.

3-6.—DISSOLVED ORGANIC MATTER

We have seen that by excretion and diffusion, in vivo and poste mortem changes, marine organisms give more or less complex organic compounds to the sea water and also that in principle the nitrogen of these compounds can be remineralized by bacteria. The question arises, however, whether this remineralization is rapid or slow. If the latter were true, considerable quantities of dissolved organic nitrogen should be found in the ocean.

Duursma (1961) investigated this problem by reducing the organic nitrogen to ammonia and estimating it. Before this the acidified sea water was stored for 24 h for sedimentation in order to remove the nitrogen of particles in suspension. He obtained values for the North Atlantic ranging between 40 and 400 µg/l of dissolved nitrogen thus confirming the order of magnitude given by several other authors. Duursma also studied over a year the amount in the surface, at the light vessel "Texel" in the southern North Sea; it ranged between 80 and 550 µg/l with a maximum in April–May (Fig. 3-12). In general, the quantity of organic nitrogen in sea water is of the same order as that of the inorganic nitrogen. This was confirmed by Armstrong & Tibbits (1968) using photochemical "combustion" by irradiation with ultraviolet light in order to transform the nitrogenous organic matter to nitrite and nitrate. In the English Channel at 10 m depth the values were from 2 to 9 µg-at. organic nitrogen/l, (28–126 µg/l) with few spatial or seasonal variations; the values are close to the total amount of inorganic nitrogen, which, however is much more variable (Table 3-1).

In deep water, 1000 m, the amount of dissolved organic nitrogen is far from being negligible, and Duursma found it, with few exceptions, to lie

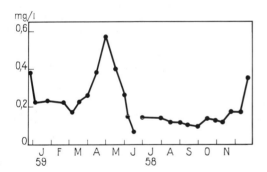

Fig. 3-12.—Annual variations in the amount of dissolved organic nitrogen (mg N/l) in the surface water near the light vessel Texel, July 1958–June 1959 (from data of Duursma, 1961).

TABLE 3-1

Variations in amount of dissolved organic and inorganic nitrogen, at 10 m at the Station E (50°02′ N: 04°22′ W) at the entrance to the Channel (after Fig. 2 in Armstrong & Tibbits, 1968).

µg-at.N/l	March	Apr.	June	1966 Aug.–Sept.	October	Dec.	1967 Jan.–Feb.	March	April
Inorganic N	5.0	1.8	0.2	0.7	2.8	4.6	7.0	7.7	3.8
Organic N	5.2	5.7	2.6	5.8	5.0	2.6	2.8	6.7	7.2

between 100 and 200 µg/l; this is due, not to the particularly refractory character of this organic nitrogen, but to the absence of factors which are favourable to its decomposition, such bacteria or particles giving available substrata (Duursma, 1961). There is then a notable quantity of organic nitrogen in solution in sea water. We are beginning to get some data on the nature of this dissolved material which includes amino acids, nucleic acids, polypeptides, and proteins. In general, part of this organic matter is more or less refractory and stable but, other fractions are more transitory and probably of greater biological importance e.g., particularly those containing amino acids and enzymes (phosphatases).

3-7.—ESTIMATION OF NITROGEN

The fact that the nitrogen utilizable by the phytoplankton exists in different forms markedly complicates its estimation in the sea. For nitrite, a simple characteristic reaction is available using sulphanilamide in acid solution: the method has been known for a long time. Unfortunately nitrites are not abundant in sea water. For nitrates, the old and very capricious methods using reduced strychine have given way to the methods depending on the reduction of nitrates to nitrite by passing through a cadmium column. For ammonia, several more or less satisfactory methods are available. In one of the best, but quite complicated, the ammonia is transformed by reaction with bispyrazolone to rubazoic acid and extracted by a solvent (Prochazkova, 1964). Recently the analytical methods have been improved by the use of automatic analysis (Technicon) which enables rapid analysis of numerous samples to be made in uniform conditions. For the analysis of nitrates, Stephanes (1970) has described a method using a cadmium column, allowing the estimation over a range from 0.2 to 40 µg-at. NO_3–N/l (standard deviation for 10 repeated samples; 0.2 µg-at. for an amount of 20 µg-at.). In spite of difficulties arising from the extraction, the ammonium method of Prochazkova was automized by MacIsaac & Olund (1971), and allowed 10 estimations/h with a precision of ±10% at a level of 0.5 µg-at. NH_4–N/l, and covering a range of 0.1–35 µg-at. NH_4–N/l.

The amounts are expressed in different ways and the equivalent are as follows:

1 microgram-atom/litre (µg-at./l) = 1 milligram-atom/cubic meter (mg-at./m^3).

1 microgram/litre (µg/l) = 1 milligram/cubic meter (mg/m^3). 1 µg-at./l = 14 µg/l.

3-8.—QUANTITATIVE VARIATIONS IN INORGANIC NITROGEN

Because of their biological importance, the amount of nitrate, nitrite and ammonia in the sea and their variation have long been estimated. The earliest investigations were made at Plymouth by Atkins, Harvey and Cooper.

Fig. 3-13 represent the seasonal variations observed in the sea off Plymouth in 1931 for the three forms of inorganic nitrogen.

In the winter: at the surface and at the bottom the inorganic nitrogen is abundant and is over 100 µg/l: 4/5 corresponds to nitrate and the rest is made up of ammonia and a small amount of nitrate.

In the spring: there is a sudden fall in the nitrogen; the ammoniacal nitrogen is taken up more rapidly than the nitrate nitrogen by the phytoplankton bloom which develops. In June nitrate almost disappears from the surface, and the only detectable nitrogen is mainly in the form of ammonia present in small quantities and probably produced by direct regeneration; near the bottom the amount of nitrogen is higher since nitrate is not consumed.

In the summer: at the beginning of July, nitrate is absent from the warmer superficial zone which is separated from the deeper water by a thermocline; only ammonia produced by direct regeneration is utilizable by the phytoplankton. When the stratification is interrupted in August mixing of the water brings a certain amount of nitrate to the surface which enables a new massive outburst of phytoplankton, which in turn causes a decrease in the amount of nitrogen.

In the autumn: the phytoplankton multiplies less and less and the mixing of the water becomes more important: nitrate begins to accumulate from the bottom to the surface forming the spring stock. During this period the nitrite, never has a maximum.

Such a quantitative development is essentially the same in all temperate zones that have been studied, with the following principle outline;

(a) A preponderance of the nitrate over the ammonia and low importance of nitrites. While nitrate may reach or even surpass 200 µg/l, ammonium nitrogen is rarely above 60 µg/l; high concentrations near the coast may be due to continental water (up to 200 µg/l); values of nitrite-nitrogen higher than 10 µg/l have only rarely been observed (Raymont, 1963).

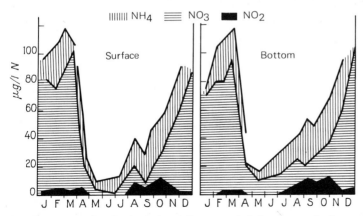

Fig. 3-13.—Variation in ammonium-, nitrate-, and nitrite nitrogen in the upper layers and near the bottom in the English Channel, 20 miles south west of Plymouth during 1931; depth 70 m (after Cooper, 1933, in Harvey 1955).

(b) At the surface there is an abundance of nitrogen in the winter, with decreases during the spring and summer.

This phenomenon appears clearly at the entrance to the English Channel (Fig. 3-14), as well as in the Gulf of Maine on the American Atlantic coast (Fig. 3-15). In the Pacific, off Oregon, the superficial layer of more than 50 m depth is virtually denuded of nitrate during the summer (Fig. 3-16). This

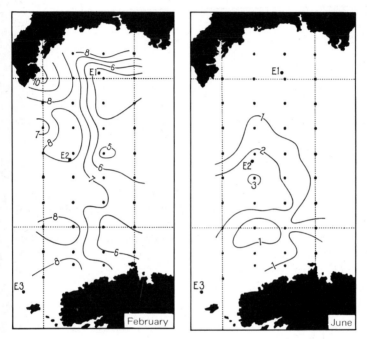

Fig. 3-14.—Distribution of nitrate-nitrogen (in µg-at. NO_3–N/l) in February and June 1961 at the entrance to the English Channel (after Armstrong, Butler & Boalch, 1970).

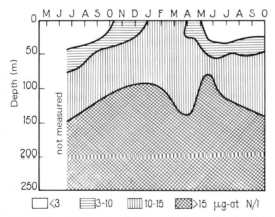

Fig. 3-15.—Seasonal variations of nitrates in the Gulf of Maine (after Rakestraw, 1936 in Barnes 1957).

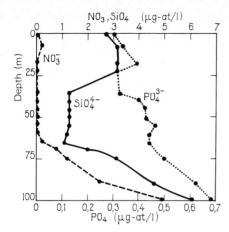

Fig. 3-16.—Vertical distribution (in μg-at./l) of nitrate nitrogen (NO$_3$–N) on the 26th July 1968 in the Pacific Ocean off the coast off Oregon (45°10' N: 126°56' W): distribution of phosphate (μg-at. PO$_4$–P/l) and silicate (μg-at. SiO$_4$–Si/l) are also shown (after Anderson, 1969).

decrease of the amount in the surface water is generally found in all temperate and cold zones. The seasonal rhythm is reduced only at low latitudes in tropical waters.

(c) Reduction of the seasonal variations in the depth, up to its complete disappearance. This may be seen in Fig. 3-5 where, from 150 m the amount of nitrate–nitrogen is constantly higher than 210 μg/l (15 μm-at.). The depth at which seasonal variations disappear is essentially a function of the transparency of the water and of currents.

Apart from this general occurrence of the seasonal and vertical distribution of the inorganic nitrogen, investigations have shown the existence of important differences, according to the region. Fig. 3-17, after Deacon (1933), shows that in the southern half of the Atlantic Ocean at depth of above 1000 m, there is a gradient of increasing amounts, from 200 μg/l in the tropical zone to more than 500 μg/l in the Antarctic zone. This large amount is again found in the deep water of the central zone of the Indian and Pacific Oceans (Fig. 3-18). In the Mediterranean, on the contrary, the amount at 1000 m depth is less than 100 μg/l (Vaccaro, 1965). Only in shallow zones, near the entrance to rivers are higher values found (Fig. 3-19).

3-9.—THE NITROGEN BALANCE

The main part of the phytoplanktonic nitrogen arising from either ammonia or nitrate is taken up by herbivores of the zooplankton or by herbivorous fishes; a fraction is removed by them with subsequent remineralization (excretion, death). The nitrogen of the herbivores is consumed by the first level carnivores, which again remove a fraction of it and

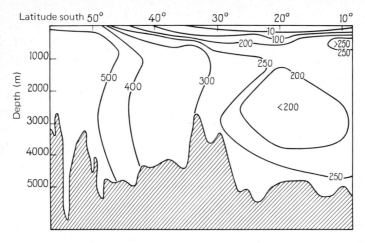

Fig. 3-17.—Distribution of nitrate (+ nitrites) (in µg/l) along 30°W in April–May 1931 in the South Atlantic: Antarctic Convergence is at about 50°S and the Subtropical Convergence at about 41°S (after Deacon, 1933).

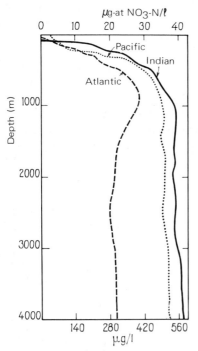

Fig. 3-18.—Distribution of nitrates as a function of depth in the central zone of the three major oceans: upper, in µg-at. NO_3–N/l: lower, in µg/l (after Sverdrup et al., 1946).

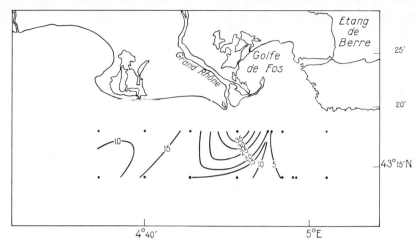

Fig. 3-19.—Distribution of nitrates (µg-at. NO$_3$–N/l) at the sea surface at the entrance to the Grand Rhône, in autumn (after Blanc, Leveau & Szekielda, 1969).

so at each level we have a loss of organic nitrogen, which is transformed to ammoniacal nitrogen; at higher trophic levels the absolute quantity of nitrogen integrated into the living matter decreases. We have, therefore, a classical pyramid; at its base is photosynthetic production and at its apex the highest link of the food chain. As regards the ammoniacal nitrogen, it is partly re-cycled in the phytoplankton and partly transformed to nitrate nitrogen. This simplified cycle is schematized in Fig. 3-20 which gives the reservoir of nitrate–nitrogen, the pyramid of the nitrogen integrated into living matter, and the re-cycling of the ammoniacal nitrogen. An arrow near the top of the pyramid represents the "uptake" by fisheries.

We shall try to estimate the values in this scheme by considering a surface of 1 m^2 of the global ocean. The average depth of the sea is 3800 m (Sverdrup et al., 1946) and considering in general an average amount of 450 mg/m^3 of

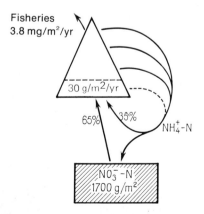

Fig. 3-20.—Scheme of the nitrogen balance for 1 m^2 of the global ocean (see text).

nitrogen (Vaccaro, 1962) the reserves per m² is of the order of 1700 g. Accepting a mean annual photosynthetic production in the ocean of 150 g/C/m² (Ryther, 1959) and an average ratio of 16 atoms nitrogen to 100 atoms of carbon, an average annual amount of about 30 g of nitrogen are integrated into the phytoplankton for each square metre which represents about 1/60 of the stock, a relatively high value. Vaccaro estimated the direct re-mineralization of ammonia by re-cycling as 35%. We have tried to estimate the uptake by fisheries. In 1971, 55×10^6 tons of marine fish (including salmon and shad which grow in the sea) were collected (annual statistics of fisheries of the F.A.O.) for a surface of the oceans and seas of 361×10^{12} m². Taking as a basis for calculation an amount of 2.5% of nitrogen (as a function of wet weight) in the whole fish (Vinogradov, 1953, Osterhout, 1970) this corresponds to nitrogen uptake of 3.8 mg/m²/year. It appears, therefore, that the annual incorporation of nitrogen in the vegetal matter is high while the annual uptake by fisheries is relatively small—of the order of one thousandth of the production. If, however, we make the same calculation for a zone with an active fishery such as the North Sea, supposing it to be arbitrarily separated from the rest of the ocean the balance is markedly different; for an average reserve of 14000 mg N/m² (an average depth of 100 m: spring amount 10 μg-at./l) the annual loss of nitrogen (calculated in page 46) is 77.5 mg/m² i.e., 1/180 of the stock which is substantial.

3-10.—MODES AND SIMULATIONS OF THE NITROGEN CYCLE

The rapid improvement in our knowledge of the nitrogen cycle and the rapid development of new analytical methods will presumably enable us to make studies on the nitrogen cycle using the techniques of models and the methods of simulation. Walsh & Dugdale (1971) investigated the Peruvian

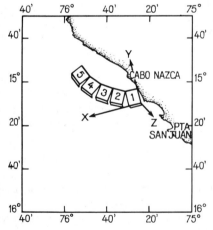

Fig. 3-21.—Location of the five spatial blocks used for simulating the development of the phytoplankton in the upwelling plume off Peru: each block is 11 km wide and they are situated along the axis of the plume formed by the upwelling of the water (after Walsh & Dugdale, 1971).

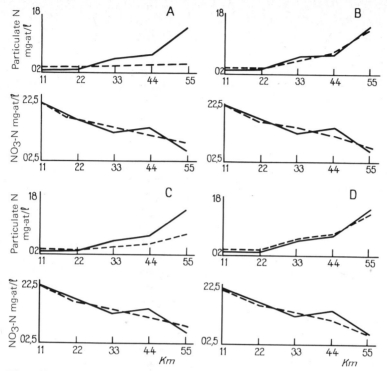

Fig. 3-22.—Comparison of observed curves (———) and simulated curves (----) for models with interference of different factors during day time: A, advection, diffusion and NO_3 uptake; B, to the factors in A gradient NH_4 uptake is added; C, sinking is added to the factors in B; D, grazing is added to the factor in B. On the abscissa kilometers from the origin of the plume; on the ordinate nitrate-nitrogen and phytoplankton nitrogen (after Walsh & Dugdale, 1971).

upwelling system. In a preliminary study they divided the upwelled water plume into five blocks 11 km wide and 10 m deep, (Fig. 3-21, compare with Fig. 4-16). Fig. 3-22 illustrates the results obtained. Taking into account advection and diffusion and the absorption of nitrate-nitrogen by phytoplankton gives only a poor agreement between the particulate nitrogen of the phytoplankton calculated by simulation and that observed. Adding to the model an absorption of ammoniacal nitrogen according to a gradient which increasing from the first block to the fifth as a function of the decreasing of amount of nitrate nitrogen, considerably improves the fit of the observed curve to that calculated. This enables us to consider the gradual absorption of ammonia as a working hypothesis. In the same way the addition to the simulation of the consumption of the phytoplankton by herbivores (grazing) seems to be interesting but a consideration of the sinking of the phytoplankton by gravity reduces the agreement between the curves. This mode of research is probably very expensive since it requires a lot of equipment for analysis and calculations but it seems to be very promising.

CHAPTER 4

PHOSPHORUS AND ITS CYCLE

4-1.—THE NEED FOR PHOSPHORUS

Phosphorus is an essential constituent of living matter, particularly as a component of adenosine triphosphate (ATP), of nucleic acids, and of phospholipids. The phytoplankton must find a source of phosphorus in the marine environment; since this element is found only in small quantities in the sea, its study is of great importance.

The experiments of Kain & Fogg (1960) using bacteria-free cultures of *Prorocentrum micans* give us a detailed example of the requirements for phosphorus. Phosphorus was added as dipotassium hydrogen phosphate and the number of cells (no/mm^3), after 23 days was found to be proportional to the initial quantity of phosphate (Fig. 4-1). This experiment also allows us to calculate the quantity of phosphorus needed for a cell, namely, 0.94 pg-at. (= 29 pg) and knowing the volume of the cell the quantity of phosphorus/mm^3 may be calculated. Some examples (in μg/mm^3) concerning different species are as follows;

Prorocentrum micans 0.055
Phaeodactylum tricornatum 0.011
Peridinium sp. 0.05
Isochrysis galbana 0.03
Asterionella japonica 0.05
Asterionella formosa 0.001 (fresh water diatom)

It should be noted that the amount of phosphorus in the cells may decrease when the source is inadequate as is shown by the values at the left of Fig. 4-1. Experiments with *Phaeodactylum tricornatum* enable us to obtain cells in which the ratio between carbon and phosphorus (in atoms) is 0.4% instead of the usual proportion of 1.8%. At a certain limit, cell division stops and at this point the amount of cellular phosphorus decreases to 0.002 pg-at. per cell (Kuenzler & Ketchum, 1962). Long before growth stops it is possible to distinguish a threshold in the concentration of phosphorus in the environment below which the growth rate begins to decrease. For *Chaetoceros gracilis*, Thomas & Dodson (1968) showed that this threshold is about 0.22 μg-at./l (about 7 μg/l). It is of the same order, 0.25 μg-at./l, for *Asterionella japonica* (Goldberg, Walker & Whisenand, 1951) and twice as high for *Phaeodactylum*, 0.55 μg-at./l (Ketchum, 1939).

With regard to the proportion of the phosphorus relative to nitrogen (the N/P ratio) it may vary markedly as shown in Table 4-1 where the extreme values are 5.4 and 53.0. Variations of the same order may occur within the

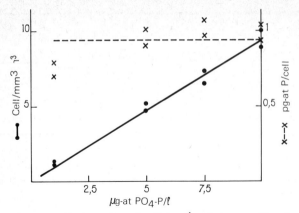

Fig. 4-1.—Final number of cells/mm^3 and the quantity of phosphorus/cell in a culture of *Prorocentrum micans* as a function of the initial quantity of PO$_4$–P added as dipotassium hydrogen phosphate (after Kain & Fogg, 1960).

TABLE 4-1

N/P ratios (in atoms) of phytoplanktonic organisms: the species, source of nitrogen, temperature in °C, and the reference are indicated.

	°C		N	P	N/P
Asterionella japonica$^{(+)}$		Kain & Fogg, 1960	0.27*	0.05*	5.4
Ditylum brightwellii$^{(\times)}$ (NH$_4$)	14°5	Strickland et al., 1969	19.3	1.8	10.7
Ditylum brightwellii$^{(\times)}$ (NO$_3$)	14°5	Strickland et al., 1969	17.5	1.9	9.2
Ditylum brightwellii$^{(+)}$	20°	Strickland et al., 1969	15.4	1.3	11.8
Prorocentrum micans$^{(+)}$		Kain & Fogg, 1960	1.1*	0.05*	20.0
Peridinium sp$^{(+)}$		After Kain & Fogg, 1960	0.47*	0.05*	9.2
Gonyaulax polyedra$^{(\times)}$ (NH$_4$)	22°	Strickland et al., 1969	11.1	0.7	15.9
Gonyaulax polyedra$^{(+)}$	22°	Strickland et al., 1969	12.0	0.29	41.4
Cochonina niei$^{(\times)}$ (NH$_4$)	20°	Strickland et al., 1969	13.9	1.7	8.1
Cochonina niei$^{(\times)}$ (NO$_3$)	20°	Strickland et al., 1969	12.4	1.2	10.3
Cochonina niei$^{(\times)}$ (deficiency)	20°	Strickland et al., 1969	7.0	0.73	9.6
Isochrysis galbana$^{(+)}$		Kain & Fogg, 1960	1.6*	0.03*	53
Natural phytoplankton		Harris & Riley, 1956			16.7

*Values in µg-at./mm^3 of algae: other values in µg-at./100 µg-at. C.
$^+$Laboratory experiments, small volumes and high concentrations of nutrient minerals.
$^\times$Experiments in deep tanks, the inorganic nutrient concentrations low and similar to natural conditions.

same species; for *Chlorella pyrenoidesa* Ketchum & Redfield (1949) obtained a N/P ratio of 5.6 for normal cultures, 30.9 for cultures with a phosphorus deficiency and 2.9 for cultures with deficiency in nitrogen.

4-2.—ABSORPTION OF PHOSPHORUS

The phytoplankton absorbs its necessary phosphorus essentially in the form of orthophosphate ion; anions of orthophosphoric acid H$_3$PO$_4$, namely,

H_2PO_4 and HPO_4^{2-} are almost exclusively the forms found in the sea*. Nevertheless, experiments have shown that besides inorganic phosphate, organic phosphates and particularly glycerophosphates can be used as a source of phosphorus by numerous species of unicellular algae, e.g., *Phaeodactylum tricornatum, Tabellaria flocculosa, Asterionella formosa*, (Provasoli, McLaughlin & Droop, 1957). According to more recent work, many planktonic forms are able to produce phosphatases associated with the cell surface and these enzymes gives rise to a phosphate ion which is assimilated and an organic fraction which remains in the environment (Kuenzler & Perras, 1965). Such a phosphatase is found in many species when the environment becomes poor in inorganic phosphate and conversely the addition of the latter blocks the production of phosphatase as shown for *Coccolithus huxleyi* in Fig. 4-2. Although it seems as if there is direct utilization of organic phosphate under these conditions it will only be inorganic phosphate which is absorbed.

The extracellular inorganic phosphate of the environment should be incorporated into the cells in the form of glucose-1-phosphate, which would be transformed into glycogen and polyphosphates. We should remember that volutine, a compound found in some vegetable cells, and for a long time enigmatic, is now considered as a form of polyphosphate.

Solorzano & Strickland (1969) using the diatom *Skeletonema costatum* and the peridinian *Amphidinium carteri* (both in bacteria-free culture) found that with excess of phosphate, 30% of the cellular phosphate was present in the form of polyphosphates. A deficiency of phosphate in the environment

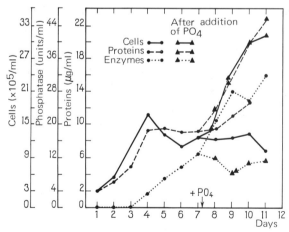

Fig. 4-2.—Variation of cell density, amount of protein, and enzymatic activity (phosphatase) during the growth of a culture of *Coccolithus huxleyi*: 1-l culture agitated several times during the day, containing initially 3.7 μg-at. PO_4–P at 20°C, under an irradiance of 0.02 cal/cm²/min: PO_4–P added at the 7th day to give 25 μg-at. PO_4–P: enzyme measured at pH 9.7 (after Kuenzler & Perras, 1965).

---

*In the following discussion the term phosphate unless otherwise specified will mean orthophosphate.

leads to the utilization of these polyphosphates. These two species are capable of using polyphosphates with chains of 70 atoms in bacteria-free cultures. In *Amphidinium*, the polyphosphates may be hydrolysed quicker than they are used and give rise to phosphates which accumulate in the medium.

4-3.—REMINERALIZATION OF PHOSPHORUS

In most of the organic compounds the phosphorus is present as PO_4 and, a priori, it may be thought that it could be more easily returned than nitrogen, often present in organic forms, to the marine circuit (Barnes, 1957). On the other hand, as we have just seen, the last stage of the remineralization may be carried out by the phosphatases of the phytoplankton just before absorption of the inorganic phosphate; in this case, for some of the phosphorus the remineralization may be largely hidden.

After its uptake by phytoplankton a part of the organic phosphorus may pass into the medium as indicated recently by Kuenzler (1970). For the diatom *Cyclotella cryptica*, Fig. 4-3 shows the fall in organic phosphate which follows the proliferation of the culture and the appearance in the medium of dissolved organic phosphorus; this then re-assimilated, partly

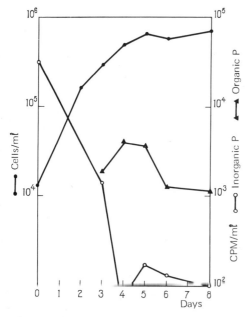

Fig. 4-3.—Growth, phosphate uptake, and dissolved organic phosphate excretion of *Cyclotella cryptica*; salinity 20‰; temperature 20–24°C; light intensity 0.2 cal/cm²/min: phosphate evaluated by measuring the quantity of radioactive phosphate, initially introduced as a tracer, in counts/min/ml (CPM/ml) (after Kuenzler, 1970).

due to the secretion of phosphatases, and this explains why the amount reaches a maximum and then decreases.

After passing through zooplankton the organic phosphorus is excreted and the importance of excretion in the rapid regeneration of phosphates has been demonstrated notably by the use of ^{32}P. The speed of the return is particularly striking since, according to Pomeroy, Mathews & Min (1963), 35–60% of the phosphorus of planktonic organisms is excreted each day in the form of inorganic phosphorus; these authors do not give details on the material used. Conover (1961) estimated the excretion in *Calanus finmarchicus* to be 10% of the stock of phosphorus of this copepod and suggested the rates would be higher in smaller planktonic organisms. This was confirmed by Johannes (1964) who show the speed of excretion, measured as the time required to excrete a quantity of phosphorus equal to the total amount in the body, increased considerably with a decrease in the size of the animal (Fig. 4-4): the protozoa renew their phosphorus 10 to 30 times during the day.

Pomeroy, Mathews & Min (1963) also found that in addition to the inorganic phosphorus a significant fraction of the phosphorus was excreted in organic form by the zooplankton.

When the organisms die an important part of their organic phosphorus, dephosphorylated by phosphatases, passes to the environment with production of orthophosphate. Hoffman (1956) observed, with organisms killed by

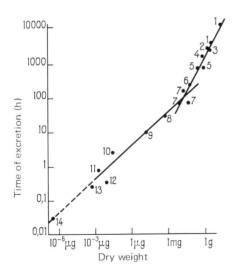

Fig. 4-4.—Relation between the time necessary for excretion of a quantity of phosphate equivalent to that found in the body, expressed in h (ordinate), and the dry weight of different marine animals (abscissa). 1, *Tridacna crocea*; 2, *Penaeus setiferus*; 3, *Crassostrea viginica*; 4, *Modiolus demissus*; 5, *Uca pugnax*; 6, *Salpa fusiformis*; 7, *Littorina irrorota*; 8, *Lembos intermedis*; 9, *Artemia salina* (nauplii); 10, *Euplotes crassus*; 11, *Euplotes trisulcata*; 12, *Euplotes vaunus*; 13, *Uronema* sp; 14, hypothetical phagotrophic flagellate of $1\,\mu^3$, volume (2.5×10^{-7} µg dry weight) (after Johannes, 1964).

heat or chloroform, that non-bacterial remineralization may account for 70% of the phosphorus in two days. Under natural conditions with bacteria present, it does not seem that the return of phosphorus to dissolved phosphate is so rapid. Skopintsev & Bruck (1940) found that in a sample of plankton at 18°C 60–70% of the organic phosphorus was re-mineralized after 54 days; the processes were more rapid in the first twenty days. This confirms the results of Cooper (1935) on phytoplanktonic diatoms, who found mineralization of the order of 50–60% in 3 to 4 weeks.

In the organic detritus there is a phosphorus fraction which is remineralized only after a long time and under bacterial action. It is agreed that bacterial action is necessary in order to complete remineralization but remembering the production of dissolved organic phosphorus (by diffusion from phytoplankton or excretion of zooplankton) as a result of the action of phosphatases and the excretion of inorganic phosphate, the rôle of bacteria does not seem dominant. We still need further investigations on the ecological importance of the different ways of remineralization.

It has long been considered that the degradation of plankton and marine organisms gives a more or less important quantity of organic phosphorus in solution. In 1955 Harvey insisted (Fig. 4-5) on the relative importance of this fraction in relation to inorganic phosphate, the phosphorus in detritus, and the living organisms. Fig. 4-5 show a seasonal variation increasing considerably after the spring bloom of the phytoplankton and decreasing in the

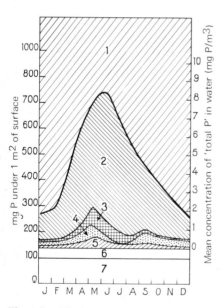

Fig. 4-5.—Distribution during the year of the phosphorus in a 70 m column of water in the English Channel: 1, dissolved inorganic phosphate; 2, dissolved organic phosphorus; 3, phosphorus in the detritus; 4, phytoplankton phosphorus; 5, zooplankton phosphorus; 6, fish phosphorus; 7, benthic fauna phosphorus (after Harvey, 1955).

autumn. Strickland & Austin (1960) found the same development in Canadian Pacific waters.

Some further details were added by Solorzano & Strickland (1969) on this dissolved organic phosphorus. First, as regards polyphosphates (measured as organic phosphate by the methods now available for analysis) they showed that only a small amount is present in this fraction except in the case of pollution. For the main part the dissolved organic phosphorus is true organic phosphorus and of this fraction only a small part is easily hydrolyzed by enzymes (Strickland & Solorzano, 1966): the dissolved organic phosphate given off, as already seen, by the phytoplankton (Kuenzler, 1970) might, therefore, be transient and rapidly transformed by phosphatases. The major part of the dissolved organic phosphorus is more difficult to hydrolyze; Solorzano & Strickland (1969) suggested that it is constituted largely of nucleic acid material.

4-4.—THE PHOSPHORUS CYCLE

Fig. 4-6 shows the phosphorus cycle in the sea as it appears according to recent work. The pathways which are in doubt are shown by a broken line, and each pathway is distinguished by a number relating to the following discussion.

1. This is the absorption by the phytoplankton of dissolved phosphates. In order to complete it we must remember the part played by benthic algae whose consumption of phosphate is significant in shallow water.

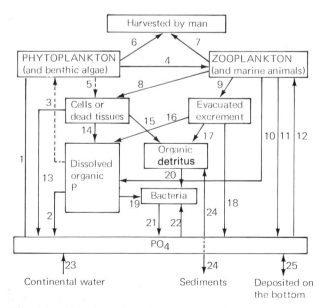

Fig. 4-6.—Phosphorus cycle in the sea; numbers refer to the explanation in the text.

2. The dissolved organic phosphate is able to be hydrolyzed giving inorganic phosphate. This last stage may be extremely small when the hydrolysis is by the phosphatases of the phytoplankton with the immediate absorption of the phosphates.

3. Under experimental conditions phytoplankton cells can absorb dissolved organic phosphorus, but this absorption may be attributed to the action of the above mentioned phosphatases and corresponds to the pathways 1 + 2.

4. Consumption of the phytoplankton by zooplankton. In order to simplify the scheme the benthic and nectonic marine animals are included together with the zooplankton.

5. It is agreed that in general all the cells of the phytoplankton are, at a certain time, removed by the zooplankton. Pathway 5 is, therefore, considered as purely theoretical.

6–7. We have mentioned the fisheries which do not directly concern plankton, but as a result of the activity of the algae, crustaceans, molluscs, fishes, and in certain limited areas, whales, this loss of phosphate may be quantitatively important.

8. Death and destruction of the zooplankton and of other marine animals.

9. This pathway might under certain conditions be of great importance. During the period of high proliferation of the phytoplankton the zooplankton and particularly the copepods consume more phytoplankton than they can digest and assimilate, and so it is evacuated as faecal pellets rich in diatoms or peridinians, ground but undigested or even still alive; however such superfluous feeding is in dispute (see Chapter 11-3).

10–11. Excretion of inorganic phosphate and dissolved organic phosphate.

12. There is a good reason to think that a certain amount of direct absorption of inorganic phosphate by the marine animals can occur; Marshall & Orr (1961) observed that *Calanus finmarchicus* placed in sterile water containing water labelled with ^{32}P produced radioactive eggs. This had already been observed in 1930 by embryologists working with fertilized eggs and the larvae of sea urchins. It is difficult to estimate the quantitative importance of this absorption which balances the pathway 11.

13, 14, 15. On the death of animals a part of the organic phosphorus is remineralized to orthophosphate and may be present as dissolved phosphorus in the environment (13) as in a dissolved organic form (14), and the rest remains in the organic detritus (15).

16, 17, 18. It is the same for the phosphorus from excretion or in faecal pellets; in these, orthophosphates are rapidly formed under the action of phosphatases of the cells.

19, 20. Dissolved organic phosphorus and organic phosphorus of the organic detritus are consumed by bacteria under conditions which are still inadequately known.

21. The phosphorus so transformed again enters into solution in the form of orthophosphate, or by bacterial phosphatase is transferred to the environment, either before or after being integrated into bacterial cells.

22. The importance of this pathway under natural conditions is still uncertain but it is well known under experimental conditions. In sea water stored in bottles there is a rapid increase in the bacterial population together with a decrease in phosphate and increase in oxygen consumed (Fig. 4-7). This indicates how necessary it is to 'fix' a sample when it is to be used for phosphate analysis when it cannot be estimated at once. The consumption of phosphate is a function of the organic matter available for the bacteria in the medium, as shown in the results of the experiment given in Fig. 4-8.

23. The continental water (rivers, drains) contribute to the enrichment of the sea in phosphate, at least in the littoral zone.

24. In shallow waters the sedimentation of organic detritus leads to a loss

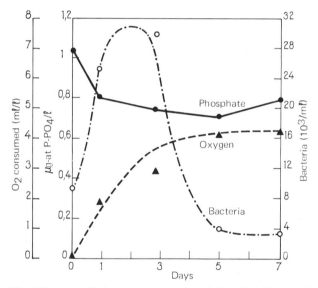

Fig. 4-7.—Assimilation and regeneration of phosphate by bacteria in sea water kept in bottles after enrichment with phosphate: phosphate determinations are in triplicate, others in duplicate (after Renn, 1937).

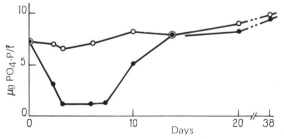

Fig. 4-8.—Changes in phosphate concentration during storage in a sample of sea water filtered through Whatman No. 3 paper: open circles without addition; solid circles with the addition of 2.5 mg glucose and 0.25 mg of $(NH_4)_2 SO_4/l$: this increased the bacterial growth due to the addition of phosphorus free nutrient (after Harvey, 1955).

of phosphorus; such sinking may only be temporary, an increase in turbulence being again able to circulate the detritus, so that the mechanism can be reversed.

25. The sediment and interstitial water contain phosphorus, part of which is soluble. Arrhenius (1952) found at a depth of several thousand metres in the pacific there was 0.2–0.9 g of phosphorus/kg in the upper layer of the sediment. The shells found at moderate depths contain about 0.3% of phosphorus (Harvey, 1955). There is no doubt that the phosphorus of the sediment can be taken up again to the overlying water, but we have no data on the importance of this phenomenon under natural conditions (Barnes, 1957).

4-5.—ESTIMATION OF PHOSPHORUS

The estimation of phosphorus is based on a reaction studied by Deniges, adapted for sea water by Atkins and improved particularly by Harvey, Cooper & Armstrong. It is a colorimetric method, in which a phosphomolybdate complex is formed by the addition of ammonium molybdate in acid solution; on reduction this phosphomolybdate complex gives blue colour and ascorbic acid is used for this reaction. This molybdenum blue is developed even when only traces of phosphate ions are present.

For more details on this technique and for the method for dissolved organic phosphate analysis in sea water see Strickland & Parsons (1968). Stephens (1970) automated this method enabling a range of 0.5 to 5.0 μg-at. PO_4–P/l (15.5 to 155 μg PO_4–P/l) with a standard deviation of 0.06 μg-at. PO_4–P/l on 10 samples at a level of 2.0 μg-at. PO_4–P/l to be determined.

Although the method of estimation of inorganic phosphate is satisfactory and widely used there is a suspicion that this method is able to estimate certain organic phosphates. Strickland & Parsons (1968) distinguished, therefore, between the dissolved inorganic phosphate (D.I.P.) and soluble reactive phosphate (S.R.P.) defined as the phosphate present in a sample of water which reacts in 5 min with the acidified molybdate. Observations suggest that the dissolved organic phosphate and soluble reactive phosphate are not always equivalent. This lead Chamberlain & Shapiro (1969) to compare different procedures of analysis and they established a bioassay for the estimation of the biologically utilizable phosphate. Their work concerns, however, only lake water.

4.6—DISTRIBUTION OF PHOSPHORUS IN THE SEA

Due to the ease of analysis there are many investigations on the distribution of phosphorus and the general outline of its distribution in the sea is clearer than for nitrates.

In cold or temperate zones where the phytoplankton bloom is clearly limited, a very marked difference is observed in the distribution in the depth according to the season. Fig. 4-9 gives a typical example of the conditions in which near the bottom the amount is almost the same in the two seasons. From year to year depending on the meteorological and in turn hydrological conditions the amount may vary to a certain degree but two principal features remain, namely large variations during the year in the superficial water with a minimum in the summer and a maximum in the winter and the amount is always higher in the depths where the variations decrease in amplitude. Fig. 4-10 shows these two phenomena. At a great depth there are no seasonal variations in phosphate and as for nitrate the deep water constitutes an important reserve. This is seen in Fig. 4-11 which represents a transect made in the south Atlantic along 30°W from 35°S: below 1000 m the water almost always has >50 µg/l, whereas at the surface in the same area the amount is zero over the same period.

If now we consider the central zone of the three oceans we reach the same conclusion as for nitrates: the phosphorus in the deep water of the

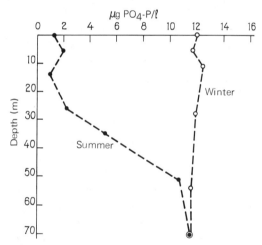

Fig. 4-9.—Distribution of inorganic phosphate (PO_4–P) as a function of depth 20 miles off Plymouth in the winter and summer (5th January, 1949, and 31st August, 1948) (after Armstrong & Harvey, in Harvey, 1955).

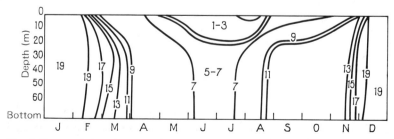

Fig. 4-10.—Variations in inorganic phosphate during 1924 at a station in the English Channel: numbers indicate the amount in µg PO_4–P/l (after Harvey, 1955).

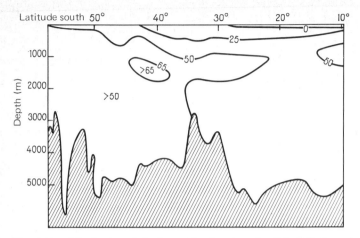

Fig. 4-11.—Distribution of the inorganic phosphate (μg PO_4–P/l) along the meridian 30°W in April–May 1931, in the South Atlantic: the Antarctic Convergence is situated at about 50°S and the Subtropical Convergence at about 41°S (after Deacon, 1933).

Indian and Pacific reaches 3 μg-at. PO_4–P/l (93 μg/l) and are clearly richer than those of the Atlantic where it does not exceed 15 μg-at. PO_4–P/l in its central part (Fig. 4-12); but even this latter shows a marked contrast relative to the poverty of the deep water of the Mediterranean (Fig. 4-13).

At certain periods the superficial water may be almost so completely depleted of phosphate that it becomes unfavourable to the multiplication of phytoplankton even though the deep layers still contain important reserves. If, for some reason, the deep water is carried to the surface, into the photic zone where photosynthesis is active, there will be considerable increase in the production in such an upwelling area, due to nitrate and phosphate being made available in abundance.

Fig. 4-14 shows the distribution of phosphates for the central Pacific on a south-north transect down to 300 m. Two divergences are evident, one at the equator and the other at the northern limit of the Countercurrent; they coincide clearly with an increase of phosphates in the sub-superficial layers. In the Mediterranean the deep waters are poor in phosphorus (0.3–0.4 μg/at./l) and in the superficial water it is almost always at the limit of detection. A mixing of the deep water with shallow water is seen in the rising of the isopleths as evident in the transect in Fig. 4-15.

Without doubt the clearest example of the transport of water rich in phosphate to the surface is seen in the upwelling off Peru. Fig. 4-16 shows amounts >2.5 μg-at. P–PO_4/l at 3 m depth the same as found at about 1000 m depth in Fig. 4-12. The Peruvian upwelling corresponds to a considerable production of anchovy (*Engraulis rigens*).

Solorzano & Strickland (1969) have given some indication of the amount of polyphosphates which, as has been seen, may be used by cells. It is only on the occasion of rapid multiplication of phytoplankton or of pollution that these polyphosphates have any importance (from 0.06 to 0.13 μg-at. P in the port of San Diego, California).

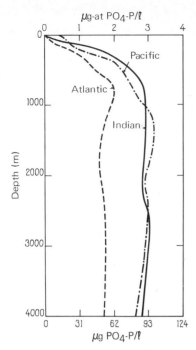

Fig. 4-12.—Vertical distribution of inorganic phosphate in the three big oceans (after Sverdrup et al., 1946).

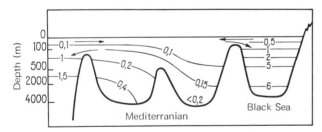

Fig. 4-13.—Diagram showing the distribution of inorganic phosphorus (μg-at. PO_4–P/l) in the Black Sea, the Mediterranean, and in the coastal Atlantic: Arrows indicate current directions (after Redfield, Ketchum & Richards, 1963).

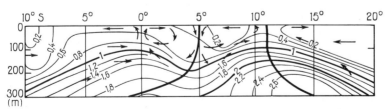

Fig. 4-14.—Distribution of inorganic phosphate (μg-at. PO_4–P/l) in a vertical section between 10°S and 20°N along the meridian 140°W, in the Pacific Ocean during October 1929 (Carnegie observations): heavy line indicates currents directed to the west on both sides of the lines; in the centre between the lines is the equatorial counter-current, directed to the east: arrows indicate the flow of water superposed on the general currents (after Sverdrup et al., 1946).

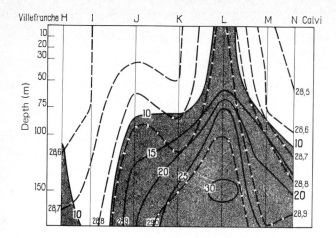

Fig. 4-15.—Vertical section of the superficial water from stations H to N, between Villefranche and Calvi (Corsica) in February 1964 showing the parallelism of the isopycnes (in broken lines) and the distribution of dissolved inorganic phosphate (μg-at. PO_4–P $\times 10^2$, solid line) indicating a mixture, at point L, of superficial water with deeper water by vertical turbulence (after Gostan, 1968).

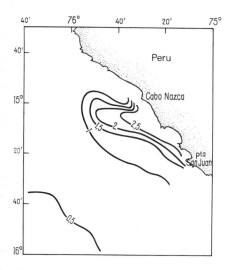

Fig. 4-16.—Distribution of phosphate (μg-at. PO_4–P/l), at 3 m, at the level of upwelling in Peru in April 1969: day observations (after Walsh, Kelley, Dugdale & Frost, 1971).

4-7.—LONG TERM VARIATIONS IN AMOUNT OF PHOSPHORUS

The methods for the measurement of phosphorus were mainly developed in the marine laboratory at Plymouth by Atkins, Harvey, Cooper and Armstrong and, since 1921, therefore, we have much data on the water off the coast of Plymouth. From 1921 to 1928 the mean of spring maximum was established at ≈20 µg/l. Since 1930 this value has shown a clear decrease to ≈14 µg/l (Fig. 4-17). Due to turbulence in the superficial water the amount there is a fraction of the quantity of phosphorus at a depth of about 400 m at the entrance to the Celtic Sea. In order to find water containing about 20 µg/l of phosphate after 1930 it is necessary to go down to 500 m depth, and this explains the impoverishment of the superficial water in phosphorus.

This impoverishment was considered by Cooper (1955) with respect to the decrease in the dog-fish (*Squalus acanthias*) fishery in ports of the English Channel and the Celtic Sea and also with the decrease since 1930, of the herring fishery, which was originally very rich during the New Year period near Land's End. These two changes may be considered as an indication of the decrease in the general production at the entrance to the English Channel and to be correlated with the decrease in available phosphorus. Cooper suggested there was a relation between this decrease in production and the milder temperatures observed since 1930 in the Arctic. Cooper's argument is as follows. The extreme temperatures lead to a greater formation of cold, heavy water in the Norwegian Sea and this overflows to the Atlantic through the shelf which exists between Island and Greenland. This overflow should be discontinuous and should develop internal waves which spread in the North Atlantic: the turbulence produced by the arrival of this wave near the Celtic Sea should cause a rise to less than 400 m of layers rich in phosphate and in consequence the enrichment of the water of the continental shelf. According to this hypothesis, therefore, a period of very low temperatures in the Arctic should be related to a period of higher production at the entrance of the English Channel.

4-8.—REGENERATION OF NITRATE AND PHOSPHATE

The degradation of the organic matter in the sea with the regeneration of inorganic carbon, nitrogen, and phosphorus may be compared with a continuous combustion which requires a certain amount of oxygen used

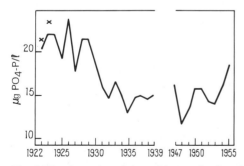

Fig. 4-17.—Annual variation of inorganic phosphate 22 miles S–W of Plymouth during the winter maximum: in 1922 and 1923 the maxima have probably been missed and were without doubt greater than the values indicated (after Cooper, 1955).

in the respiration of the organisms at the different trophic levels. Although we have seen that as regards the absorption of nitrogen and of phosphorus, there are differences as there are in the amounts of carbon, nitrogen and phosphorus of different organisms of the phytoplankton, and that there is a somewhat great variation depending upon the available material, it is possible to determine a mean value for the elementary composition of phytoplankton. Fleming (1940) published the following values expressed in atom ratios as a summary of much data:

	C	N	P
Phytoplankton	108	15.5	1
Zooplankton	103	16.5	1
Mean	106	16	1

This agrees with the values of the ratios N/P given by Harris & Riley in Table 4-1. Adopting these mean values, and taking 2 atoms of oxygen per atom of carbon and 4 atoms of oxygen per atom of nitrogen for combustion, 276 atoms of oxygen are necessary completely to oxidize the organic matter containing one atom of phosphorus (or 16 atoms of nitrogen). These are only mean values; taking into account variations in the composition of the plankton Riley (1951, 1956), considered that this ratio may normally vary between 250 and 300, but the variations may even be greater.

Below a certain depth, when exchange with the atmosphere and photosynthesis are absent, it should be possible to find a relation between the oxygen used for respiration of the organisms and the mineral regeneration of carbon, nitrogen, and phosphorus. A relatively easy estimation of the oxygen used in a given case may be obtained by comparing the measured amount of oxygen and the calculated amount for the water, assuming that the water is saturated with oxygen at its given salinity and temperature. This difference is called apparent oxygen utilization (AOU). Figure 4-18 gives the results of some analyses of water from the western Atlantic. On the left it appears that the N/P ratio (in atoms) is very close to that found for the elementary composition of plankton. On the right the ratio of the apparent used oxygen (AOU) and the nitrate (138 μM O_2/2.22 μg-at. NO_3–N) above a level of 1000 m is rather far from the theoretical value of 276/16 (138 μM O_2/16 μg-at. NO_3–N) given above. In reality, two fractions of nitrogen and phosphorus in the water must be considered, namely, a preformed fraction present in the water when it left the surface and a fraction regenerated by oxidation. In Fig. 4-18B the solid circles correspond to water taken at more than 1000 m depth where the preformed fraction is higher than at the other levels (the deviation from the line which represents the ratio 138 μM O_2/16 μg-at. N is higher).

Fig. 4-19 shows the variations of the two fractions of phosphorus for two stations in the North Atlantic, one tropical and the other at a high latitude. The importance of differences in the relative proportions of regenerated phosphorus and preformed phosphorus is dependent on the region. In the upper part (A) of Table 4-2 it is supposed that the total inorganic phosphorus

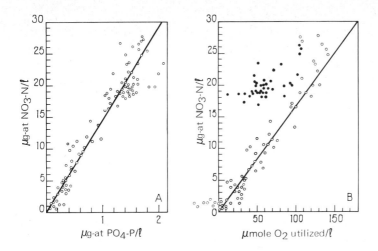

Fig. 4-18.—A, correlation between nitrate-nitrogen and phosphate-phosphorus in waters of Western Atlantic: B, correlation between nitrate-nitrogen and apparent oxygen utilization (AOU) in same sample: solid circles, samples below 1000 m-level; open circles, samples from above 1000 m: in A the lines correspond to 1 μg-at. PO_4–P/15 μg-at. NO_3–N and in B to 138 μmoles O_2/22.2 μg-at. NO_3–N (after Redfield 1934, in Redfield, Ketchum & Richards, 1963).

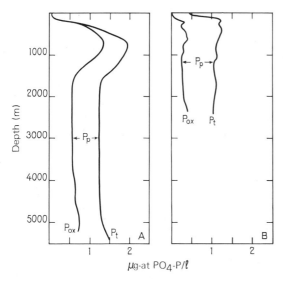

Fig. 4-19.—Distribution of phosphates at two stations in the North Atlantic: P_t, total phosphate; P_p, preformed phosphate; P_{ox}, phosphate regenerated by oxidation; A, Crawford Station 364, 16°16′ N: 54°44′ W; B, Discovery Station 3849, 55°25′ N: 33°12′ W (after Redfield, Ketchum & Richards, 1963).

TABLE 4-2

Comparison of the amount of oxygen in the deep water of the North Atlantic and in the North Pacific, assuming an absence of preformed phosphate (A) or (B) allowing for the preformed phosphate present, in estimating the equivalent oxygen utilization: values in μg-at./l (after Redfield, Ketchum & Richards, 1963).

	North Atlantic	North Pacific
A. 1. Total phosphate	1.25	3.00
2. Equivalent O_2 utilization ($A_1 \times 276$)	345	828
3. O_2 saturation value	735	735
4. Excess $O_2(A_3-A_2)$	390	−93
5. O_2 saturation	53%	−13%
B. 1. Preformed phosphate	0.75	1.50
2. Regenerated phosphate (A_1-B_1)	0.50	1.50
3. Equivalent O_2 utilization	138	414
4. Excess O_2	597	321
5. O_2 saturation	81%	44%

observed in the depths comes from regeneration; this leads in the north Pacific to a total disappearance of oxygen. In the lower part of the Table (B) the preformed phosphorus is considered and the calculated percentage in saturation of oxygen corresponds effectively to the observations. With only 44%, the north Pacific appears, therefore, much more vulnerable to an approach to an anoxic condition than the North Atlantic. Effectively, in the northern part of the Pacific there are wide zones where the amount of oxygen is very low.

This kind of oceanic biochemistry seems to have made only little progress over the past few years. It requires the cooperation of biologists and of physical chemists and its understanding will influence in a large measure the future exploration of the world ocean. It certainly gives rise to many problems (Rotschi, 1961; Pytkowitcz, 1971). For more details on the cycle of nitrogen and phosphorus see the recent review by Corner & Davis (1971).

CHAPTER 5

SILICON, OLIGOELEMENTS, AND GROWTH FACTORS

5-1.—SILICON AND SILICATE

In order to make their silicates frustules, diatoms must have sufficient quantity of available silica since silica may constitute more than 60% of their inorganic fraction and 15–20% of the dry weight (Parsons, Stephens & Strickland, 1961). The silica also forms the skeleton of silicoflagellates. Apart from phytoplankton, silica is also found in the skeleton of Radiolaria and in the spicules of numerous sponges.

In spite of the abundance of silicates in the earths crust it is remarkable that sea water is notably undersaturated in silica. The solubility of silica is of the order of 100 mg/l while the highest values found in the sea are 8–9 mg/l of SiO_2 (4 mg Si/l).

The estimation is a colorimetric one with the production of silico-molybdate blue by the addition of acid molybdate. It is agreed that besides silica in solution a small fraction of colloidal silica is also measured (Armstrong, 1965) but this is, however, utilizable by diatoms.

Apart from dissolved silica more or less high quantities of suspended silica in particulate form are found in the sea; these are of the order of mg/l (0.5 to 2.0 mg/l in the English Channel); the particles contain 17–55% of SiO_2, representing 37 to 410 μg/l of silicon, i.e., a quantity equal or higher than that found in solution (Armstrong, 1965). In deep water the proportion of silica in suspension decreases considerably and represents only 2 to 10% of the dissolved silica because of the decrease in particulate matter and the considerable increase of silica in solution. Indeed, it is in the depths that the maximum amounts are found, exceeding 4000 μg Si/l in the North Pacific below 2000 m. Fig. 5-1 shows the differences between various regions; the North Atlantic appears to be relatively poor in silica.

In shallow water the amount may decrease considerably because of the utilization of silica by diatoms and a seasonal cycle rather similar to that of the phosphorus is found. Fig. 5-2 shows two examples of this for a station at the entrance to the English Channel opposite Plymouth, one of the best studied regions as regards silica. The winter values vary from 70 to 200 μg Si/l, which in summer decrease to few tens μg/l and low values of 1–2 μg Si/l have sometimes been observed (Armstrong, 1965), it is possible, therefore, that silica can limit the production of the diatoms.

At Friday Harbor, near Seattle, on the Pacific coast of the United States seasonal variations were observed by Phifer & Thompson (1937) but the summer minimum remained above 100 μg Si/l (winter maximum 145–190 μg

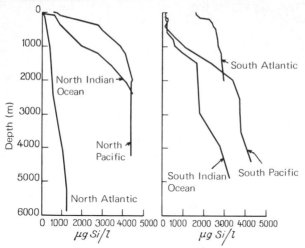

Fig. 5-1.—Vertical distribution of silica in the oceans (after Armstrong, 1965).

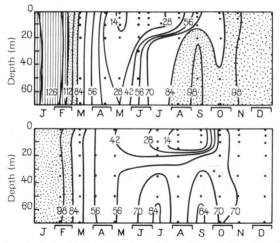

Fig. 5-2.—Vertical distribution of silica (μg Si/l) 22 miles off Plymouth, in 1959 (upper) and in 1960 (lower) (after Armstrong & Butler in Armstrong, 1965).

Si/l); this high value, which is also found for nutrients, is attributed to upwelling and to a high degree of turbulence which redissolves silica from the sediments.

In coastal regions besides such solution there is also the interference of transport by continental water. The rivers of the Gulf of Bothnia contain 1400 to 2800 μg Si/l and a decreasing quantity is found from north to the south in the Baltic Sea (Voipio, 1961 in Armstrong, 1965). At the entrance to the English Channel there is a clear inverse relation between salinity and silica as shown for February 1961 in Fig. 5-3.

As is the case for nitrogen and phosphorus the silicon taken up by the diatoms is returned to the environment, but the consumers of phytoplankton (with the exception of the radiolarians and silicious sponges which are

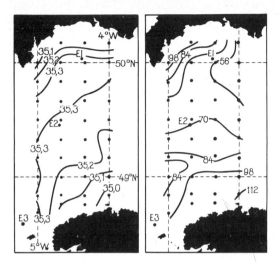

Fig. 5-3.—Comparison of the distribution of salinity (left) and of silica (μg Si/l), (right) at 10 m depth at the entrance to the English Channel in February 1961 (after Armstrong & Butler, in Armstrong, 1965).

benthic) need only a very small quantity of silica, far different from the need for nitrogen and phosphorus. Certain authors relying on laboratory experiments consider, therefore, that after its utilization the silica rapidly returns to the water. Kamatani (1971) showed that simple heating of the water to 100°C, in 2 h dissolves 80% of the silica in *Skeletonema costatum*; however, other experiments showed a much slower re-solution of silica which may be explained by specific differences other than observed by Kamatani. It seems that our knowledge on this subject is far from complete (Armstrong, 1965), and this makes it difficult to estimate the role of silicon as a limiting factor in diatom growth. We may note that in certain cases the frustules of the diatoms are able to pass through a large water column without being dissolved and reach the sediments in the depths where they constitute the diatom ooze. It might be that we have here a phenomenon of the same kind as was considered by Wangerski (1969) for carbonate particles; these, covered with an pellicule of organic matter do not dissolve in waters which are undersaturated with respect to carbonate.

Hamilton (1969) has studied the problem of silicon deficiency in lake waters. After adding sodium silicate to water with a low amount of silica (close to 0 μg-at./l) he obtained a stimulation of photosynthesis (measured by ^{14}C).

5-2.—IRON

We shall now consider the oligoelements, namely, iron manganese, and copper.

Iron is found in small quantities in cells and it is a necessary oligoelement for cultures of unicellular algae as shown by the experiments of Ryther & Kramer (1961), where the maximum development of various cultures was obtained on the addition of certain quantity of iron (Fig. 5-4). It appears that the requirement is not the same for all species and it is possible to distinguish coastal species such as *Skeletonema costatum* which require a considerable quantity of iron, and oceanic species such as *Chaetoceros lorenzianus* which need much less. This goes hand in hand with the existence in coastal waters of larger amounts of iron than those found in the open sea.

The experiments reported in Fig. 5-5 are also important. They were made in water from the Peru upwelling which is rich in phosphates, nitrates, and silicates; after adding to filtered water from 75 m depth 10% of unfiltered surface water, with its natural phytoplankton the reserves of the deep water were rapidly used; however, the addition of deferriferrioxamine B, which specifically blocks the iron and makes it unavailable, considerably reduced the growth.

Any study of the distribution of iron, in order to understand its rôle in marine waters, encounters difficulties of two kinds; the first, is analytical problems and one of methods. Armstrong (1957) uses quartz containers with the addition of HCl and digestion for 5 h at 140°C in an autoclave; secondly, is the high variability of the results. At Station E, near the Eddystone Lighthouse at the entrance to the English Channel 10 successive samples of water collected in 15 min gave a mean value of 65 µg/l with a variation of 22 µg/l. In a station in the Gasconian Gulf the amounts obtained at the different levels were as follows, expressed in µg of iron per litre: 0 m,

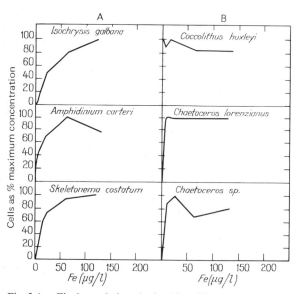

Fig. 5-4.—Final population obtained for different species of phytoplanktonic algae, coastal (A) and oceanic (B), as a percentage of the maximum concentration obtained with different concentrations of iron (added as Fe–EDTA) (after Ryther & Kramer, 1961).

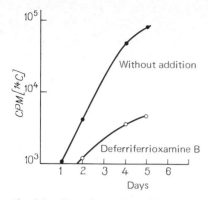

Fig. 5-5.—The effect of blocking of iron by deferriferrioxamine-B on the development of natural phytoplankton (estimated by the absorption of ^{14}C in CPM) in the water of the Peru upwelling: 90% of filtered water from 75 m were added to 10% of non-filtered water from the surface: 75 m-water had 26.08 µg-at. NO_3–N/l, 2.5 µg-at. PO_4–P/l, and 23.41 µg-at. SO_4–Si/l: 0 m-water had 21.99 µg-at. NO_3–N/l, 2.09 µg-at. PO_4–P/l, and 12.16 µg-at. SO_4–Si/l (after Barber, Dugdale, MacIsaac & Smith, 1971).

23; 10 m, 7; 100 m, 7; 200 m, 146; 300 m, 15; from 400 m to 4600 m the values varied between 4 and 27. At 200 m the amount was of a different order of magnitude. This variability may be due to the fact that unlike nitrates, phosphates and silicates only a small part of the iron is in solution. According to Cooper (1948) iron is found in the following forms:

1) ionic iron; Fe^{2+}, Fe^{2+}, $FeOH^{2+}$, in true solution and this does not exceed 10^{-8} µg-at./l.

2) colloidal inorganic iron; ferric hydroxide ($Fe(OH)_3$) and ferric phosphate ($FePO_4$) in colloidal micelles or larger agglomerates, associated with mucilaginous colloids which maintain it in this temporary state; this iron has a tendency to precipitate. The colloidal inorganic iron should form a unicellular film of ferric hydroxide on the sea surface, where it is often found in higher amounts and it may cover the surface of the living diatoms and other elements of the phytoplankton. It is also found in faecal pellets.

3) Organic iron; this is the iron which is associated with stable organic compounds.

4) Terrigenous iron; the iron which is found in the clay and inorganic matter coming from the land.

The main part of the iron which is estimated is found in the form of colloidal inorganic iron and, therefore, in the form of particles; this explains the variability of the analyses and also the necessity to dissolve the iron by a vigorous treatment during the analytical procedure at a low pH. In practice, current work separates arbitrarily the "particulate iron" retained by a 0.5 µm filter from the "dissolved iron" which passes through this filter and contains, besides true dissolved iron, very fine particles. In the water of Southampton Harbour, Head (1971) found 100–1000 µg/l of particulate iron and values of dissolved iron of the order of 10 µg/l. In the Northeastern

Atlantic he found 2.5 to 50 µg/l of particulate iron. It is still very difficult to estimate the relation between iron and primary production under natural conditions.

5-3.—MANGANESE

Harvey (1937) showed the importance of manganese for certain marine organisms of the phytoplankton in particular for *Dunaliella tertiolecta*.

The experiment which is shown in Fig. 5-6 illustrates this phenomenon. A strain of *Dunaliella tertiolecta*, kept for a long time in artificial water under the laboratory conditions, was transferred to coastal water to which phosphate, nitrate, and iron (in the form of citrate) were added. The most vigorous cultures were those which were cultured in water with an initially high quantity of manganese. The effect is very sensitive; with the addition of 0.1 µg/l, using this stock of *Dunaliella tertiolecta* Harvey could prove experimentally the adsorption of manganese on the organic detritus. The same results were obtained by Harvey with a marine *Chlorella*; however, with *Coscinodiscus excentricus* he could not show any significant effect.

The estimation of manganese in sea water is difficult and various methods have been proposed (Riley, 1965). The values obtained are generally between 0.5 to 3 µg/l, sometimes with values up to 10 µg/l. According to Hood et al. (in Riley, 1965) an important part of the manganese in sea water should exist in the form of organic complexes. Notable quantities are often associated with particles in suspension and the plankton, and the distribution of this element is often as erratic as that of iron.

In open waters McAllister, Parsons & Strickland (1960) found low values, 0.3 µg/l; referring to the results given in Fig. 5-6, it is possible to believe that in these waters deficiencies in manganese could limit the growth of certain elements of the phytoplankton.

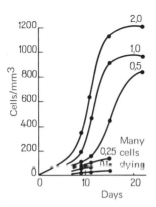

Fig. 5-6.—Growth of *Dunaliella tertiolecta* in coastal sea water enriched with nitrate, phosphate, and iron, and addition of 0.1, 0.25, 0.5, 1.0 and 2.0 µg/l of manganese (after Harvey, 1947).

According to Wangersky & Gordon (1965) the dissolved Mn^{2+} ions could be incorporated onto organic aggregates and when these sediment Mn^{2+} ions sink to the bottom where the organic matter is consumed and the Mn^{2+} ion is liberated afterwards to be oxydized to give nodules.

5-4.—COPPER

Copper is often added to algal cultures in order to ensure their development. On the other hand, the toxicity at high concentration to many organisms is well known as is the capacity of numerous organic substances to complex with it or to chelate it and so mask its biological effects (Bougis, 1962). It is accepted that in cultures it is available to the cells only in minor quantities because of the complexing substances found in the medium which prevent it from being toxic.

Recently Steemann–Nielsen & Wium–Andersen (1970) showed that in cultures in which copper was in the ionic form, it could exercise a toxic effect on photosynthesis at very low levels, of the order of several $\mu g/l$ (Fig. 5-7). These two authors suggested that the observation made by Barber & Ryther (1969) in the region of Cromwell Current (in the Pacific) that the upwelling waters are not at the first very productive, may possibly be explained by the necessity first of all for the complexing of copper in this water.

5-5.—VITAMIN B_{12}

Certain vitamins, mainly Vitamin B_{12} and Vitamin B_1 are found in the sea and can interfere with the growth of the phytoplankton. Vitamin B_{12}, also called cobalamine since it contains a small proportion of cobalt, is synthesized by certain micro-organisms and has to be supplied to others, which allows its biological estimation. In 1955 Droop showed that this vitamin was

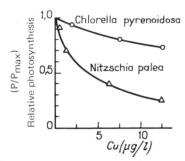

Fig. 5-7.—Relative photosynthesis as a function of the quantity of ionic copper added to the medium for *Chlorella pyrenoidosa* and *Nitzschia palea* (after Steemann–Nielsen & Wium–Andersen, 1970).

necessary for a typical planktonic diatom *Skeletonema costatum*. About 70% of the phytoplankton cultures have to have an external supply of vitamin B_{12}. To estimate vitamin B_{12} in the sea different organisms such as *Lactobacillus leichmannii*, *Euglena gracilis* from clone 3 and now the marine diatom *Cyclotella nana*, clone 3H, which is sensitive in the range of 0 to 2 ng vit. B_{12}/l (Ryther & Guillard, 1962) are used. In 1956, Cowey showed, largely in the North Sea, that the amount of vitamin B_{12} may vary during the year from 0.1 to 2 ng/l; the amount decreases in the spring and increases in the autumn.

Menzel & Spaeth (1962) found that although absolute values are small variations of the same order are present in Sargasso Sea and these go hand by hand with the development of primary production (Fig. 5-8). They also found that the values increase towards the bottom with a maximum at about 500 m. This increase with depth was also noted in the North Pacific where, however, between 100 and 200 m (Fig. 5-9), the amounts are much higher

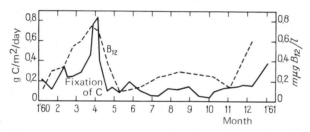

Fig. 5-8.—Comparison of the primary production and the mean concentration of vitamin B_{12} in the photic zone in the Sargasso Sea off Bermuda (after Menzel & Spaeth, 1962).

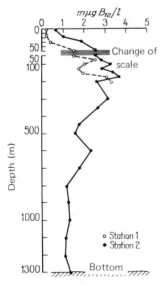

Fig. 5-9.—Vertical distribution of vitamin B_{12} in two stations south of Los Angeles in May 1965 (after Carlucci & Silbernagel, 1966).

than 2 ng/l. In this region, this vitamin may be undetectable in the surface waters.

It may be concluded that in its distribution vitamin B_{12} shows seasonal cycles increasing at moderate depths; however, it seems that the attempts to show a stimulation of the production by enrichment of water with their contained phytoplankton have not been successful. The question, posed by Droop in 1961, whether vitamin B_{12} is a limiting factor for the multiplication of phytoplankton remains open.

5-6.—VITAMIN B_1

Vitamin B_1, or thiamin chloride, constitutes in the form of its pyrophosphoric ester a coenzyme, carboxylase, which is important in metabolism. McLaughlin and then Droop (1958a) showed that vitamin B_1 was necessary for several marine protists. Among those which showed a requirement for thiamin are Haptophyceae such as *Prymnesium parvum*, Crysophyceae such as *Monochrysis lutheri* and Cryptophyceae such as *Hemiselmis virescens*. Other organisms including the diatoms *Phaeodactylum tricornutum* or *Skeletonema costatum* and peridinians *Peridinium trochoideum*, do not show this requirement.

The thiamin molecule has a thiazol moiety which contains nitrogen and sulphur and a pyrimidine moiety containing nitrogen (Fig. 5-10). Droop tried the separate effect of these two parts of the thiamin molecule on organisms which require a supply of vitamin B_1. Fig. 5-11 shows his results; addition of pyrimidine alone did not improve the growth of a culture of *Hemiselmis virescens*, but the addition of thiazol had, on the contrary, a positive effect. It is concluded that in this case the thiazol part of the molecule is necessary and is supplied by thiamin. In *Monochrysis lutheri* and *Prymnesium parvum* the results were the opposite, i.e., the pyrimidine nucleus of the thiamin was necessary for their growth.

Different organisms have been proposed for the biological estimation of thiamin in sea water; recently the marine yeast, *Cryptococcus albidus*, in which the growth is almost linearly related to the vitamin content in the range 10 to 300 ng/l, has been used and this enabled Natarajan & Dugdale (1966) to measure the amount of his vitamin in the water of the Northeastern Pacific. They found surface concentrations from 0 to 500 ng/l, with the

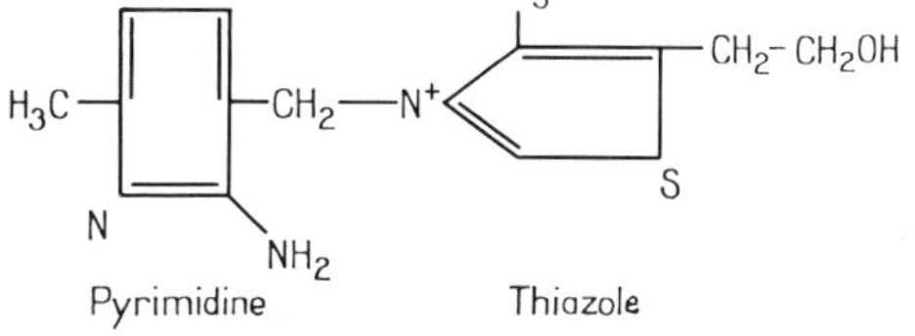

Fig. 5-10.—Structure of thiamine.

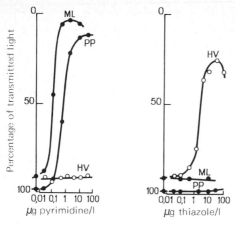

Fig. 5-11.—Response to pyrimidine and to thiazole of cultures of *Monochrysis lutheri* (ML), *Prymnesium parvum* (PP) and *Hemiselmis vurescens* (HV): density of culture measured optically (after Droop, 1958a).

coastal water being generally richer. The amounts usually decreased with depth up to 50 or 75 m and then remained at the limits of the detection.

Seasonal variations of thiamin are found which are generally positively related to high primary production which explains its superficial maximum. On the other hand, near the Aleutian Islands Natarajan (1970) observed, a significant negative correlation with the amount of NO_3–N, PO_4–P, and silicates. It seems quite certain that the thiamine is related to the primary production of the phytoplankton, but whether the relation is causal is unknown. It should, however, be mentioned that in water relatively poor in thiamin Natarajan obtained in vitro, by enrichment with this vitamin, a doubling of the production (measured by absorption of ^{14}C). In the same series of experiments positive results were also obtained with niacin, pantothenic acid, *p*-amino-benzoic acid, but rather paradoxically not with biotin. Using marine bacteria for a bioassay Belser (1959) found that this compound improved the growth in culture of certain species of sea water organisms. This worker has also found the presence of uracil and iso-leucine to improve growth.

5-7.—OTHER FACTORS

Many of the preceding results on elements or substances necessary for phytoplankton growth were obtained using cultures. Certain other data also obtained by culture methods are concerned with the condition of development of unicellular algae and we shall now review these relevant facts.

(a) pH.—Experimentation with pH is difficult because of the absorption of CO_2 which rapidly modifies the pH. An old experiment by Bachrach & Lucciardi using a mixed culture of *Nitzschia* and *Navicula* is reported in

Fig. 5-12. At about pH 7.4–8.8 the pH does not interfere with growth and this interval largely covers the pH variations experienced in the open sea.

Kain & Fogg (1958a, b; 1960) more recently carried out experiments on cultures using Tris buffer (trishydroxymethylaminomethane). Their results and the preceding data (Fig. 5-13) show the negative influence of the pH on the growth at pH 8.5 for *Asterionella japonica*. *Prorocentrum micans* seems to be more tolerant and its growth is effected only above 8.75.

(b) Salinity.—Between 20 and 40‰ *Asterionella japonica* and *Prorocentrum micans* develop satisfactorily (Fig. 5-14) but in culture planktonic algae can withstand conditions far from that of sea water. This phenomenon is still more marked for organisms of littoral pools (Fig. 5-15). The diagram shown in this figure illustrates clearly the small range for *Skeletonema costatum* which, although usually considered as an euryhaline species, appears distinctly stenohaline in comparison with supralittoral species such as *Monochrysis lutheri*. This species can also multiply in a medium with a very low amount of Ca^{2+} and Mg^{2+}, yet it does not withstand a simultaneous deficiency of both ions (Droop, 1958b).

(c) Sulphur compounds.—Interesting results have been obtained, in cultures of diatoms with the addition of sulphur compounds or inorganic sulphur as shown in an experiment with *Skeletonema costatum* (Table 5-1).

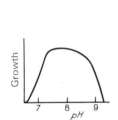
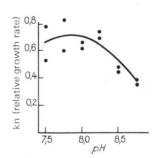

Fig. 5-12.—The effect of pH on the growth of a mixed culture of *Nitzschia* and *Navicula* (after Bachrach & Lucciardi, in Harvey, 1955).

Fig. 5-13.—The effect of pH on the growth of the diatom *Asterionella japonica* (after Kain & Fogg, 1958).

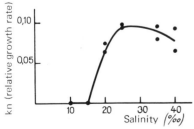

Fig. 5-14.—The relative growth constants of *Prorocentrum micans* in media of different salinities: cultures inoculated with cells grown at 35‰ (after Kain & Fogg, 1960).

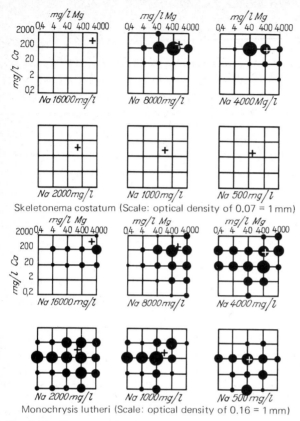

Fig. 5-15.—Response of the planktonic diatom *Skeletonema costatum* and a chrysophycean from a littoral pool (*Monochrysis lutheri*) to varying concentrations of Na, Ca, and Mg: diameter of the circles is proportional to the optical density of the yield in cultures: crosses (+) indicate the position in the diagram where Ca, Mg, and Na have the proportions of sea water, for which the concentrations (in mg/l) are as follows, Na, 10720; Ca, 420; Mg, 1320 (after Droop, 1958b).

TABLE 5-1

Effect of the addition of inorganic sulphur and of cystine on growth in culture of the diatom *Skeletonema costatum* (after Harvey, 1955)

Condition of culture	Number of cells after 60 h in the light
Surface water enriched with nutrient minerals	2658
Same water + 1 mg/l of $Na_2S \cdot 9H_2O$	7258
+ 1 mg/l of cystine	4244
+ 0.5 mg/l of $Na_2S \cdot 9H_2O$ and 0.5 mg/l of cystine	9204

Similar results were obtained with other sulphur compounds—glutathione, thiourea and thiamin. Their positive effects are due to their interference in the metabolism of silica necessary for the building of the frustule (Harvey, 1955).

(d) Molybdenum.—Although it has so far been only investigated on a fresh water algae, *Scenedesmus*, it is interesting to mention the experiments of Arnon (1958) which showed the rôle of molybdenum (Figs 5-16 and 5-17). The molybdenum seems to be an obligatory element for the utilization of nitrates, but it has no effect when the nitrogen is supplied as ammonium or urea. It seems that this element interferes at the level of nitrate reduction. It also plays a rôle in the fixation of molecular nitrogen, so that molybdenum is very important in the nitrogen cycle. According to the available data its amount in the sea, from 2–18 µg/l, seems at first sight adequate but this needs to be examined again as emphasized by Sournia & Citean (1972).

(e) Antibiotic substances.—In experiments on fresh water algae it has

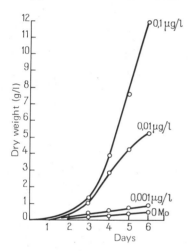

Fig. 5-16.—The effect of molybdenum on the dry weight (g) of *Scenedesmus* obtained per litre of nutrient solution (after Arnon, 1958).

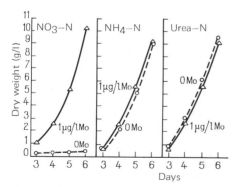

Fig. 5-17.—The effect of molybdenum on the growth of *Scenedesmus*: nitrogen added in three forms, nitrate, ammonium, urea: on the ordinate dry weight/l (after Arnon, 1958).

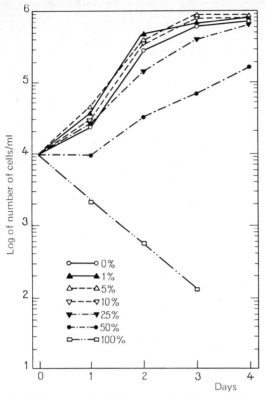

Fig. 5-18.—Growth of *Skeletonema* in different dilutions of a filtrate of *Olisthodiscus* (100‰ to 0‰) with added mineral nutrients (after Pratt, 1966).

been found that excretion products could interfere and make impossible the co-existence of two algal species in the same medium. Pratt showed that a similar phenomenon exists for the two abundant species of the phytoplankton of Narragansett Bay (Rhode Island, U.S.A.), namely, the diatom *Skeletonema costatum* and the flagellate *Olisthodiscus lutens*. Fig. 5-18 shows the growth of an inoculum of *Skeletonema* in a filtrate from a culture of *Olisthodiscus* enriched with mineral nutrients. In a filtrate diluted to 50% growth is markedly delayed. The active substance seems to be an ectocrine produced by *Olisthodiscus*, yellow brown in colour and similar to tannin. Similar substances were found by Craigie & MacLachlan (1964) to be excreted by *Fucus vesiculosus* and which caused a lysis of cells.

Another kind of action was suggested by Timmis & Epstein (1959): pteridines can act as antimetabolites. Pteridines were isolated from cells of Cyanophyceae and the impossibility of co-existence of cyanophycean and euglenids in fresh water may be explained on this basis.

We may mention that an antibiotic effect of phytoplanktonic algae on bacteria, either purely marine or of terrestrial origin has been observed (Sieburth, 1968). Sieburth & Pratt (1962) also showed in the littoral zone an anticoliform factor liberated into the sea during a bloom of *Skeletonema costatum*.

CHAPTER 6

QUANTITATIVE STUDY OF THE PHYTOPLANKTON

The importance of the phytoplankton as the starting point for organic production in the seas has long been appreciated and many investigations have been directed to determine quantitative estimates. After giving an outline of the methods used we shall examine the principal results so far obtained.

6-1.—ENUMERATION

The ideal would be to obtain, for a given volume of water, the number and the characteristics of the different species. This entails the separation and concentration of the cells, followed by their identification and estimation of their numbers.

For the phytoplankton which are retained on a net (micro-phytoplankton) separation and concentration are relatively easy; for quantitative studies it is enough to make a series of net tows and to analyse and count the contents. There are, however some difficulties:

1. The volume filtered varies with the net used in spite of the adoption of standard nets with known characteristics. Important variations of the volume filtered may still arise from differences in the towing velocity and the more or less rapid clogging of the meshes. The use of a propeller-type counter to estimate the volume of water effectively filtered (as for instance in Clarke-Bumpus net), eliminates this cause of error.

2. A list of species with their number related to a unit volume of water enables one to treat certain aspects of phytoplankton ecology but it is inadequate for studies requiring a knowledge of the biomass. The volume of the cells vary even within the same species and even more from one species to another. A relatively simple approach is to take into account the volume of the cells (calculated previously for each species as a function of a reference length) and then to express the results in mm^3 or μm^3/volume of water.

Smayda (1965) used the plasma volume in order to eliminate the vacuoles which are often very big in the diatoms; he used the equation:

$$V_p = S \cdot e + 0.10 V$$

where V_p is the plasma volume, S the cell surface, V the cell volume and e the thickness of the parietal cytoplasm, calculated between 1 and 2 μm as a function of the ratio S/V ($S/V < 0.35$, $e = 2$; $0.35 < S/V < 0.50$, $e = 1.5$; $0.51 < S/V < 0.89$, $e = 1.0$; $S/V \geq 0.90$ the equation is simplified to $V_p = V$).

A further attempt to estimate the carbon of the cells from their volume was made by Strathman (1967): he obtained the best results using the plasma volume.

For nanoplankton the collection of cells is much more difficult and several techniques have been used, namely, (a) centrifugation followed by identification and counting under an ordinary microscope (b) sedimentation in a vessel with a thin bottom and counting with an inversed microscope (Utermohl method), and (c) concentration by filtration and counting the concentrated suspension. An indirect method adapted from bacteriology has been used in which cultures are seeded from more or less diluted samples.

Ballantine (1953) has made a detailed comparative study of these methods and she concluded that the centrifugation method, which was for a long time discredited, could be used if certain precautions are taken; namely, the centrifugation should be gradually finished, and the water should be removed with a bent pipette connected to a vacuum pump (Table 6-1). The method has the advantage of being able to use an ordinary microscope which enables better derminations to be made of the forms present. The dilution method, which is very long, gives mediocre results. These methods, used for nanoplankton, give more or less errors but have the advantage of being able to relate the results to a precisely measured volume of water, which is not the case with net collections.

6-2.—MEASUREMENT OF PIGMENTS

The biomass as deduced from the volume of the phytoplankton is not a very precise estimate of the quantity of total organic substance because of the differences in cell structure depending upon the group and species, particularly as regards the vacuoles. On the other hand, counting is often a long and laborious procedure, and much effort has been made to estimate the phytoplanktonic organic mass more directly or to obtain an index which is adequately representative. The estimation of pigments and especially chlorophyll answer this requirement. Introduced by Kreps & Verbinskaia (1930) for the Barents Sea, and adopted by Harvey (1934) for the English Channel, this method was notably improved by Richards & Thompson (1952). An acetone extract of phytoplankton gives an absorption spectrum

TABLE 6-1

Nanoplankton, separated by centrifugation, in a sample of water from the west end of the Breakwater at Plymouth, collected on the 10th December 1952; values are mean number of cells/m^3 (after Ballantine, 1953).

Total number	<2 μm	2–5 μm	5–10 μm	>10 μm
17.5	2.9	4.5	2.3	0.9

similar to that of Fig. 6-1. There are two absorption maxima, one at 400–500 nm and the other at about 650 nm. The second one is due only to chlorophylls as shown in Table 6-2 and this is used for the determination.

By measuring within the upper maximum at three different wave lengths, it is possible to calculate the quantity of each chlorophyll. The equations recommended by an international working group of Unesco (Anonymous, 1966) are as follows:

$$Chl_a (\mu g/ml) = 11.64 e_{663} - 2.16 e_{645} + 0.10 e_{630}$$
$$Chl_b (\mu g/ml) = -3.94 e_{663} + 20.97 e_{645} - 3.66 e_{630}$$
$$Chl_c (\mu g/ml) = -5.53 e_{663} - 14.81 e_{645} + 54.22 e_{630}$$

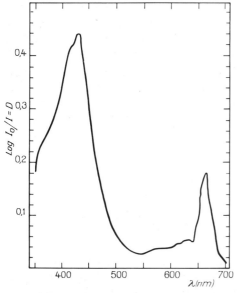

Fig. 6-1.—Absorption spectrum of a sample of natural phytoplankton extracted with 90% acetone (after Richards with Thomson, 1952).

TABLE 6-2

Specific absorption coefficients of some plankton pigments (chlorophyll and carotenoids) in 90% acetone solution (after Richards with Thompson, 1952).

Pigment λ (nm)	665	645	630	510	480	450	420
Chlorophyll a	66.7	16.4	11.9	2.6	1.9	8.9	70.7
Chlorophyll b	6.5	45.6	12.7	3.5	13.6	54.0	26.8
Chlorophyll c	1.1	4.4	10.4	2.1	5.4	78.5	37.3
β-carotene	0	0	0	45.3	293	224	148
Fucoxanthin	0	0	0	56.6	203	249	169
Astacin-type pigment	0	0	0	169	249	221	147
Non-astacin type pigment	0	0	0	45	203	246	171

These concern the concentration of the chlorophylls in acetone extracts. e_{663} is the extinction (optical density or absorbance $= \log I_0/I$) measured at 663 nm, subtracted from the extinction at 750 nm (corrected for turbidity) and brought to cell length of 1 cm. Under natural conditions the amounts of chlorophyll c and even more of chlorophyll b (Table 6-3) are small but Parsons (in Anonymous, 1966) insisted on the validity of their estimations in oceanic water; most workers are satisfied with the determination of chlorophyll a.

This tricolorimetric method has been considerably extended so as to replace the net method by collecting a sample of water and filtering it through several μm filters; it enables a rapid estimation of almost the total phytoplankton including most of the nanoplankton to be made.

This method has been further improved by measuring the fluorescence rather than the absorption of the chlorophyll, i.e. its capacity to emit light when excited by luminous energy at 430 nm. An acetone solution of chlorophyll emits at 670 nm (Holm-Hansen et al. 1965). With the development of extremely sensitive fluorimeters it has become possible to use this method for the estimation of chlorophyll in situ (Lorenzen, 1966), i.e., directly on the water, without any extraction; continuous analyses are, therefore, possible.

An important source of error in the use of chlorophyll for the estimation of the phytoplanktonic biomass is the presence of phaeophytin, a degradated chlorophyll, whose spectrum is only little different (Bogorad, 1962) and which, therefore, interferes with the chlorophyll determination even in in situ measurements. Acidification of chlorophyll converts it to phaeophytin and this occurs particularly in faecal pellets excreted in large quantities mainly by herbivorous copepods. Based on this fact a correction can be made by making two measurements of fluorescence, one before and one after addition of acid (Saijo, Ijzuka & Asaoka, 1969).

Many workers have tried to deduce the quantity of carbon in the phytoplankton from the quantity of chlorophyll, using a direct relation:

TABLE 6-3

Results of measurements of chlorophyll on different cultures and on sea water samples (collected at Port-Hacking, Australia): MgCO$_3$ added before filtration, filtrate ground and extracted for 10 min (after Humphrey & Wooton, in Anonymous 1966).

Chlorophyll μg/l	a	b	c
Dunaliella	1.51	0.30	—
Skeletonema	0.89	—	0.47
Nannochloris	1.24	0.65	—
Sea water	0.72	0.08	0.13
Sea water	1.77	0.02	0.19

mg C = $F \times$ mg chlorophyll. Unfortunately, the coefficient F varies considerably, between 10 and 100, according to the population concerned (Strickland, 1960) and there are slight variations in the amount of chlorophyll in the cell even during a 24-h cycle (Yentsch & Scagel, 1958). The coefficient F has to be used with precaution. Ryther & Menzel (1965) accepted a value of 35 as representative of a healthy and in actively growing phytoplankton population. Ryther et al. (1971) found a value of F close to 50 in the upwelling off Peru where the phytoplanktonic development is considerable.

6-3.—OTHER METHODS

A third group of methods unlike those just considered which relate specifically to the phytoplankton, are those which estimate the small-sized particles in the sea water (seston). Among such particles the phytoplankton is very often the dominant part but it is associated in various proportions with other living particles, elements of the zooplankton, and inorganic and organic non-living particles (tripton).

The first method of this type is the counting of particles with a Coulter counter, which gives in addition to the numbers the mean volume of the particles; it is the least selective method since it does not distinguish living from non-living particles.

A second method consists in the analysis of certain components of the living matter and so, after filtering a given volume of water, the particulate organic carbon, nitrogen, protein or glucose are estimated. If the inorganic particles are then eliminated the non-living organic particles will still interfere in the results. In order to remove this disadvantage the estimation of ATP (adenosine triphosphate), which relates to the energy reactions of living matter and which disappears rapidly when the cells are dying, has been proposed; the reaction for ATP is very sensitive and depends on a photochemical reaction carried out on an enzymatic system extracted from fireflies (Holm-Hansen & Booth, 1966; Danmas & Fiala, 1969). There are some difficulties of interpretation; the amount of ATP may vary in different species (for instance in *Skeletonema costatum* it is 0.031% of the dry weight while in *Ditylum brightwellii* it is 0.12).

Finally, these different non-selective methods should be used together with the preceding methods.

6-4.—VERTICAL DISTRIBUTION

A study of the different layers of water shows much variation in the distribution of the phytoplankton and it becomes scarce in the deep layers. This is not surprising since plant plankton must have adequate light for photosynthesis and can develop only in the upper layers of the oceans. The numbers given in Table 6-4, according to Lohmann (in Steuer, 1970),

TABLE 6-4

Quantitative distribution of coccolithophorids as a function of depth, in May 1901, in front of Syracuse (Mediterranean) (after Lohmann, in Steuer, 1910).

Depth in m	0	20	50	77	155	230	431	631
No individuals/l	176	308	2980	368	16	50	2	0

concerning the coccolithophores in 1 l of water from different depths illustrates this distribution. In this example the coccolithophores, which become rare below 100 m, disappear completely at more than 500 m depth. Another general character of the vertical distribution is that, in spite of a direct dependence on light for photosynthesis, phytoplankton is not most abundant in the superficial water, where light intensity is greatest but at a lower level. Two factors are responsible for this phenomenon:

1. At high light intensity the photosynthesis of the algae decreases and, as we have already seen, may even be seriously disturbed.
2. Many of the phytoplanktonic organisms, particularly diatoms cannot move independently and so tend to sink. Those with a high density will, therefore, remain below the level of high production.

To maintain phytoplanktonic populations near the surface a certain amount of turbulence is essential to renew the superficial zone with passively carried plant cells. The level of the maximum of abundance varies according to the geographical region. Riley, Stommel & Bumpus (1949) found a maximum at 10 m depth over Georges Bank but at 100 m depth in the Sargasso Sea as well as in the Florida Strait. The conditions in these two latter areas are markedly different: the turbulence in the Florida Strait, located on the Gulf Stream, is clearly higher. According to Davis (1955) it may be that the phenomena responsible for the vertical distribution are much less simple than "envisaged" in the classical interpretation. It is possible to find a confirmation in the observations of Ryther & Hulburt (1960), at a station about 100 miles off New-York, on the 27th March 1957. According to all the measurements, temperature, salinity, oxygen, phosphate, nitrate and nitrite the water at this station was absolutely homogenous. The distribution of five principle species of the phytoplankton show, on the contrary, a considerable variation (Fig. 6-2).

The vertical distribution of the chlorophyll is very often also characterized by a sub-surface maximum (Anderson, 1969; Saijo, Ijzuka & Asaoka, 1969) at a depth of 50 to 150 m, i.e., in the lower part of the photic zone or even below it. This maximum contains an important fraction of phaeophytin (Fig. 6-3). Finally, we may note that the peridinians because of their motility are able to present a specific vertical distribution and several samples of vertical migration are known particularly for *Gonyaulax polyedra*. This species is often very abundant in the coastal water near La

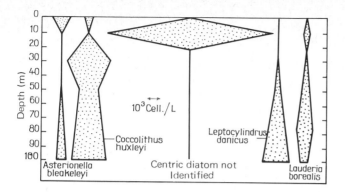

Fig. 6-2.—Vertical distribution of the five most abundant species of phytoplankton, at Crawford, Station 136 on 27th March 1957, about 100 miles off New-York (after Ryther & Hulburt, 1960).

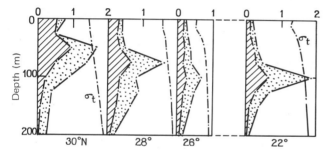

Fig. 6-3.—Vertical distribution of chlorophyll a (shaded) of phaeophytin (dashed) and of the density (σ_t), on a north south transect, on the meridian 132°E, south of Japan in May–June 1968 (after Saijo, Ijzuka & Asaoka, 1969).

Jolla (California) swimming to the surface during the day and sinking to 10–15 m during the night (Eppley, Holm-Hansen & Strickland, 1967).

6-5.—SEASONAL FLUCTUATIONS

Regular observations at the same place shows large phytoplanktonic variations during the year and these are generally repeated according to the same cycle from one year to another. The data obtained at the Isle of Man at Port Erin Bay and reported by Johnstone, Scott & Chadwick (1927) are an excellent example. From 1907 to 1920, many weekly surface samples were taken with a 35 cm diameter net, gauze number 20 towed at the surface for a distance of half a mile. The number of cells were related to one collection considered as a unit. The results of the monthly mean number over 14 years of observations for *Chaetoceros*, which represents a major part of the phytoplankton collected by nets, are shown in Fig. 6-7. From March to June there is a considerable development of *Chaetoceros*, a maximum in May with

8×10^6 cells/tow. In July and August the density remained low, but, in September and October there was a new outburst, which was followed by a period of 4 months in which the number of cells did not exceed 5×10^4/tow. The ratio of the extreme densities (6.5×10^3 and 7.95×10^6) is of the order of 1:1000. But the variations are still higher, if, instead of looking at the mean over 14 years the numbers obtained during one year are considered, e.g., those of 1912, (Fig. 6-4); the spring maximum in this year reaches 18×10^6 cells/sample, the autumn maximum 7.7×10^5, and the ratio of the extreme values is 1:18000 (1100 in December and 18×10^6 in April). The development of the population of *Chaetoceros* in Port Erin Bay show their two successive outbursts during a year, one in the spring and the other in the autumn: the spring outburst in March is the most important. This example illustrates a general character of the seasonal distribution of the phytoplankton: namely, the existence of a cycle marked by considerable differences in its density during the year. There may be two maxima, a large one in the spring and a smaller one in the autumn (Fig. 6-5) or only one maximum may exist (Fig. 6-6). Since this seasonal cycle is related a priori to important seasonal variations in physico-chemical factors in the cold or temperate zones it would be expected that in tropical regions the cycle would be blurred and even disappear. Saurnia (1969) has considered this question recently and came to the conclusion that in tropical neritic waters (for the oceanic water

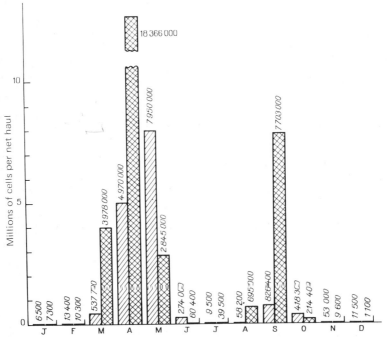

Fig. 6-4.—Mean number of *Chaetoceros* cells in a net-tow: left (shaded) means of the years 1901–1920; right (diagonals) 1912 results; Port Erin Bay (after Johnstone, Scott & Chadwick, 1927).

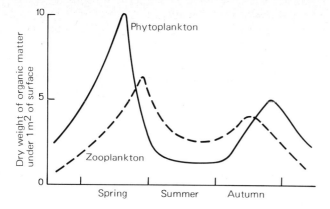

Fig. 6-5.—Seasonal variations in the phytoplankton biomass (composed mainly of diatoms) and of zooplankton biomass, in the English Channel 4 miles off the coast (after Harvey, 1955).

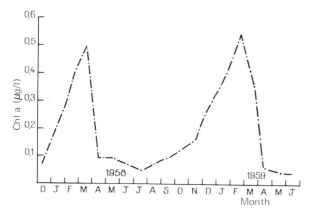

Fig. 6-6.—Variations of the phytoplankton, (µg chlorophyll a/l) at Castellon (Mediterranean Sea, Spain) (after Herrera & Margalef, 1961).

the data are inadequate) there is also a cycle, related to variations in hydrographical (currents, upwelling) or climatic (winds, rains, monsoons) conditions. Fig. 6-7 gives an example of a cycle in the waters off Abidjan (Ivory Coast, Gulf of Guinea): from June to November there is a large amount of chlorophyll a which coincides with a period of upwelling related to the winds regime.

6-6.—ECOLOGICAL ANALYSIS OF THE PHYTOPLANKTONIC BLOOM

We have to credit Marshall & Orr (1930) for the excellent and now classical, detailed analysis of the ecology of the spring bloom. The work was done in Loch Striven, near Millport, in the Clyde Sea Area (Scotland). In this

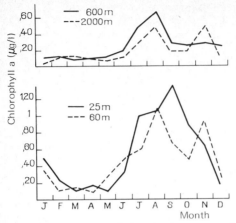

Fig. 6-7.—Seasonal variations in the amount of chlorophyll *a* in the surface water at four stations located off Abidjan, above bottom depths of 25, 60, 600 and 200 m; monthly means from June 1964 to June 1967 (after Reyssac, 1970).

loch the conditions are more stable than in the open sea, a fact which facilitates the analysis and, furthermore, the spring outburst is almost exclusively due to one species of diatom, *Skeletonema costatum*. Besides couts *Skeletonema*, measurements of temperature, salinity (which enabled the calculation of the density) oxygen, pH and phosphate were made. In 1927 observations (Fig. 6-8) began on the 18th of March when the outburst had already started; this reached its maximum at the surface from the 14th of March with more than 12 500 cells/cm^3. From the 18th of March the maximum density of diatoms was found deeper in the water and the deepening isopleth in their diagram is very characteristic being due to passive sinking of the cells, not compensated by turbulence. Indeed, the density shows that from the 18th to the 24th of March there was a clear stratification of the water, resulting from the slightly lower surface salinity and so giving stability. The following phenomena accompanies the multiplication of *Skeletonema costatum*.

1. An increase in the pH of several tenths of a unit as a result of the absorption of CO_2 during photosynthesis.
2. A supersaturation of oxygen which reached 130%, also due to the photosynthesis.
3. A fall in phosphate from 20 μg to few μg, this being absorbed and incorporated in the new organic matter.

Two comments are relevant, the considerable increase in the number of cells with depth is not accompanied by concomitant variations of the oxygen and pH which proves that the cells do not multiply at lower levels but are carried from the upper level and take little part in photosynthesis. These cells will contribute to the sediment of the bottom (the planktonic "rain"). On the other hand the proliferation of the cells in the spring outburst stops

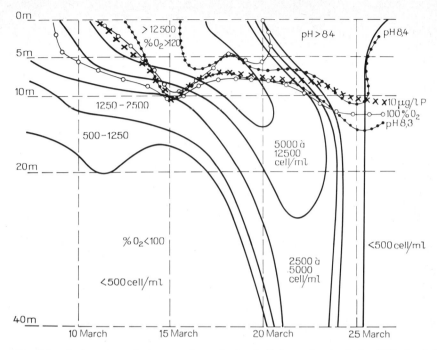

Fig. 6-8.—Spring phytoplankton outburst in Loch Striven, Scotland, in 1927: isopleths correspond to the density of cells (in ml) of *Skeletonema costatum*, which essentially comprises the phytoplankton at this time of the year, to the amount of PO_4–P (μg/l) to pH, and to percentage of oxygen saturation (after Marshall & Orr, 1930).

before all the phosphorus is utilized which suggests that factors other than phosphate are limiting. This latter was not found in the open sea, and may be due to the special conditions in this loch where the spring outburst corresponds to a development of an almost pure culture of *Skeletonema*. This progression of the spring outburst with an oblique deepening of the isopleths as a function of time was confirmed in 1927 by similar observations to those of 1926. In 1928, on the contrary, the phenomenon was different, the outburst, late in the month being only moderate and interrupted by a mixing of the water marked by uniform density, pH, oxygen and an increase in phosphate.

Marshall & Orr (1924–1929) give the date of the spring outburst, observed during successive years as determined by the appearance of cell densities higher than 200 cells of *Skeletonema* cm^3, and they tried to relate the observations to the increase in light, as estimated in day-hours. In general, the threshold of the outburst goes hand in hand with a tendency to an increase in hours of light, but a detailed correlation is much less satisfactory (Fig. 6-9).

Another important factor also interferes as the preceding example showed, namely, the degree of stability of the water. We shall see how it is manifested by an examination of the factors which control it.

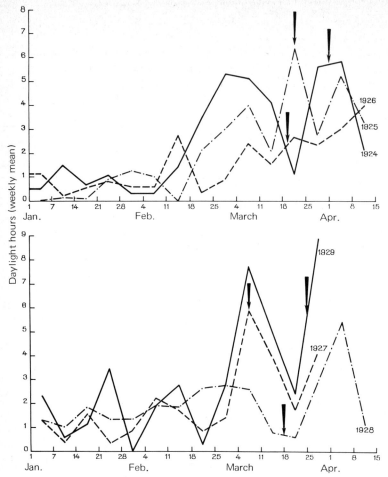

Fig. 6-9.—Development of the weekly daylight hours in Loch Striven during the first months of the years 1924 to 1929 and the threshold of the spring outburst (arrows) (according to data of Marshall & Orr, 1930).

6-7.—CAUSES OF THE PHYTOPLANKTONIC BLOOMING; SVERDRUP THEORY

In order that the phytoplanktonic population should multiply its gain of material must be higher than its losses. In other words, the total production of organic matter by photosynthesis must be greater than that used in respiration, i.e., $P > R$ where P is the production and R the consumption by respiration.

$$P > R \tag{1}$$

These conditions occur above the compensation depth, D_c, since at this level, by definition, photosynthesis is equal to respiration. But since the luminous radiation varies, the inequality (1) has to be verified over a mean 24 h cycle; this is the mean compensation depth, calculated for 24 h. This depth should be always considered and it may be symbolized by $\bar{D}_c$. The respiration is not, however, the only cause of the reduction of the mass of the phytoplanktonic population in a given layer. The density of many species is higher than that of the water and they sink slowly; this sinking is sometimes balanced in flagellates and other motile species by searching for an optimal level but this is made at the cost of loss of material used in the movement, and we lack data on the energy so spent. On the other hand, the marine environment is far from being still and particularly the swell is the origin of a more or less important turbulence which has the tendency to mix and homogenize the superficial layers of the sea to a depth which, in the Norwegian Sea, reaches 400 m (Sverdrup, 1959). The increase in phytoplankton, during the day in the illuminated layers will, therefore, be opposed by the turbulence which tends to equalize the distribution to more or less moderate depth. We should remember that turbulence will oppose the sinking due to gravity of the inert cells and disturb any directed movements of mobile cells. All these factors are represented in Fig. 6-10, to which is added the grazing of phytoplankton by the zooplankton and eventually the return to the environment of the undigested phytoplankton (superfluous feeding). In summary, it is not only for the short period of 24 h that the inequality ($P > R$) must be realized in order that a phytoplanktonic outburst will occur but also for all the thickness of the superficial layer mixed by turbulence.

Assuming that production is not limited by the lack of nutrients, the principal factor on which the quantity of phytoplankton produced depends is the available luminous energy. At high latitudes, for instance in the Norwegian Sea, this may be at the beginning of the winter. Since the nitrates and phosphates are then abundant in the superficial layers the development

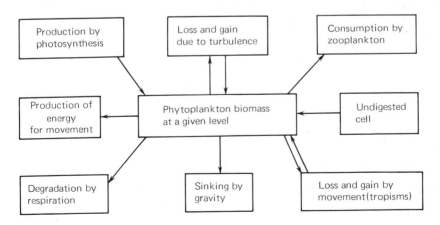

Fig. 6-10.—Diagrammatic representation of the factors interfering in the equilibrium at a given depth of the phytoplanktonic biomass.

of the phytoplankton will go hand in hand with the increasing hours of light as the day becomes longer; this happens at the end of the winter and the beginning of the spring. The luminous energy increases from the end of February with some sudden changes, differing from one year to another and depending on the cloud cover. A progressive delay in the start of phytoplankton production would be expected on proceeding towards the poles. But it is not so, the multiplication of the phytoplanktonic cells very often takes the form of a sudden spectacular spring outburst, and at the same latitude and the same light conditions the phenomenon is not simultaneous.

We have to credit Sverdrup (1953) with an attempt to explain and put into equations the required conditions for this outburst. Sverdrup defined as the critical depth, D_{cr}, the depth at which $P = R$, and assumed the existence in the surface of a well mixed layer of depth D_m, where the phytoplankton has a homogenous distribution independent of the adjacent water. As this homogenous superficial layer becomes thinner relative to the critical depth more of the phytoplankton will multiply rapidly. When this increase becomes very rapid the outburst will take place. Sverdrup gives the following approximate equation for this critical depth.

$$D_{cr} = \frac{\bar{I}_e}{KI_c}$$

where K is the extinction coefficient (Chapter 2-2c), I_c is the irradiance of the compensation depth, and $\bar{I}_e$ the algebraic mean of the luminous energy which passes the surface (Sverdrup assumes in contrast to Strickland (Chapter 2-3), only the radiation from 420 nm to 560 nm to be active). The method of deducing this equation is given in the Appendix (p. 317). Sverdrup applied his theory to the results obtained in 1949 by a weather ship located at the point M in the Norwegian Sea. Fig. 6-11 shows the results obtained. It is only at the beginning of April that the lower limit of the homogenous superficial water passes the critical depth and this coincides with the first important development of phytoplankton, which exceeds over 20000 cells/l. During April the bottom of the mixed superficial layer oscillates from one side of the critical depth to the other and the phytoplankton population is almost stable, consumed more and more by the zooplankton which appears in quantity after the first phytoplanktonic bloom. From the 10th of May the lower limit of the homogenous layer exceeds the critical depth and phytoplankton shows a new outburst, doubling its density. The agreement between the theory and the facts seems to be striking. Marshall (1958) confirmed this agreement for the arctic zone. His data are given in Table 6-5. For the Arctic water the critical depth exceeds the depth of the mixed layer in March–April, and it is during this period that the spring outburst begins. In Atlantic water, the production only begins later in May or June, but it is only in May and mainly in June that the depth of the homogenous layer becomes less than the critical depth. Finally, the observation that the retreat of the ice is followed by a phytoplanktonic blooming is also explained by this theory,

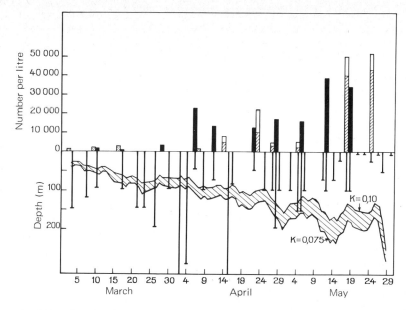

Fig. 6-11.—Results of observations at Weather Ship M (66°N, 2°E) from March to April 1949: upper, solid, phytoplankton, number of cells/l; shaded, copepods; clear, nauplii for these two groups using the same scale, at the left, which corresponds in these cases to number in a vertical sample from 0 to 100 m: lower, vertical lines represent the thickness of the superficial mixed layer (from 0 to D_m) and the curves fringing the shaded zone represent the critical depth, D_{cr}, calculated for two values of the extinction coefficient ($K = 0.10$ and $K = 0.075$) between which the probable value is found (after Sverdrup, 1953).

TABLE 6-5

Comparison of the critical depth and the depth of the homogenous surface layer (after Marshall, 1958).

Month		November–February	March–April	May	June	July–October
Critical depth		0.5 m	30–170 m	140–190 m	190–240 m	270–300 m
Depth of homogenous layer	Arctic water station	75 m	50 m	25 m	25 m	30–60 m
	Atlantic water station	>200 m	>200 m	>150 m	25–75 m	40–80 m

since the melt water forms a light homogenous layer of 10–25 m thick, much less than the critical depth.

Recently Parsons & LeBrasseur (1968) applied Sverdrup's equation on a large scale to the North East Pacific (North to 40°N, from 125°W to 160°W). Their results are shown in Fig. 6-12, where, for the different zones, the depth

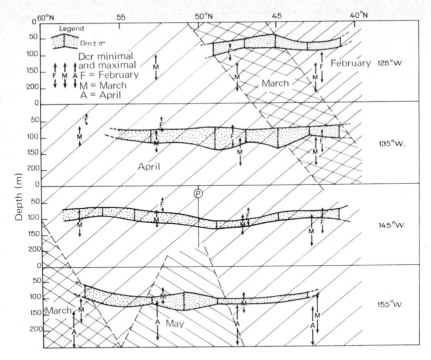

Fig. 6-12.—Comparison of the depth of the superficial mixing layer ($D_m \pm \sigma$) and the calculated critical depth (D_{cr} minimum and maximum) in February, March and April, in four north-south transects in the north east Pacific: meteorological Station P is indicated: critical depth in April shown only along the 155°W line; in the rest of the area it is below the mixing depth ($D_m \pm \sigma$); shaded the zone where the conditions required for the phytoplanktonic outburst are established in February, March, April & May (after Parsons & LeBrasseur, 1968).

of the mixing layer ($D_m \pm \sigma$) and the critical depth (D_{cr}), calculated for the months February, March, April and May, are given. It appears that the conditions for the initiation of the spring outburst are present from February along the coast of the United States (40–45°N, 125°W) while for the meteorological Station P (50°N, 145°W) and the water situated more westward, it is delayed until May.

6-8.—SUCCESSION OF THE PHYTOPLANKTONIC POPULATIONS

One of the remarkable results of the counting method was the demonstration of the existence during a given year, of a succession of species in phytoplanktonic populations. Considerable changes in the specific composition of the population are superposed on the important seasonal variations of the total phytoplankton which have just been considered. During their observations Johnstone, Scott & Chadwick (1924) established the succession represented in Fig. 6-13. Grall & Jacques (1964), have given detailed specific

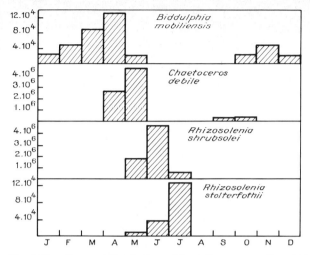

Fig. 6-13.—Seasonal distribution of four diatoms near Port Erin: ordinate, mean number of cells per net tow, according to the observations in the years 1907 to 1920 (after Johnstone, Scott & Chadwick, 1924).

succession of the diatoms in Roscoff (Fig. 6-14) from 1962 to 1963. The phytoplankton is, therefore, composed of a series of dominant species which succeed each other during the season, the order of this succession hardly changing in successive years.

There are several causes of this phenomenon. Temperature is certainly one of the first factors involved. Sverdrup, Johnson & Fleming (1946) cited the following succession for the Gulf of Maine:

April, 3°C: *Thalassiosira nordenskioldii, Porosira glacialis, Chaetoceros diadema*;
May, 6°C: *Chaetoceros debilis*;
June, 9°C: *Chaetoceros compressus*;
August, 12°C: *Chaetoceros constrictus, Chaetoceros cinctis, Skeletonema costatum*.

Without doubt the temperature plays a role in this phenomenon since cold and warm water species are known, but a comparison of optimal temperatures obtained in vitro and in situ for various species does not give good agreement with the observed succession and a replacement of one dominant species by another may take place rapidly and without any marked change in temperature (Fig. 6-15). It appears quite clearly, that with a low amount of nitrogen either as ammonium or nitrate, the coccolithophore *Coccolithus huxleyi* has a higher multiplication rate than the diatoms *Ditylum brightwellii* and *Skeletonema costatum*. Eppley (unpublished data) has observed, off the coast of southern California, that *Coccolithus huxleyi*, is dominant relative to the diatoms except when the upwelling brings water richer in nitrates to the surface.

The fourth factor, and probably the most important, concerns what may

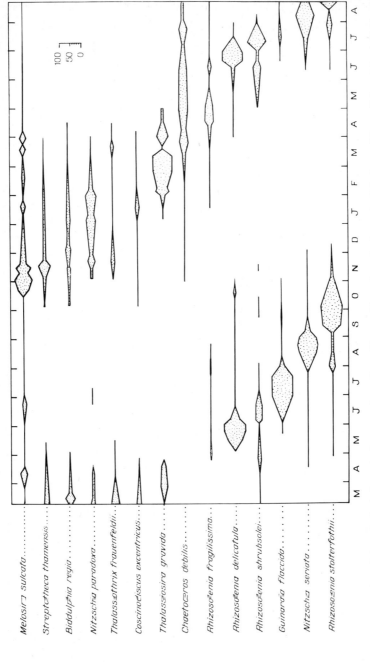

Fig. 6-14.— Succession of the principal phytoplankton species collected by net (% total number) at point B in front of Roscoff from March 1962 to August 1963 (after Grall & Jacques, 1964).

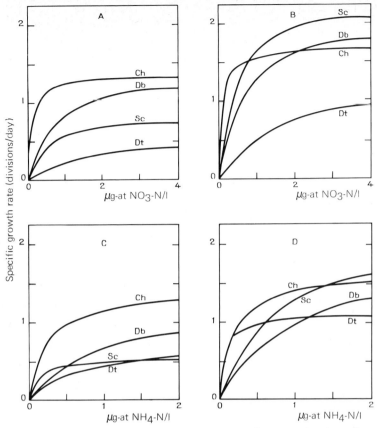

Fig. 6-15.—Specific growth rate (number of divisions/day) calculated as a function of nitrate and ammonium concentrations at two different irradiances; A, C, 5% (0.025 cal/cm²/min) of the superficial irradiance (400–700 nm); B, D, 20% 0.1 cal/cm²/min: Ch, *Coccolithus huxleyi*; Db, *Ditylum brightwellii*; Sc, *Skeletonema costatum*; Dt, *Dunaliella tertiolecta* (after Eppley, Rogers & McCarthy, 1969).

be called the conditioning of the environment by secreted or excreted organic matter (the external metabolites of Lucas, vitamins, antibiotic substances), and coming from bacteria and the phytoplankton. The irradiance also seems to be involved, since the optimal values, for maximum photosynthesis, varies according to the group of algae considered (Fig. 2-12). The richness in nutrient salts is the third factor, since, according to Barker (1935), certain peridinians may be satisfied with an extremely low amount of nitrogen and probably also of phosphorus; this could explain why they often replace diatoms after the spring outburst which had lead to a marked decrease in the amount of these nutrients. The interpretation of this factor was clarified by Eppley, Rogers & McCarthy (1969) who estimated the specific growth rate (or relative, see Appendix p. 315) of different species as a function of the amount of nitrogen. In this conditioning we may also include the establishment of satisfactory equilibria between metallic ions and chelating substances, as we have seen in our discussion on copper. On the other hand, the loss of organic phosphate, by phytoplankton, which is utilizable only immediately by species secreting the required phosphatases may also be one of the ways of this conditioning.

The relative rates of multiplication of different species may play a remarkable role in the establishment of the seasonal succession. Margalef (1958), from his study on the Ria de Vigo, gives a scheme for the succession during the phytoplanktonic bloom in which the rate of multiplication is involved at the same time as there is a decrease in the quantity of nutrient salts and probably a conditioning of the water: the following sequence is suggested:

1. Small diatoms with a rapid multiplication (one to two divisions per day) proliferate (*Skeletonema, Leptocylindrus, Chaetoceros fragilis*);
2. Large diatoms with slower multiplication rates develop, (*Schroederella, Bacteriastrum*);
3. Finally, a higher proportion of motile cells appear, mainly dinoflagellates (*Prorocentrum, Ceratium . . .*) with a low growth rate, ending, in quiet water, with red water.

Recently, Pratt (1966) has given an interesting example of specific succession in Narragansett Bay on the Atlantic coast of the United States, and with the help of experimental work, tried to explain the phenomena from May to October; blooms of *Skeletonema costatum* and of *Olisthodiscus luteus* (Xanthophyceae) are observed, and most often seem to exclude each other (Fig. 6-16). Experimentally, as we have seen *Skeletonema* is inhibited by high concentrations of substances secreted by *Olisthodiscus*, but is stimulated by low concentrations. On the other hand its rate of multiplication is higher. Pratt proposed the following scheme: *Olisthodiscus* succeeds *Skeletonema costatum* by the production of an ectocrine substance; *Skeletonema* assures its dominance by its higher rate of multiplication, and is also more favoured by the secretion of *Olisthodiscus* when present in small quantities.

In summary, five principal factors are concerned in phytoplankton succession: temperature, irradiance, the amount of nutrient salts, conditioning of the water, and the rate of multiplication. To these factors, sometimes selective consumption by zooplankton organism must be added. This is of course only valid for the same mass of water and the replacement of one mass of water by another (different in its characteristics, in particular salinity) may rapidly and considerably modify phytoplankton populations. Finally, besides the preceding factors the problem of "seeding" must be considered. In order that a species will develop it must be introduced to any given zone. It seems that certain species even though neglected (or unobserved) in a survey may persist all the time but only in small numbers yet still enable the initiation of a new multiplication when the conditions become favourable (Riley, 1966). On the contrary, Braarud (1966) cited the example of *Coccolithus huxleyi* which in certain years pollutes the fjord of Oslo but

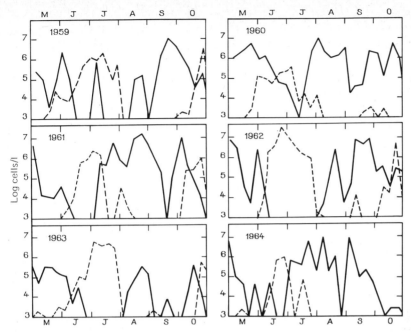

Fig. 6-16.—Abundance in the surface of *Skeletonema* (solid lines) and *Olisthodiscus* (broken line) in May to October during 6 years of weekly observations in Narragansett Bay (U.S.A.) (after Pratt, 1966).

in the other years its abundance is low; this may be attributed to variable introduction of the initial population. There are species which form resting spores and these sink into the deeper waters or to the bottom where they remain until suitable conditions return and so ensure the "re-seeding" (meroplanktonic species). We cited the example of *Biddulphia mobilensis*. According to recent work this also may be the case for certain coccolithophores.

Even when a succession at a specific level cannot be proved in the phytoplanktonic population it is frequently possible to recognize a succession in the variation of the character of the phytoplankton. Margalef (1963) studied this problem by a consideration of the specific diversity for which he elaborated the following equation, derived from an expression due to Brillouin from information theory, and representing the entropy of a system of particles:

$$\bar{D} = \frac{1}{N} \log_2 \frac{N!}{n_1! \, n_2! \ldots n_s!}$$

when $\bar{D}$ is the diversity index, N is the total number of individuals observed in a sample and $n_1, n_2 \ldots n_s$ the number of individuals of the species $1, 2, \ldots s$ (for a further study of different expressions of diversity see Travers (1971). Fig. 6-17 gives the development in time of this diversity index for two stations in Ria de Vigo (Spain). This varied from 1.27 to 3.50

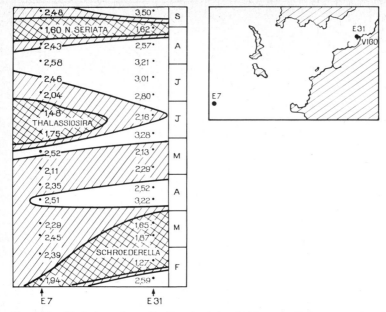

Fig. 6-17.—Variations of specific diversity index (in bits per cell) for two stations in Ria de Vigo from February to September 1962; phytoplankton samples collected by net by F. Fraga: dominant species indicated, *Schroederella delicatula*, *Thalassiosira* sp., *Nitzschia seriata* (after Margalef, 1963).

bits per cell from February to September when the diversity is low and this corresponds to the dominance of one particular species *Schroederella delicatula*, *Thalassiosira* sp. or *Nitzschia seriata*; when the index is high it means that more species are present without a clear dominance of any one of them. When we represent the data of Station E31 on a graph (Fig. 6-18) with the time on the abscissa and diversity index on the ordinate we find a

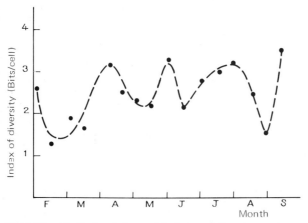

Fig. 6-18.—Cyclical variations of the index of diversity at the Station E31, according to the data of Fig. 6-17.

sinusoidal variation of the index. In the same place we have a cyclical development of the phytoplankton population as a function of time. Margalef introduced here the idea of maturity; the population with little diversity is considered as young, developing to very diverse population as it becomes more mature. The calculation of the specific diversity requires a detailed systematic analysis by a qualified specialist. The pigment diversity, also introduced by Margalef, is much more rapidly estimated. It depends upon the fact that in a population with low maturity, during rapid growth, a relatively high proportion of chlorophyll a is found, while in very mature populations the composition of the pigments is much more complex, with a very high proportion of carotenoids. It corresponds to the ratio:

e_{430}/e_{665}

estimated on an acetone extract of phytoplankton (it is possible to use the value of e_{663} instead of e_{665}). By comparing this index to the index of specific diversity Margalef found a significant correlation of $+0.405$ ($P = 0.01$). Travers (1971) in Marseille also found a satisfactory correlation ($+0.54$ significant at $P = 0.01$) for the early months of the year, but subsequently the correlation became poor when the carotenoids measured at e_{430} come mainly from organic detritus and not from phytoplankton. Although this method of pigment diversity is simple, precautions are needed in its use.

6-9.—SPATIAL VARIATIONS

After considering the vertical distribution and the seasonal variations of the phytoplankton at a given geographical location, we shall now discuss variations in space.

The observations of Margalef (1963) on the distribution of the diatom *Skeletonema costatum* will be used as an example. Margalef studied three series of 27 samples, with one month difference between them (Fig. 6-19). We can see that not only the maximum densities but also their location varies markedly from one month to another and that instead of having a homogenous distribution of the species there is often a rapid change, notably in the 30th of July, from high densities. If we represent graphically the variation, with density on the abscissa and frequencies on the ordinate, we see that the range of densities is extremely wide and that the distribution is very far from normal. It is only by transformation of the densities to their logarithmic values that a normal distribution is obtained, as seen in Fig. 6-19 (upper left).

This example illustrates a very general character of the spatial distribution of the phytoplankton, which is found for the specific distribution as well as for the general distribution, as estimated by the amount of chlorophyll a; the distribution is heterogeneous. The use of statistical methods will always need a transformation of the data in order to normalize their distribution; this will often be obtained by a logarithmic transformation.

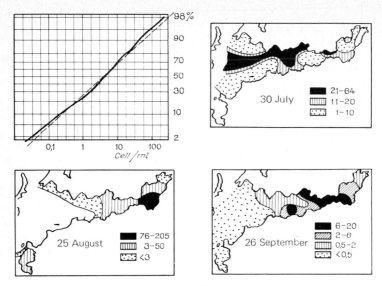

Fig. 6-19.—Distribution of the diatom *Skeletonema costatum* in the surface waters of Ria de Vigo, on three different days in 1955 (27 stations, length of the Ria about 20 miles): no. cells/ml, obtained by filtration of 200 l of water with a fine gauze, is the minimal estimation (the ultimate loss supposed to be proportional to the density of the cells): towards left, cumulative distribution of the whole samples (logarithmic abscissa, ordinate in probability scale); continuous line corresponds to the distribution of the samples (24%, for example, is lower or equal to 1 cell/ml; 70% to 10 cells/ml); broken line is a straight line interpolated by eye; near coincidence of these two lines verify that logarithmic transformation normalizes the data (after Margalef, 1963).

The spatial heterogeneity of the phytoplankton may also be observed on a very small scale as illustrated by a transect made by Platt, Dickie & Trites (1970) in Saint-Margaret's Bay, near Halifax. The phytoplankton was estimated as chlorophyll *a* at successive stations, a distance of 0.1 nautical miles apart, and the results show a series of maxima alternating with low values (Fig. 6-20). The study of the correlation of a given station with those stations more or less near to one another shows (Fig. 6-21), a positive correlation for stations which are separated by less than 0.2 nautical miles or more than 0.7 nautical miles and a negative correlation for stations between these values. Clearly, the transect passes through areas richer and poorer in phytoplankton or, more simply, the distribution is patchy and in this case the patches are of about half a nautical mile across.

Bainbridge (1957) gathered together much data on the spatial distribution of the phytoplankton obtained by different techniques and concluded that there were patches or zones of concentration which, in general, do not have a massive or irregular form but are elongated, elliptical or strip-like, their length being from 3 or 4 to 100 times their width. The dimensions of these zones varies considerably, from several metres to tenths of a nautical mile. Two more examples will be discussed. On a profile made across the North Sea with Hardy's plankton recorder the zones of concentration form a

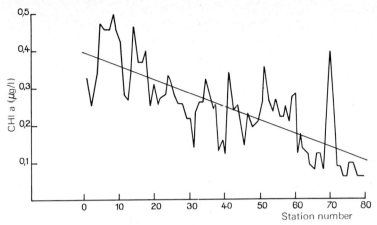

Fig. 6-20.—Variation of chlorophyll concentration at 10 m depth along an 8 mile transect in St. Margaret's Bay (Canada, N.E.) on the 27th June 1968: distance between stations is 0.1 mile: straight line regression line of chlorophyll as function of station number has been drawn in (after Platt, Dickie & Trites, 1970).

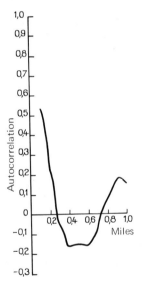

Fig. 6-21.—Serial autocorrelation of the amount of chlorophyll a expressed as a function of distance between stations (only distances less than or equal to 1 mile): transect of the 27th June 1968 (after Platt, Dickie & Trites, 1970).

succession of saw like peaks (Fig. 6-22). In the Mediterranean, along the south coast of Spain, Margalef & Estrada also observed a series of peaks (Fig. 6-23); the peaks corresponding to the highest densities (Section 1) also show a specific diversity lower than that of the closer spaced stations.

The existence of a very marked spatial heterogeneity in very small distances has lead to the questioning of the validity of the sampling method

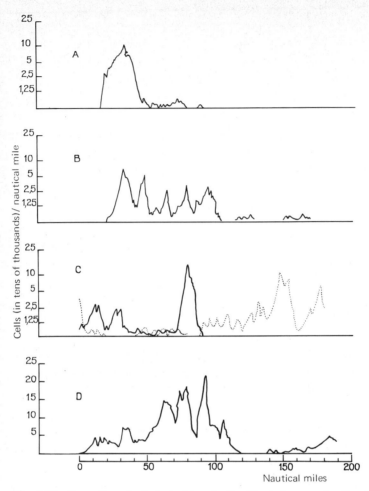

Fig. 6-22.—Recording of phytoplankton in the North-Sea on the line Hull-Bremen: A, *Rhizosolenia styliformis* (26th October, 1932); B, *Biddulphia sinensis* (27th November, 1935); C, *Rhizosolenia styliformis* (continuous line) and *Biddulphia sinensis* (dotted line) (20th October, 1935); D, dinoflagellates (20th June 1937) (after Lucas, 1940 in Bainbridge 1957).

itself. Platt, Dickie & Trites (1970) studied this question, making a series of 10 stations distributed in a square of more or less large area and, therefore, more or less near to one another. The total variance σ_T^2 is analysed by separating it into three components:

$$\sigma_T^2 = \sigma_0^2 + \sigma_1^2 + \sigma_2^2$$

when σ_2^2 is the variance due to the true differences between the stations: σ_1^2 is the variance due to the different samples of the same station; σ_0^2 the variance due to errors (sub sampling, manipulation, technique). Fig. 6-24 shows the results of the calculation of s_2^2 (estimates of σ_2^2). Even for ten stations distributed in 1/16 of a square mile (400 m × 400 m) and about 200 m

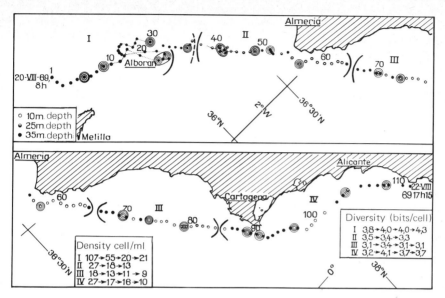

Fig. 6-23.—Maxima of density of phytoplankton (multiple concentric circles) along a section in western Mediterranean to the south of Spain: whole transect has been divided into 4 sections (I, II, III, IV) according to the density of phytoplankton and its specific composition: in the insets, average cell densities (cells/ml) and average diversities (bits/cell) are given for the four sections, as follows; the first number is the average corresponding to the maxima, the second the average of the stations one place nearer to the maxima, the third, the average of the stations two places nearer to the maxima and the fourth when given, the average of the stations three places distant from any maximum (after Margalef & Estrada, 1971).

apart a significant variance can exist between stations. This variance increases and reaches a plateau when an area of a quarter of a mile (800 m × 800 m) is reached. This plateau may be interpreted by assuming that when the area has reached a minimum which contains a peak and a lowering of the heterogeneous distribution, the variance is stabilized. As regards s_0^2 and s_1^2 their mean values are similar: the error due to the variability between successive samples at the same point (different bottles) is of the same order as that of the analytical error. By expressing these results in terms of the coefficient of variations (on the logarithmic values of the data) conclusions of the same kind are reached (Fig. 6-25). After a rapid increase this coefficient became stabilized at a mean value of 50–60%. For the determination of chlorophyll in the laboratory it is 8%.

Taking into account the movement of the water which defines the patches, the significance of successive measurements at the same point, when used to obtain development with the time, will be, a priori, small. When the variations are notable it seems that more certain conclusions can be drawn. Platt, Dickie & Trites have examined nine stations in St. Margaret's Bay (over an area of about 6 × 4 nautical miles) during the autumnal outburst (Fig. 6-26). Expressed in mg of chlorophyll *a* under 1 m², the development of the amount at different stations is very similar except

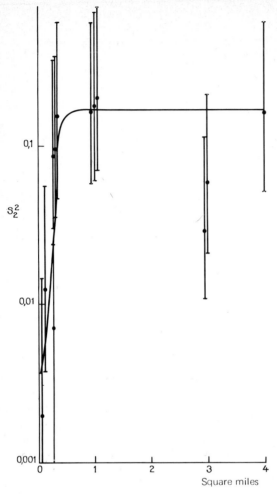

Fig. 6-24.—Variation of the between station variance (s_2^2) of amount of chlorophyll a, measured at a constant number of stations (10), as a function of the increase in total sampling area (1 square mile = 1610 × 1610 m): vertical bars represent confidence intervals (after Platt, Dickie & Trites, 1970).

during the fifth week at Stations 3, 6 and 8, which continued to increase while the others decreased. During the period of these observations the logarithmic coefficient of variation calculated for each week and each level (1, 5, 10 and 15 m) oscillated between 11 to 111% with an overall mean of 42%.

In concluding this study of the spatial variations of the phytoplankton we shall accept that this distribution is heterogeneous, presenting zones of concentration, or patches, alternating with zones of low densities and that detailed quantitative studies will be necessary to determine the causes of this heterogeneity and in order to obtain a satisfactory sampling procedure.

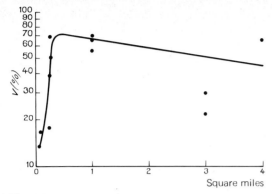

Fig. 6-25.—Change in coefficient of variation ($V\%$) of a single chlorophyll a measurement (transformed to natural logarithms) as a function of the increasing of sampling area (1 square mile = 1610 m × 1610 m); the number of stations (10) kept constant: line drawn by eye schematizes the change (after Platt, Dickie, & Trites, 1970).

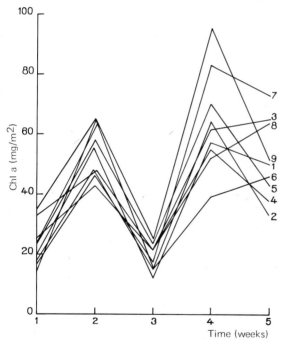

Fig. 6-26.—Variation in chlorophyll *a* concentration under a surface of 1 m² at nine stations in St. Margaret's Bay (near Halifax) during the autumn bloom of 1967 (after Platt, Dickie & Trites, 1970).

Platt (1972) from his continuous series of chlorophyll estimations suggested that the concentration of the phytoplankton at a given point is in a large part related to the turbulence. We shall examine the influence of this on the distribution of zooplankton later (Chapter 9-4).

6-10.—REGIONAL VARIATIONS: GEOGRAPHICAL VARIATIONS

Returning to Fig. 6-23 from Margalef and Estrada (1971) we see that the whole transect is divided into 4 sections, which take into account the density of phytoplankton and its specific composition; in other words, superposed on the spatial variation which we have considered above, there are variations of higher order which enable one to distinguish more or less vast regions. An example of such regional variations is given by Margalef (1967b) for the coast of Venezuela (Fig. 6-27) where, taking account of cell density as

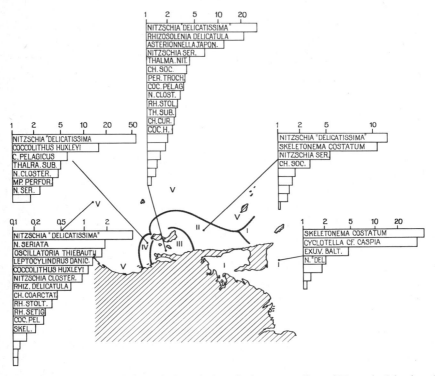

Fig. 6-27.—Regional variations of phytoplankton in the surroundings of Margarita Island on the N.E. coast of Venezuela: for each region the mean densities (from 0 to 50 m) of the most frequent species are indicated (in cells/ml, log scale): specific diversity, not calculated, appears to be low in region I and high in regions III and VI (after Margalef, 1967b).

well as the floristic composition and specific diversity, five regions may be distinguished. These regional variations are by no means fixed but their limits may be labile in time and in space.

Finally, after dealing with the spatial and regional variations it is logical to consider quantitative variations according to different geographical areas. Since, however, these may be more or less masked by seasonal variations, they are not always easy to determine. It is only by a comparison of well studied areas over several years that these can be determined. Marshall

(1933) tried to compare different zones, for which some data were available, by calculating the daily mean number of diatoms/l. She gave the following results:

Loch Striven (the Clyde sea) 1926: 1.009 400/1 ($1000/cm^3$)
(slight variation from one year to another).
La Jolla (California Coast). 1922: 40 000/1 ($40/cm^3$).
(average year chosen).
Plymouth 1915: 14 300/1 ($14.3/cm^3$).
Low Isles (Australia). 1928: 5800/1 ($5.8/cm^3$).

These numbers are in accordance with the conclusions of Lohmann (1920) who, after the "Deutschland" cruise, estimated that in the tropics the number of phytoplanktonic organisms per litre may be expressed in thousands, in temperate oceanic water in tenths of thousands, and in temperate coastal waters in hundreds of thousands.

6-11.—THE COLOURED WATER

We have already mentioned that massive "pollution" by phytoplankton sometimes occurs in the sea, one of the most famous being that observed on the 23rd July, 1947 off Venice (West coast of Florida) where *Gymnodinium breve* represented 98.99% of the phytoplankton with a density of 60×10^6 cells/l. Such "pollution" is more frequent in some regions than in others; most often the appearance and disappearance is sudden and unpredictable. The abundance of pigmented cells colour the sea water intensively and this phenomenon is often called "red-water" or "red-tide"; it corresponds to what the limnologists call a bloom. We shall use a more general term "coloured water" since other cells, green or yellow-brown may also be responsible for this phenomenon. Besides Florida the west coast of equatorial Africa is a zone where coloured water is frequent. Nuemann (1957) observed this phenomenon along 600 km of the coast of Angola. Coastal waters are not the only place where this phenomenon occurs. The "Marine Observer" contains regular observations made by commercial ships often in the open water; Fig. 6-28 is an example of this. Holmes, Williams & Eppley (1967) have studied in detail the red water produced by peridinians in the bay of La Jolla. Each time one or two species was largely dominant; *Gymnodinium* sp. and *Cochlodinium* sp. in May 1964, *Prorocentrum micans* in April, 1965, *Gonyaulax polyedra* in June and July 1965. The density of the cells may reach several millions or several tens of millions per litre and the amount of chlorophyll has reached several hundred µg/litre. Even with 4 or 5 µg/l (equivalent, for example, to 0.2×10^6 cells of *Gonyaulax polyedra*/l the water appeared coloured. These red waters were in the form of elongated patches often parallel to the coast, and variable in dimensions, the width between several tens to several hundreds of meters and the length from several hundred meters to several kilometers. The patches appeared in

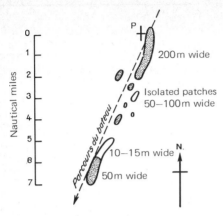

Fig. 6-28.—Distribution of patches of coloured water (deep brown-red) observed by S. S. Hupeh on the 17th July 1950, about 200 miles north of Formosa: position of point P: 28°25′N: 122°06′E (after Suffren, 1951, in Bainbridge, 1957).

a particular cycle, formed at about 10.00 h in the morning and disappearing in the evening before 16.00–19.00 h; this is explained by a cyclical behaviour in the positive phototaxis of this dinoflagellate (cf., Chapter 6-4).

A certain stability of the water is necessary for the establishment of this phenomenon, as is the case for the initiation of the spring diatom outburst. This water is found in periods of quiet seas and disappears when the sea becomes rough. Nitrogen and inorganic phosphate whose amount decreases inversely as density of the cells as shown in Table 6-6, are also needed. This nutrient supply takes place in the zones where the coloured water occurs either by upwelling or by the arrival of water from the continent. Moreover, it seems that, at least for the dinoflagellates, the alternation of concentration and of phototactic dispersion enables the cells to use the nutrient resources of a larger volume of water than that in which they are concentrated during daytime (Holmes et al, 1967). This may be also apply to growth factors like vitamin B_{12}.

The dinoflagellates which are most frequently the cause of coloured water, are not, however, the only group which can be responsible for this phenomenon. "Pollution" by the tropical cyanophycean *Oscillatoria* (*Trichodesmium*) may cover very large areas Goering et al. (1966), and Subba Rao (1969) has observed blooms and colouration of the water by the diatom *Asterionella japonica*, with over 90×10^6 cells/l ($36 \mu g/l$ of chlorophyll a) in the waters of the Bay of Bengal.

Finally, coloured water may have a disastrous effect on the environment. The species which constitute this may excrete toxins to the environment which may kill marine populations or the decomposition of the mass of organic matter so produced may reduce the quantity of available oxygen to a very low level and cause high fish mortality. Table 6-7 shows an example of a fall in oxygen concentration due to coloured water in the Bay of Loanda (West South Africa).

TABLE 6-6

Data on the red water of *Gonyaulax polyedra*, La Jolla, 30th June 1965 (after Holmes, Williams & Eppley, 1967).

Depth m	Chl. a (µg/l)	Density of Gon. (10^3 cell/l)	Particul. C (µg/l)	Org. dis. C (µg/l)	Particul. N (µg/l)	Org. dis. N (µg/l)	NO_3–N plus NO_2–N (µg/l)	Org. dis. P (µg/l)	PO_4–P (µg/l)
1	23.4	713	4000	1200	652	84	1	9	5
7	8.6	125	2100	980	194	78	2	8	17
12	3.6	11	370	850	63	59	107	—	42

TABLE 6-7

Variations in the oxygen (mg/l) in the Bay of Loanda, 5 miles off the coast (after Nuemann, 1957).

Depth (m)	0	10	20	30	50
27th July, 1951	9.36	3.85	3.76	3.53	3.80
20th September, 1951	9.31	0.59	0.60	0.24	0.24

Brongersma-Sanders (1948) has attributed to the coloured water a geological role: the accumulation of fossil fishes may be due to massive mortality of fishes caused by such water in zones of upwelling.

Particular mention must be made concerning red water, observed in different points of the globe and produced by the ciliate *Mesodinium rubrum* (40–50 μm). This contains chloroplasts and is capable of autotrophic nutrition; it belongs, therefore, to the phytoplankton, but the status of its chloroplasts is not very clear (Taylor, Blackbourn & Blackbourn, 1971).

CHAPTER 7

PRIMARY PRODUCTION

7-1.—REMARKS ON PRIMARY PRODUCTION: METHODS OF MEASUREMENT

There has long been criticism of all the quantitative techniques involved in primary production including the counting of cells, the estimation of chlorophyll, the estimation of plant matter in the water; all this gives only the quantity of phytoplankton at the moment of sampling, i.e., the standing crop. From an ecological point of view the biomass at any given time is much less interesting than is the production of new organic matter.

Since photosynthetic production is the basis of all the marine production it is termed primary production. Parallel to pelagic primary production of phytoplankton, there is also benthic primary production which has its origin in the algae living on the bottom.

More generally, the primary production in an aquatic medium is the whole organic matter produced by photosynthetic or chemosynthetic organisms. In the sea, the pelagic primary production is attributed classically to phytoplankton i.e., essentially to unicellular algae. In certain lakes it may be completely different. In the Green lake at Fayetteville (U.S.A.), studied by Culver & Brunskill (1969), more than 80% of the primary production is due to sulphur-oxidizing photosynthetic bacteria. This is a meromictic lake where the waters are poorly mixed and the deep layers which are not renewed contain hydrogen sulphide.

Theoretically, by repeated analysis over a period of time it is possible to estimate the primary production which is none other than the growth of the phytoplankton (the rate of change of the biomass with time). An example of this is given later (Fig. 7-16). The long time required for this method and the possible consequent errors due to movement and replacement of the water during the period of sampling as well as the consumption by herbivores, has lead to a search for other methods for the estimation and calculation of production. (For a general account of the methods for measuring the primary production see Vollenweider, 1969). It is possible to classify the methods into three groups.

In the first, samples of sea water containing its natural phytoplanktonic population are incubated in transparent vessels. In the older methods the oxygen production was measured, but most commonly the ^{14}C method or most recently the Coulter counter method is used. The results are then extrapolated to the mass of water under investigation. In the second group, instead of working with small volumes over a short time, the methods are similar but are integrated over considerable masses of water over a period of days, weeks or even months. We shall discuss the phosphate method, the CO_2 variation method and the oxygen variation method.

Finally, a third group of methods uses calculations relying on the relation between photosynthesis and luminous energy; the chlorophyll method or mathematical models are more or less complex.

Whatever method is used it is necessary to define the nature of the production measured as either gross or net production. Gross production corresponds to the entire synthesis of organic matter over a given time, and net production to the new organic matter available at the end of this period. For a given interval of time the net production will be equal to the gross production less the matter "lost" during this time, by respiration. But the expression of the results on a day-time basis meets with some difficulties e.g., the necessity, in order to estimate correctly the daily net production, for the measurements made during daytime hours to take into account also the respiration during the night. For data obtained over a long period, information on the consumption by herbivores may be lacking.

We must note that there may be some confusion between the two terms "production" and "productivity". We shall use the terms according to the definition of Wood (1967), namely, production is the quantity of matter produced in unit time, and productivity is the capacity to produce per unit time. Production will, therefore, have an absolute value, while productivity is only relative (see Chapter 7-12).

7-2.—THE OXYGEN METHOD

The general outlines of this method have already been given in connection with Jenkin's experiments (Chapter 2-4). A sample of sea water is collected and the dissolved oxygen determined. Bottles containing 200 to 400 ml are filled with the water and immersed at the depth at which they were collected. One bottle is transparent and receives light while the other is darkened and does not receive light. At the end of a period of 24 or 48 h, the oxygen content of the water in each of the bottles is estimated. The increase in oxygen in the "light" bottle is interpreted as a measure of the carbon assimilated by the phytoplankton in suspension in the sample of water and not lost by respiration; it is net photosynthesis (P_n). The decrease of oxygen in the dark bottle is interpreted as a measure of the carbon lost by the respiration of the phytoplankton (R_p). The sum of these two values is considered as equivalent to the total carbon assimilation i.e., gross photosynthesis.

$$P_g = P_n + R_p$$

In reality, it is not so simple, and as shown by Pratt & Berkson (1959), certain corrections must be applied.

(a) The first cause of error lies in the fact that in sea water enclosed in bottles bacteria develop quite rapidly and consume oxygen (see Fig. 4-7).

The respiration of the phytoplankton measured by the decrease in oxygen in the dark bottle, is, therefore, overestimated since bacterial respiration may reach 40–60% of the total. The net photosynthesis, measured by the increase of the oxygen content in the "light" bottle, is underestimated by a value equal to the quantity of oxygen used in the respiration of bacteria (R_b). The respiration of the zooplankton (R_z), which cannot be completely eliminated, increases this underestimate. Under these conditions only the gross photosynthesis is correctly measured.

$$P_g = P_{na} + R_p + R_b + R_z$$

when P_{na} is the apparent net photosynthesis (Fig. 7-1).

(b) The second cause of error is due to the fact that the phytoplanktonic population in vitro does not remain similar in the two vessels. While it is almost stable in the dark bottle in the "light" bottle it can increase from 50 to 100% since the experiments are sometimes of a somewhat long duration in order to obtain measurable changes in the amount of oxygen. As a consequence, the values obtained do not correspond exactly to the production in the sea water from which the sample was taken; this is a serious cause of error. The bacteria in the bottle may also interfere; as a result of their development, accelerated by contact with the glass walls of the bottle, they can produce material (nitrogen, phosphorus, vitamins) available to the phytoplankton and able to favour the growth.

(c) The photosynthetic ratio i.e., the ratio between the molecules of oxygen liberated to the CO_2 absorbed is not necessarily equal to unity. This is true for the synthesis of hexoses but it is about 1.4 for lipid synthesis, and for proteins it may vary depending on the source of nitrogen e.g., 1.05 if it is NH_3–N, 1.6 if it is NO_3–N which is rich in oxygen. Under natural conditions it is not easy to determine this ratio and usually a value of 1.2 is accepted (Strickland, 1960). This uncertainty should be taken in account when comparing results obtained by other methods.

(d) Due to the limits of accuracy of the estimation of oxygen the method can only be satisfactory when photosynthetic activity is adequately high. The prolongation of the experimental period in order to overcome this difficulty gives rise to other errors noted above.

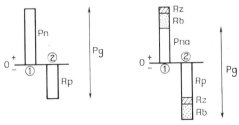

Fig. 7-1.—Scheme illustrating the significance of measurements by the oxygen method: left, theoretical measurement; right, real measurement: Pn, net production; Rp, phytoplankton respiration; Pg, gross production; Rz, zooplankton respiration; Rb, bacterial respiration; Pna, apparent net production: 1 = light bottle; 2 = dark bottle.

Finally, the oxygen method largely gives an approximation to the gross photosynthesis and is not adequate for waters whose production is poor.

In order to obtain the daily net production from measurements limited to daytime periods Crouzet (1972), in agreement with Brown (1953) and Brown and Webster (1953), consider that night-time respiration is the same as day-time respiration, and he corrects the net production by taking into account the respiration. This net production, so corrected, corresponds to about 60% of the original net production. Even so, errors in this method due to bacterial respiration remain difficult to estimate.

7-3.—CARBON-14 METHOD

The method was first developed by Steemann-Nielsen (1952), and depends on the following principal. A given quantity of $^{14}CO_2$ is added to sea water of which the concentration of CO_2 is known. If it is accepted that the $^{14}CO_2$ is assimilated by the phytoplankton only by photosynthesis, which is almost true, and that $^{14}CO_2$ is assimilated at the same rate as $^{12}CO_2$, then by determination of the ^{14}C in the phytoplankton at the end of the experiment, it is possible to estimate the total quantity of carbon fixed by multiplying the quantity of assimilated ^{14}C by a factor corresponding to the ratio $^{14}C/^{12}C$ at the beginning of the experiment. In practice, water samples are collected at different levels, generally chosen by taking into account the different levels of illumination. The water is transferred to bottles, ^{14}C added bicarbonate, and the bottles are immersed at their original depth. At the end of the experiment the water is filtered and the quantity of ^{14}C assimilated by the phytoplankton is determined by measuring the β-radiation emitted by the filter. For more details on this technique see Brouardel & Rinck (1963) and Strickland & Parsons (1968).

The production is expressed for each depth as mg of fixed $C/m^3/h$. It varies between 0 and several tens of mg (for instance Platt & Subba Rao (1970) obtained 81.55 $C/m^3/h$ during a phytoplanktonic outburst). Generally the measurements done over half daylight time enable one to obtain the daily production ($C/m^3/day$) by multiplying the results by 2. Yet the existence, under certain conditions of a decrease in the photosynthetic capacity of the phytoplankton during the afternoon (afternoon depression) may bias this extrapolation (Eppley & Strickland, 1968).

Platt (1971) in contrast to most other authors made his measurements from 10.00 to 17.00 h, but developed a method of calculation to convert the results to daily production.

The ^{14}C method, which seems to be very attractive, in practice, requires, nevertheless, important corrections (Steemann-Nielsen & Hansen, 1959; Strickland, 1960; Yentsch, 1963; Saurnia, 1968a).

(a) The $^{14}CO_2$ may be incorporated into the organic matter of the phytoplankton in ways other than by photosynthesis and parallel experiments in the dark show a fixation of the order of 1–5% of the maximal photosynthesis, at saturation, and a correction for this is generally made. It seems that this correction is greater for populations of algae deprived of nutrient salts, probably by a decrease in photosynthesis at saturation. In this way, dark fixation in oligotrophic tropical waters may be over 50% of that in light and sometimes even equal to it. Relevant to this are the results of Brouardel (1973) who showed that the dark fixation increased considerably with the temperature and much care is required in the interpretation of production measurements made at high temperatures and of Morris, Yentsch & Yentsch (1971) who observed that the ratio between dark and light fixation increases when the density of cells is decreased. Indeed, our knowledge of how CO_2 is integrated in the biochemical cycles is still insufficient for a correct interpretation of the measurements.

(b) The $^{14}CO_2$ is not assimilated at exactly the same rate as $^{12}CO_2$; there is a difference of 5 to 6%. This isotopic effect is compensated by multiplying by a correction factor of 1.05 or 1.06.

(c) The ^{14}C assimilated in the organic matter may be returned by respiration; according to Steemann-Nielsen 60% of the CO_2 liberated by respiration during the experiment should come from CO_2 which was previously fixed by photosynthesis and should contain, therefore, $^{14}CO_2$ which is so returned. It is necessary, in order to correct for this to know the effect of respiration. Steemann-Nielsen estimated it as 10% of the maximal photosynthesis at saturation ($R/P_{max} = 0.1$) which gives a correction of 6% (60% of 10%). This has been contradicted by Ryther (1955) who in experiments on a culture of *Dunaliella* (known previously as *Chlamydomonas*), showed that respiration, when the nutrient salts began to disappear and the multiplication of the cells becomes slower, is of the same order as fixation by photosynthesis ($R/P_{max} = 1$). A difficulty in the employment of this method arises; according to Steemann-Nielsen & Hansen (1959) deficient cells do not persist in the sea, but sink rapidly or are consumed; however, this is not generally accepted and any correction for respiration is still a subject of argument.

Recent work has confirmed the difficulty of this problem in a study of R/P ratio i.e., respiration to photosynthesis during the spring outburst of phytoplankton. Platt & Subba Rao (1970) found that it varied from 0.1 to 5 in their observations, but instead of increasing regularly with the age, as in cultures, in the ageing of natural populations it passed through a minimum before a further increase.

(d) In order to obtain satisfactory estimations, the organic matter which was incorporated into ^{14}C must not diffuse out of the cells and so has to be effectively collected on the filter. This is not always the case and, as has long been known, a significant part of the organic matter produced may pass to the milieu, Anderson & Zeutschel (1970) after completing the usual determination of production, measured the ^{14}C of the organic matter in the filtrates with a liquid scintillation spectrometer; after eliminating residual inorganic ^{14}C, they obtained for the "production" passing into solution values reaching 10 to 20% of the particulate production by the phytoplankton. This is an important source of error, even though it is relatively much smaller for very productive eutrophic than for oligotrophic waters.

(e) Finally, Arthur & Rigler (1967) draw attention to the loss of ^{14}C which may result from the breakage of cells during filtration. The importance of this loss depends on the conditions of filtration. By determining and applying such a correction these authors obtained good agreement with simultaneous measurements using oxygen method (Table 7-1).

TABLE 7-1

Fixation of carbon calculated from simultaneous measurement of production; classical measurement of ^{14}C; values corrected for the loss on filtration; oxygen measured (after Arthur & Rigler, 1967).

	Fixed carbon (^{14}C) (mg/l/h)		Fixed carbon (O_2)(mg/l/h)	Ratio	
	(1) not corrected	(2) corrected	(3)	(1)/(3)	(2)/(3)
1st November, 1962	0.217	0.302	0.29	0.74	1.03
7th June, 1963	0.108	0.228	0.23	0.47	0.99

In spite, then, of its apparent simplicity the ^{14}C method gives rise to new problems. One of the most recent concerns the presence and determination of β-activity in the ampules of ^{14}C, used for the determinations. The biological determination of this activity has lead Steemann-Nielsen (1965) to propose to multiply previously obtained results by 1.45 when the activity of the ampule had been determined by physical methods. So corrected, the measurements should correspond to gross production. This correction has not yet been completely accepted (Brouardel, 1971).

The determination of production is generally made by immersing the experimental bottles from sunrise to sunset or during half a day, this procedure immobilizes the research vessel for many hours at the station. Such determinations are defined as in situ methods* and are sometimes replaced by a simulated in situ method. In this, the bottles are placed on board in an incubator which receives daylight more or less attenuated, or artificial light. This technique simplifies the operation at sea but the interpretation of the results again meets with difficulties (Kiefer & Strickland, 1970).

Recently, Jitts, Morel & Saijo (1973) in UNESCO-SCOR Working Group No. 15 on board the "Discovery" made a precise comparison of the in situ method, the simulated in situ method, and a third method, illustrated in Fig. 7-2. The samples were collected at a given level and, in this last method, divided into different fractions which were exposed in a xenon incubator to graduated irradiance; using ^{14}C this enables one to establish a production curve as a function of the irradiance, $P = f(I)$. At the level from which the sample of water was taken the changes in irradiance with the time were also measured ($I = f(t)$). The combination of the two results gives the instantane-

---

*Brouardel retains this term for methods where all the manipulations are effectively done in situ; the usual method is called by him the reference method.

ous value of production at the different times, and the integration of these values for the entire day gives the production. Experimentally, in rich or poor waters, these authors found an excellent agreement between the calculated production (P_c) and the simulated in situ production (P_s) with a correlation coefficient of 0.983 and a regression equation:

$$P_c = 0.979 \, P_s + 0.25$$

The ratio P_c/P_s is very close to unity. The comparison between the in situ production (P_i) and the simulated in situ production (P_s) is also very satisfactory, with a correlation coefficient of 0.939 and a regression equation:

$$P_s = 0.979 \, P_i + 1.53$$

If only, however, the samples from the surface are considered, the regression equation becomes (correlation coefficient = 0.959):

$$P_s = 1.391 \, P_i + 0.73$$

This difference is attributed to UV radiation, small in the simulated in situ method and in the xenon incubator and higher in the true in situ method, which decreases the value of P_i.

The irradiance was measured in quanta in the range 350–700 nm.

In conclusion, the ^{14}C method may reach an accuracy of $\pm 5\%$ when it is well done and it is a good method. Its sensitivity is 50 to 100 times that of the oxygen method (Strickland, 1960) and it gives a value essentially close to the net production for a daytime period. It does not give any indication of the quantity of organic matter used for respiration and so does not allow one to calculate either the gross production for 24 h, or the daily net production. The ^{14}C method represents in practice an indispensable tool in the analysis of the aquatic primary production but its results have, as far as possible, to be supplemented by other methods in order to obtain definitive data.

7-4.—PARTICLE COUNTING METHOD

This method was introduced by Cushing & Nicholson (1966) and was used by Sheldon & Parsons (1967) and Parsons, Stephens & LeBrasseur (1969). The particle-size frequency distribution is determined in a sample of water with a Coulter counter. This enables one to obtain the volume corresponding to each size class. The counter is then used to follow the development of the number and volume of the particles in the same water incubated in the light or dark bottle. It is then possible to determine the production in volumes of the different size ranges of phytoplankton and to convert them to carbon or calories according to the available data.

A priori it may be seen that it is impossible to separate the inert particles from the phytoplankton in counting. This major difficulty was solved by Cushing & Nicholson in the following way; from different measurements of volume the relative rate of growth of the phytoplankton k is calculated according to the equation (see Appendix equation (7)):

$$\frac{V_t - D}{V_{t_0} - D} = e^{k(t-t_0)}$$

where V_t is the volume of the phytoplankton at the time t, V_{t_0} is the volume

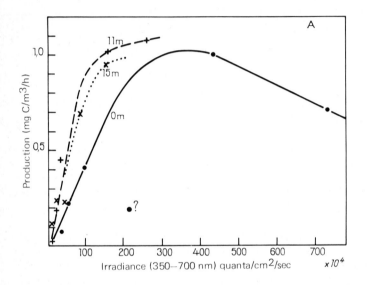

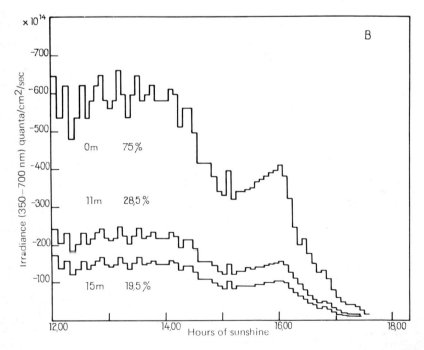

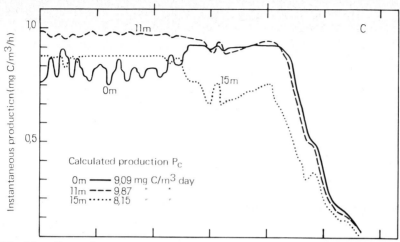

Fig. 7-2.—Changes in the primary production: A, curves showing the production measured as a function of the irradiance ($P = f(t)$), in xenon incubator, for samples collected at depths of 0, 11, and 15 m; B, changes in the irradiance at depths of 0, 11 and 15 m as a function of the day hours ($I = f(t)$); C, variation of the instantaneous production, as a function of the time of day calculated for three depths, 5, 11 and 15 m from the preceding curves $P = f(I)$ and $P = f(t)$: calculated daily production (Pc) obtained by integration (after Jitts, Morel & Saijo, 1973).

of phytoplankton at the time t_0 and D the volume of the detritus. This rate of relative growth is not exact since in the counting, the volume D, of the detritus particles interferes, and, therefore, it varies as a function of time. The artifact in the calculation consists of attributing to D, supposed to be invariable, arbitrary values: the value at which k shows no further change is considered the true value. This may be seen clearly in Fig. 7-3 where D is slightly less than 25.

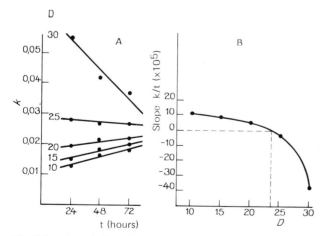

Fig. 7-3.—A, variation of relative growth rate k, as a function of time, for different arbitrary values of detritus volume ($\mu m^3 \cdot 10^4$/ml): B, range of the slope of the lines in A as a function of D: zero slope gives the real value of D (after Sheldon & Parsons, 1967).

7-5.—THE PHOSPHATE METHOD

Under normal conditions phytoplankton cells contain a relatively constant amount of phosphorus. The decrease in inorganic phosphorus in the sea, in successive estimations, is related to its absorption by the phytoplankton. Theoretically, this should allow one to calculate the quantity of new cells produced (expressed as carbon). Two assumptions must be made in order to use this method:

1. that the regeneration and re-cycling of inorganic phosphorus is low in the zone in question during the period considered,
2. the absorption of organic phosphorus compounds is negligible.

Since it is difficult to accept these assumptions, the phosphate method has not been much used (Steel, 1958; Legendre & Watt, 1970); however, the automation of the phosphate estimation enables one to consider its possible use and Fig. 7-16B, relevant to the Peruvian upwelling is a good example.

7-6.—CO_2 CHANGES METHOD

Since CO_2 is absorbed during photosynthesis it is logical to try to follow its variations in situ in order to estimate primary production. This was one of the first methods used; Atkins (1922) measured production in the English Channel using this method. Recently, Teal & Kanwisher (1966) have used it in waters near to Woods Hole, particularly in Vineyard Sound. They measured the partial pressure of the CO_2 or pCO_2 and calculated the variation of the total CO_2 in the sea water. In these measurements not only of photosynthesis, but also the respiration of the phytoplankton and of the zooplankton interfere and the results obtained correspond, therefore, to the apparent net production, i.e., $P_g - R_p - R_z$.

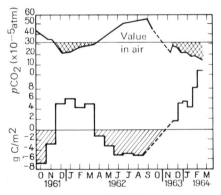

Fig. 7-4.—Changes of pCO_2 in sea water as a function of time and the values of monthly mean net production (positive or negative) deduced from these changes in the water of Vineyard Sound (North west Atlantic) (after Teal & Kanwisher, 1966).

The pCO_2 method has an advantage over the change in oxygen method, particularly since it is possible easily to make continuous recordings at the surface, during the passage of the boat.

The measurements of Teal & Kanwisher enabled them to observe patches of 2 to 20 km in diameter, with deviations of 14% from the mean pCO_2 of the region; this was during a spring outburst, in which, pCO_2 was strongly negatively correlated with the amount of chlorophyll ($r = -0.85$; $P = 0.005$) but in May the correlation was not so strong ($r = 0.57$; $P = 0.05$), and in July and December it disappeared completely. Fig. 7-4 shows an example of how the production results were obtained. During six months following the beginning of the winter or spring outburst, the apparent net production ($P_g - R_p - R_z$) was about 17 gC/m² for Vineyard Sound, 70 gC/m² for Nantucket Sound and more than 100 gC/m² for Georges Bank.

7-7.—CHANGES IN OXYGEN METHOD

Variations in the amount of oxygen in the water are a result of different opposing actions: physical on the one hand by diffusion, particularly at the sea surface, and biological reactions on the other, namely, photosynthesis and respiration. The difference between the last two give the net production which may be calculated from the changes in oxygen, on condition that they are corrected as a function of the diffusion; but the water masses are moving and one must eliminate this cause of error.

Gilmartin (1964) used this method in the relatively simple conditions of a fjord in British Columbia (Fig. 7-5). Taking into account the changes of water at the fjord inlet and diffusion at the surface, the variation of

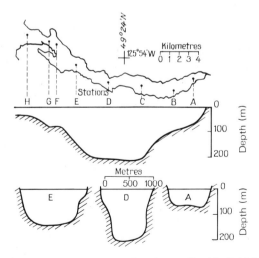

Fig. 7-5.—Scheme of Indian Arm, a fjord in British Columbia, studied by Gilmartin: below, longitudinal profile and sections at Stations A, D and E (after Gilmartin, 1964).

biological origin in the oxygen balance of this fjord may be estimated with a good degree of approximation. He obtained in this way a net production of 381 gC/m^2/yr. Using the negative variations obtained out of the photic zone and due to the oxidation of the organic matter produced, this author could extrapolate for the whole fjord an estimation of the minimal organic matter oxidized by respiration, equivalent to 290 gC/m^2/year. The gross production so estimated was 671 gC/m^2/year.

Riley (1956), and more recently Minas (1956) in the Mediterranean, have also used variations in oxygen for the estimation of the primary production.

7-8.—THE CHLOROPHYLL METHOD

Experiments show that for saturation irradiance it is possible to establish a relation between the chlorophyll *a* measured in mg, and the photosynthetic production catalysed by it, expressed in mg C/h. Table 7-2 gives some determinations of the ratio, production/chlorophyll. This ratio, number or coefficient of assimilation (which we call Q_{max} to indicate that it is established at saturation where it has maximal value), varies considerably, from 1 to 10 which in practice makes it difficult to use. It is necessary to estimate it more precisely for the population of the water studied. Ryther & Yentsch (1957), who introduced this method, adopted a coefficient of 3.7 (Woods Hole, Mass) and Grall (1972) one of 3.0 (Roscoff); more generally, a value of 3-4 is used as a first approximation.

The second factor which interferes is light. As we have already seen, production is proportional to irradiance at values lower than the saturation value. Below this value, and for a given depth, one only needs to know the amount of chlorophyll and the irradiance to calculate primary production. If the irradiance is not measured directly, its value at the surface together with the extinction coefficient of the water will enable one to estimate it. Since the maximal photosynthesis P_{max}, ($Q_{max} \times$ Chl *a*) is obtained for saturation irradiance I_s, it is possible to obtain the photosynthetic production for lower irradiances from the equation:

$$P = \frac{I}{I_s} \times Q_{max} \times \text{Chl } a$$

TABLE 7-2

Approximate values of ratio of organic carbon synthesized (in mg/h) and the quantity of chlorophyll (in mg) with light at saturation value (after Strickland, 1960).

1–2	Natural marine population in incubator (Holmes)
About 4.5	All pigments added; maximal values for the photic zone (Currie)
3	Coastal water (Riley)
6	Mean for coastal water (Ryther & Yentsch)
5–10	Various pure cultures (Ryther & Yentsch)
3	Cultures of *Dunaliella euchlora* (Ryther)

Adopting the mean curve (Fig. 7-6), established by Ryther from his various determinations (Fig. 2-17), the value of I_s is 0.1 cal/cm²/min (7×10^{-3} watt/cm²) with the coefficient Q_{max} of Ryther & Yentsch given above and I in cal/cm²/min the equation becomes:

$$P = 10 \times I \times 3.7 \times \text{Chl } a$$

By assuming over the entire daily hours and expressing the chlorophyll a in mg/m³, the daily production in mg C/m³/day is obtained for the depth concerned.

This method is limited to relatively low irradiances, lower than I_s. In order to overcome this limitation Ryther & Yentsch used the following procedure; from the curves in Fig. 2-21 they made a summation of the relative photosynthesis for the different levels of irradiance over the entire day. Using the same operation on a certain number of days scattered over the whole year they obtained the curves of Fig. 7-7 which give the daily relative photosynthesis at different levels as a function of the total daily irradiance (300–5000 nm). At an irradiance level d (in % of I_0), for a given daily total

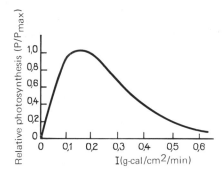

Fig. 7-6.—A, mean curve, established from data of Fig. 2-17, giving the relation between the irradiance and relative photosynthesis (after Ryther, 1956).

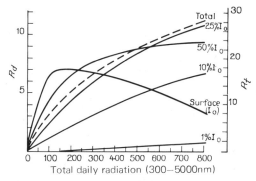

Fig. 7-7.—The relation during the year between daily total irradiance (300–5000 nm) reaching the surface continuously, and between daily relative photosynthesis ($2\Sigma_0^{12h} P/P_{max}$) at different irradiance levels (R_d) (broken line) and the total relative daily photosynthesis for all the levels R_t; total daily irradiance in cal/cm² (after Ryther & Yentsch, 1957).

irradiance, it is therefore, possible to estimate the production, P_d (in mg C/m^3/day) at this level, according to the equation:

$$P_d = R_d \cdot P_d^s$$

where R_d is the relative photosynthesis at the depth d, for the daily total irradiance considered, and P_d^s is the photosynthesis at saturation of the sample of water from depth d, expressed in mg C/m^3/h. Expressing P_d^s as function of the amount of chlorophyll/m^3 for the depth d, the equation becomes:

$$P_d = R_d \cdot Q_{max} \cdot \text{Chl } a(d)$$

and graphical integration, from the values obtained for the different depths, enables one to calculate the total production under 1 m^2.

If the phytoplankton is homogenously distributed throughout the photic zone, it is possible to simplify the above calculation by using the following equation (Ryther & Yentsch):

$$P_{eu} = \frac{R_t \cdot Q_{max} \cdot \text{Chl } a}{K}$$

where P_{eu} is the total production in the photic zone, expressed in mg of C/m^2/day, R_t is the total relative photosynthesis obtained for the value of the daily total irradiance (the broken line in Fig. 7-7), K is the extinction coefficient of sea water and Chl a the mean amount of chlorophyll a (mg/m^3) in the entire photic zone. Using this equation Ryther & Yentsch obtained results which are in a satisfactory agreement with those obtained by the ^{14}C-method.

Briefly, the chlorophyll method allows a first approximation of primary production. It is far from having the accuracy of the ^{14}C method, but it may be acceptable.

7-9.—MATHEMATICAL MODELS

It is through a consideration of the ideas of the chlorophyll method that in the recent years certain authors have suggested mathematical equations which rely on relatively simple measurements to give an estimate of the primary production. The numerous calculations involved are evidently made by computers, and this explains why these methods have only recently been developed. One of the most interesting lines of work is that of Talling (1957) which was established on the basis of Smith's equation (1936), was further developed by Vollenweider (1965) and continued by Fee (1969). Essentially it is a calculation by double integration, integration of the depth of different levels and integration during the day or half a day, of different times. As in most of the ecological models, a number of simplifications are accepted, namely:

1. the water column is optically homogeneous;

2. the water column is isothermic;
3. $P = f(I)$ can be formulated and is constant with depth and time;
4. the phytoplankton is homogeneously distributed in depth and over time.

Fee's subterminal equation is the following:

$$P = \int_{-\lambda/2}^{+\lambda/2} \int_0^\infty \frac{P_{\text{max-th}} I_{0(t)}/I_k}{\sqrt{1+(I_{0(t)}/I_k)^2}\{1+(aI_{0(t)}/I_k)^2\}^n} \, dz\,dt \tag{1}$$

where P = photosynthetic rate;
λ = length of daily irradiance;
$P_{\text{max-th}}$ = theoretical maximal photosynthesis (when $a = 0$ or $n = 0$);
$I_{0(t)}$ = irradiance on the surface at time t ($t = 0$ at noon);
a and n = model factors
I_k = theoretical parameter of luminous saturation (when $a = 0$ and $n = 0$).

I_k was introduced by Talling and is graphically defined in Fig. 7-8. Curve 1 is theoretical, and it is Curve 2 which is real; after a proportional increase of P as a function of I, it reaches a saturation value and then there is an inhibition. In order to take in account these phenomena, the arbitrary factors n and a which have no biological significance and which were introduced by Vollenweider in the following equation are used:

$$\frac{P}{P_{\text{max-th}}} = \frac{(I/I_k)}{\{[1+(I/I_k)^2][1+(aI/I_k)^2]^n\}^{1/2}} \tag{2}$$

This equation gives a family of curves the form of which depends on the parameters a and n. An increase in a gives a general lowering of the curve at all luminous intensities, while an increase of n gives a lower slope after the point of maximal curvature (Fig. 7-9). If a or $n = 0$ the underlined term in equation (2) becomes equal to 1 and the curve is asymptotic. If a and n differ from 0 we have a curve characterized by an almost linear section,

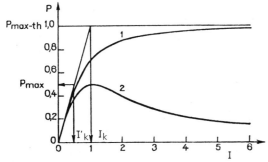

Fig. 7-8.—Diagrammatic presentation of P as a function of I, illustrating the definition of the different parameters: I_k is the irradiance for which the extrapolation of the linear part of the curve $P = f(I)$ cuts the asymptote $P = P_{\text{max-th}}$: I'_k is the irradiance at which the extrapolated straight line cuts the horizontal, $P = P_{\text{max}}$, when P_{max} is the production at saturation: 1, theoretical curve; 2, true curve (after Fee, 1969).

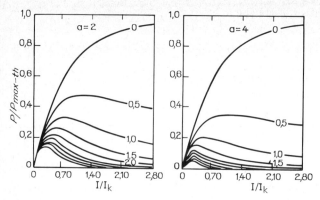

Fig. 7-9.—Theoretical curves, $P = f(I)$, for two values of the parameter a and different values of n (0.05, 1.0 etc.,) (after Fee, 1969).

followed by an optimal section and then by an inhibition section, i.e., the three characteristic sections of a real photosynthetic curve $P = f(I)$.

Equation (1) was qualified as subterminal; Fee later modified it and introduced the extinction coefficient K, so that $dI = -KId_z$, replacing d_z by this value. The necessary data are:

1. The distribution of the photosynthetic rate in situ (or in incubator, with gradiated luminous energy) as a function of depth or: $P = f(I)$;
2. the rate of photosynthesis at optimal irradiance I_s (in situ, or in incubator) or P_{max};
3. the extinction coefficient of the water, K;
4. the changes, during a day, of the irradiance at the surface, I_0.

A first programme uses as initial data the experimental data relating P to I, P_{max} and a first approximate value for I'_K ($I'_K = I_K \cdot P_{max}/P_{max\text{-}th}$). This programme gives a, n and I'_K. A second programme uses as the initial data the distribution of I_0 during the day, n, a, I'_K, P_{max}, λ and K. From relatively few data it is possible in this way to calculate production.

There are also many other types of models and Patten (1968) has reviewed them in detail. Parsons & Anderson (1970), during their investigations on the North Pacific compared the estimation of the net primary production by ^{14}C measurements (see p. 147) with that obtained by using different models. With Steele & Menzel's model (1962):

$$P_n = \frac{\alpha I_0 \text{Chl}_a}{K} \left[e^{-2e^{-Kz_m}} - e^{-2} - \frac{rKz_m}{P_m e h f(N)} \right]$$

the results were comparable. The meaning of the symbols is the following:

P_n = net production for 24 h;
α = photosynthesis per unit of luminous energy and per unit of chlorophyll;
I_0 = irradiance at the sea surface, in cal/cm²/day;
Chl a = mean concentration of chlorophyll a in the homogeneous superfi-

cial layer with a thickness z_m;
K = extinction coefficient;
r = maximal respiration rate (value adopted by Stele (1962): 0.07 gC/gC/h;
P_m = maximal photosynthetic rate: 0.066 gC/gC/h;
h = irradiance, hours in the day;
$f(N)$ = a coefficient which is a function of the limitation of photosynthesis by the availability of nutrient salts; equal to 1 when these are not limiting.

Parsons & Anderson used for α the value of 0.17 instead of 0.27 used by Steele & Menzel, justifying it by theoretical reasons, and adopted for I_0 special corrections. Under these conditions, they obtained the following relation between the production by ^{14}C corrected by night respiration and the calculated value of P_n:

$$P_{^{14}C} = 0.90(\pm 0.09)P_n + 27(\pm 22)$$

this is very close to the expected relation 1:1 (in the brackets the confidence interval with probabilities of 95% for the slope and of the intersection of the ordinate).

7-10.—RESULTS OF MEASUREMENTS

Primary production, particularly when it is estimated on sea water samples, is expressed in mg of organic carbon produced/m³ per hour, or 1 m³/day. In order to facilitate comparison, the results obtained at different depths, are generally integrated for the whole column of "productive" water and the primary production then expressed in mg or gC/m²/day.

We have numerous measurements of primary production but as already indicated, particularly as regards the ^{14}C method, they are not always easy to interpret and so to compare. First we may see that there are considerable variations in production according to the location. An example may be taken from the section worked by "Galathea" in the Atlantic (Fig. 7-10) in May and June 1952 (Steemann-Nielsen & Jensen, 1957). The lowest values 40 to 50 mg, correspond to the Sargasso Sea north of the Antilles and are incidentally the lowest values of production obtained by the "Galathea" expedition in any ocean; 85% of the results lie between 100 and 1000 mg, 70% between 100 and 500 mg. The maximum, 3800 mg, was observed in Walfish Bay, on the west coast of South-Africa and this corresponds to the presence of coloured water, where the productive layer was limited to the first 80 cm. Such geographical variations have, however, a real comparative meaning only if production is estimated over the whole year.

There are also considerable seasonal variations. In the English Channel, off Roscoff, the values of Grall (1966) range from a few mg in February to 1670 mg in August. Platt (1971) has given monthly means of the daily rate of production obtained over several years in St. Margaret's Bay, near Halifax

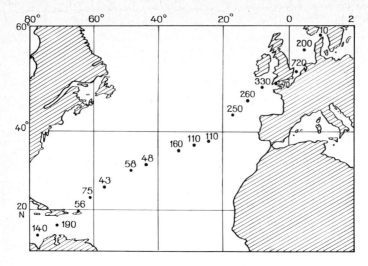

Fig. 7-10.—*Galathea* transect in the North Atlantic, in May and June 1952: for each station the primary production is indicated (mgC/m^2/day) measured by ^{14}C (after Steemann-Nielsen, 1957).

(Fig. 7-11); after a low winter production there is a considerable increase in March, corresponding to the spring outburst of phytoplankton. The level of production falls in May and rises again in August with an autumnal maximum in September, which accompanies the autumnal diatom outburst.

The estimation of the annual production requires therefore numerous measurements using, if possible, different methods. We shall examine the results obtained for different regions.

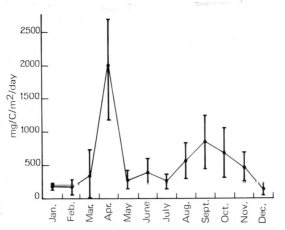

Fig. 7-11.—Monthly means of the daily primary production measured by ^{14}C (confidence interval is indicated for $P = 0.05$) in St. Margaret's Bay near Halifax (N.E. Canada) from June 1966 to June 1969 (after Platt, 1971).

North-West Atlantic

Ryther & Yentsch (1958) estimated the primary production off New York, using the chlorophyll method with the determination at each station of the ratio between photosynthesis and chlorophyll by the oxygen method in an incubator over 24 h: they adopted the mean coefficient of 3.7, in spite of its considerable variability, and so calculated gross production. For a certain number of stations, they made supplementary determinations by the ^{14}C method: out of 17 pairs of values so obtained, 5 agree to within 10%, while 4 give values for the production by ^{14}C less than half of those obtained by calculation from the chlorophyll. During the various cruises, important variations were often recorded from one station to another, with stations 10 to 15 miles apart. From these results, Ryther & Yentsch established the mean daily primary production for 5 coastal stations, 5 deep water stations, and 5 intermediate stations (Fig. 7-12). It appears on these diagrams that the annual range of variation is not very different in the coastal water (0.20–0.85 gC/m^2/day) and the open water (0.10–1.10 gC/m^2/day). For the very closed Long Island Sound using the data of Riley (1956), Ryther & Yentsch obtained an annual range of 0.23 to 1.70 gC/m^2/day for 1952–1953 and of 0.27 to 1.51 gC/m^2/day for 1953–1957; Riley's estimations were made by following the chemical variations in situ throughout the water column over a period of several weeks. From similar data (Riley, 1957) the gross production in the northern part of the centre of Sargasso Sea varies from 0.09 to 0.89 gC/m^2/day. Taking all these values and using a coefficient of 1.25 to transform the oxygen data to carbon and interpolating by an approximate curve the results during the year, Ryther and Yentsch calculated the annual primary productions shown in Table 7-3.

We recall the estimations of Teal & Kanwisher (1966) obtained by following the changes of pCO_2 namely, 100 gC/m^2 for Georges Bank, 70 gC/m^2 for Nantucket Sound and 17 gC/m^2 for Vineyard Sound. It is very

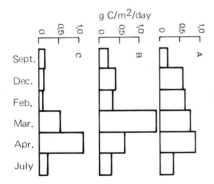

Fig. 7-12.—Daily primary production off New York, at 5 coastal stations with less than 50 m depth (A), 5 intermediate stations from 100 to 200 m depth (B) and 5 stations in the open sea with depth >1000 m (C), during six successive cruises (1956–1957) (after Ryther & Yentsch, 1958).

TABLE 7-3

Annual gross primary production in the North West Atlantic (after Ryther & Yentsch, 1958).

	Total depth of zone (m)	gC/m²/yr
Long Island Sound	25	380
Continental shelf off New York	25–50	160
Intermediate zone off New York	50–1000	135
Surroundings of the shelf off New York	1000–2000	100
Northern centre of Sargasso Sea	> 5000	78

difficult to compare them with the preceding data since they correspond only for 6 months, starting from the winter or spring outburst and are net production values, corrected for zooplankton respiration.

Platt, whose annual curve of primary production is reproduced in Fig. 7-11, estimated the annual production in St. Margaret's Bay as 190 gC/m² (±60). It is a value intermediate between the gross and net production. In Table 7-3, it is seen to lie between the production in Long Island Sound and that of the continental shelf off New-York.

Western Mediterranean

Since the early data of Brouardel & Rinck (1963), relatively numerous observations are available, mainly due to the work of Minas (1970) and Brouardel (1971). At the entrance to the bay of Villefranche, the values observed range from 70 mgC/m²/day in November to 300–350 mgC/m²/day in May and June. The annual production measured by ^{14}C is, according to these measurements, of the order of 40 to 50 g of C/m². Half way between the coast of Provence and Corsica, at the first position of the floating laboratory, the annual rate of primary production, also measured by ^{14}C method, is about 80 C/m² (Minas, 1970), which is of a comparable order of magnitude. On the contrary, between this central zone and the coast, the production should be clearly lower, which agrees with the conclusions of

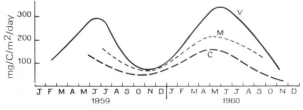

Fig. 7-13.—Variations in the primary production, measured by ^{14}C, at Villefranche-sur-Mer (V), at Monaco (M) and at Cape Martin (C) (Mediterranean) (after Brouardel & Rinck, 1963).

Brouardel & Rinck in their station off Cape Martin 10 miles from the coast (Fig. 7-13). The littoral enrichment is attributed to rain water and the central enrichment to a divergence, supplying the superficial water with mineral nutrients (see Fig. 4-15).

As for the exact significance of these ^{14}C measurements, Brouardel estimated that they could represent about half of the total production in the north of the Western Mediterranean, which is of the order of 100 g (+50) C/m^2/year, which agrees with the conclusions of Sournia (1973).

Northern Pacific

For this region we have considerable data many of them due to Parsons and his collaborators, working at the Nanaiamo Laboratory (Vancouver Island). Before giving the general results of their investigations, we shall emphasize a study particularly interesting for its methodology; it deals with the spring outburst in the Strait of Georgia which is connected to the water of Fraser River (the water in the sea is influenced by an outlet of a river forming a kind of superficial tongue, which is called plume). For estimating the primary production of the Fraser River Plume, Parsons, Stephens & LeBrasseur (1969) used a complex method which takes into account the fact that, with the exception of the summer, primary production is maximal at the surface or very close to it.

A series of determinations enabled them to establish a regression equation relating the particulate carbon of the phytoplankton and the chlorophyll *a*.

$C(\mu g/l) = 59.5(\pm 7.3) \text{Chl } a\,(\mu g/l) + 77.0(\pm 15.6)$.

By the particle counting method, with incubation of the surface water for 36 to 72 h under natural conditions of temperature and light they estimated, on the other hand, the relative growth rate expressed in generations/day; this varied from 0.41 to 2.27 between February and May (Table 7-4). Multiplying the quantity of particulate carbon by the growth rate they estimated the surface production.

For the deeper levels, they used the following methods: They measured the fixation of ^{14}C by natural populations in the laboratory during the period concerned, at different luminous energies and this enabled them to obtain the relation between the relative photosynthesis and the light intensity ($P/P_{max} = f(I)$ curve, Fig. 7-14). Knowing the light energy at any given level, they deduced from this curve the photosynthesis expressed as a percentage of the surface photosynthesis and, in proportion to this percentage, estimated the number of generations per day. This was multiplied by the amount of particulate carbon (estimated from the chlorophyll *a*, as the mean amount, in the surface mixing layer) to give the production at the level considered. Fig. 7-14 and Table 7-4 show the results of these observations during early 1967. The data obtained correspond to net production since it is the new organic matter which is estimated.

TABLE 7-4

Primary production in the Fraser River plume from February to May 1967 (after Parsons, Stephens & Le Brasseur, 1969).

Period	1 16–27th Feb.	2 28th Feb.–13th March	3 14–28th March	4 29th March–9th Apr.	5 10th Apr.–27th Apr.	6 25th Apr.–15th May	7 15th May–25th May
I_0, luminous energy utilisable for photosynthesis on the surface (cal/cm^2/min)	0.030	0.031	0.038	0.069	0.063	0.098	0.097
K, extinct coefficient for this radiation	0.52	0.43	0.35	0.35	0.35	0.43	0.73
Mixed layer depth (m)	2.5	7.5	2.5	15.0	2.5	2.5	2.5
Amount of phytoplankton carbon in this layer (mg/m^3)	71.5	131	405	255	95	136	160
Rate of maximal multiplication (at surface) (generations/day)	1.16	0.41	1.14	2.27	0.77	0.53	1.10
Calculated production in the mixed layer (mgC/m^2/day)	170	127	1,135	2,132	210	199	353

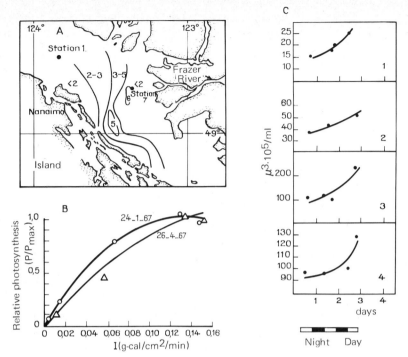

Fig. 7-14.—A, distribution of chlorophyll a (μg/l), 13th February 1967 in Strait of Georgia (British Columbia, Canada); B, relative photosynthesis and light intensity curves ($P/P_{max} = f(I)$), experimentally established; C, curves showing the increase of particulate matter on incubation, measured by Coulter counter during the first four cruises and allowing the relative growth rate of the particle population to be estimated (after Parsons, Stephens & Le Brasseur, 1969).

From this study, Parsons, Stephens & LeBrasseur estimated the total net production at about 50 gC/m^2, from February to May 1967, in the zone influenced by Fraser River Plume.

Parsons, LeBrasseur & Barraclough (1970) using the preceding data together with others from 1965 to 1968, published values of the daily production for different months for the Strait of Georgia, which are given together with certain other data in Table 7-5. For the whole year the total net production, according to these estimations, is about 120 gC/m^2. We have seen, when discussing the oxygen method, that for a fjord opening into the Strait of Georgia, Gilmartin (1967) obtained a net production of 380 gC/m^2/yr which is clearly higher.

In inshore waters the primary production may be much higher. Stephens, Sheldon & Parsons (1967) estimated the primary production in Departure Bay at Nanaimo as 200 gC/m^2/yr; this high production is probably related to the mixing of the water by the tides.

Parsons, LeBrasseur & Barraclough have remarked, that to the production of particulate carbon by phytoplankton, in the Strait of Georgia the

TABLE 7-5

Average primary production data and some factors for Strait of Georgia (British Columbia, Canada) for the period 1965–1968; S.D. in parenthesis (after Parsons, LeBrasseur & Barraclough, 1970).

	J	F	M	A	M	J	J	A	S	O	N	D	Max.	Min.
1 Irradiance cal/cm²/month	75 (9)	166 (34)	273 (57)	412 (30)	562 (54)	599 (122)	606 (37)	503 (71)	362 (29)	171 (19)	95 (14)	60 (5)	701	57
2 Secchi disc depth (m)	9.5 (3.0)	8.5 (1.7)	9.5 (3.2)	7.5 (3.8)	4.0 (2.9)	3.8 (2.9)	5.0 (4.8)	5.8 (2.6)	11.6 (2.6)	8.2 (2.6)	8.4 (2.0)	9.3 (6.3)	16	1
3 Silicate (μg-at./l)	—	56 (5)	50 (13)	22 (10)	48 (20)	33 (9)	24 (10)	25 (15)	26 (11)	38 (13)	44 (4)	52 (4)	100	8
4 NO₃–N (μg-at./l)	22 (0.8)	23 (0.8)	21 (2.7)	15 (3.3)	14 (2.9)	7 (2.8)	8 (4.2)	9 (4.4)	12 (4.9)	17 (6.0)	23 (2.3)	24 (1.9)	27	<1
5 PO₄–P (μg-at./l)	2.0 (0.18)	2.1 (0.19)	1.8 (0.29)	1.6 (0.69)	1.5 (0.32)	0.9 (0.41)	0.8 (0.29)	0.9 (0.37)	1.1 (0.48)	1.4 (0.41)	1.9 (0.23)	2.3 (0.53)	3.2	0.3
6 Seston (mg/l)	3.3 (2.6)	1.0 (0.7)	2.0 (2.1)	2.5 (1.6)	4.2 (3.0)	4.8 (2.6)	2.7 (1.4)	2.7 (1.7)	1.8 (1.0)	2.0 (1.7)	1.8 (1.3)	—	8.8	0.2
7 Particulate C (μg/l)	250 (64)	200 (57)	210 (57)	220 (110)	320 (130)	410 (95)	390 (130)	410 (160)	280 (130)	230 (86)	170 (43)	180 (87)	740	50
8 Particulate N (μg/l)	15 (3.4)	16 (13)	21 (5.3)	32 (11)	32 (26)	53 (13)	50 (17)	40 (19)	34 (15)	32 (11)	15 (2.9)	14 (5.9)	115	9
9 Chlorophyll a (μg/l)	0.9 (0.3)	1.0 (0.5)	2.1 (1.6)	3.0 (2.1)	3.2 (2.4)	4.3 (1.9)	3.0 (1.6)	2.4 (1.4)	2.0 (1.1)	2.0 (1.3)	0.8 (0.5)	1.0 (0.4)	10	<0.5
10 Soluble C (mg/l)	—	1.5 (0.6)	2.0 (1.1)	2.5 (1.1)	2.8 (0.9)	2.9 (1.4)	3.0 (1.9)	3.2 (1.9)	3.0 (1.0)	2.0 (1.0)	1.2 (0.1)	1.2 (—)	8.5	0.3
11 Primary production (mgC/m²/day)	20 (—)	250 (122)	470 (240)	550 (450)	1200 (820)	320 (300)	270 (130)	310 (280)	200 (120)	230 (85)	120 (48)	—	2562	10

1, measured at Nanaimo; 2 to 10 (inclusive) are average values for 0–20 m.

transport of particulate carbon by continental water should be added; this carbon of exterior origin reaches a value of an order comparable to that of the primary production in a year, but it is less effective as regards the level of its utilization by herbivores.

Finally, for the North Pacific, data on the production measured by ^{14}C during a series of observations on the ship "American Mail Line" between Seattle and Yokohama (Parsons & Anderson, 1970) are less complete. They show (Fig. 7-15) that production starts first near the coast and at the south of the Aleutian Islands and later between 150 and 170°W which confirms the results already obtained for this region (Fig. 6-12). These measurements by ^{14}C were made on samples from the surface incubated on board at light energies equal to 100, 54, and 9% of the surface energy. 10% of the fixed carbon was subtracted (with a correction taking into account the thickness of the mixing layer) for night respiration to obtain an estimation of the daily net production.

The upwelling off Peru

Ryther, Menzel, Hulburt, Lorenzen & Corwin (1971) carried out an extremely interesting study—both as regards methods and results—on the primary production of the upwelling off Peru. They followed the development in a mass of water over five days with a buoyed parachute drogue placed at 10 m depth (mean speed 0.37 kt) and estimated the variation of organic

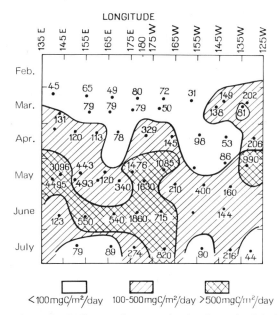

Fig. 7-15.—Daily net primary production determined during 7 crossings of the "American Mail Line", March to July, 1969, from Seattle to Tokyo (from 125°W to 180°W the transect is near 50°N, then oblique to the south cutting 40°N at about 150°E) (after Parsons & Anderson, 1970).

carbon by different methods. The results are given in Fig. 7-16. The changes in the concentrations of inorganic phosphate, nitrate-nitrogen and oxygen (Fig. 7-16B) reflect a net production of 50 to 60 gC/m^3 which is a considerable mean net production of $10 gC/m^2/day$. About a half of this production had been accumulated in the third day in the superficial layer (0.50 m) in the form of living phytoplankton, as shown by the direct measurement of particulate carbon, or by the transformation to carbon of the data on either chlorophyll or particulate phosphorus (Fig. 7-16A). On the fifth day, however, the plant biomass is not more than a fifth of the total production of the five days and, after eliminating the possibility of sinking to deeper layers or of an important diffusion of organic matter into the surrounding water, the authors concluded that there is a loss due to consumption by the phytoplanktophagous anchovy. The measurement of production by ^{14}C, made at the same time, also gave a very high rate, the highest so far observed of from $3.14 gC/m^2/day$ on the first day to $11.74 gC/m^2/day$ on the fifth day.

We have not yet enough data to determine with certainty the annual production of this upwelling region; the upwelling zones are discontinuous and fluctuating and the intensity during the year varies greatly. Relying on the anchovy catches by fishermen and sea birds and taking into account the total surface known to constitute the upwelling zone, Ryther et al. (1971) estimated, however, that a carbon production of an order of $1000 g/m^2/yr$ was necessary to account for the production of anchovy.

7-11.—LIMITS OF PRIMARY PRODUCTION

In the preceding estimations we have seen that the daily primary production, may attain, and sometime exceed gC/m^2. Under natural conditions one may consider that a certain number of factors limit this production. Can it be increased under optimal conditions? Can it increase indefinitely?

In reality since the multiplication of the phytoplankton algae itself increases the absorption of light in the environment and, as a consequence, limits the available energy for photosynthesis, it may be thought that there is a maximum value which cannot be exceeded. Steemann-Nielsen (1962) has considered this problem, supposing, for simplification, that the light in the sea is absorbed only by the pigments of the phytoplankton. He estimated experimentally the quantity of chlorophyll retaining 99% of the incident light (for 1% of the maximal incident light the photosynthesis was supposed to be zero). In this way, he obtained $800 mg Chl/m^2$ for the green algae and $400 mg Chl/m^2$ for algae such as diatoms with brown pigments. Under natural conditions, where light is absorbed not only by phytoplankton pigments but also by the water itself, by dissolved coloured substances, by detritus in suspension, and by the zooplankton, the estimated maximal values would clearly be lower and probably do not exceed $300 mg Chl/m^2$. With such a concentration of chlorophyll, the estimation of photosynthesis

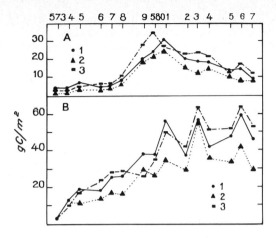

Fig. 7-16.—Changes in the biomass (gC/m^2) determined by different criteria and integrated from the surface to 50 m. A; 1, direct measurement; 2, from the amount of chlorophyll; 3, from particulate phosphorus: B; 1, from PO-P$_4$ uptake; 2, from NO$_3$-N uptake; 3, from oxygen production. Stations worked at 6h, 12h and 18h over a 5-day period; first day, Stations 573, 574, 575; second day, Stations 576, 577, 578 etc. (after Ryther, Menzel, Hulburt, Lorenzen & Corwin, 1971).

in temperate waters during the summer gives a daily gross production of 4.6 gC/m^2 (0.35 × 2.7 × 16.300 = 4.536 mgC/m^2/day; 0.35 = experimental coefficient; 2.7 = Q_{max}; 16 = the number of light hours). The maximal production in an aquatic environment will, therefore, be between 4 and 5 gC/m^2/day, under optimal conditions. In the Sölleröd Sö lake in Denmark with a good supply of nutrient salts, Steemann-Nielsen obtained values of this order, namely, 3.8 gC/m^2/day. As we have already seen, he obtained a similar value during the period of "coloured water" in Walfish Bay, corresponding to a superficial layer, 0.8 m thick, rich in phytoplankton. At this depth the incident light was reduced to 1% of its value at the surface. It seems, however, that under certain conditions, the value of 4 to 5 gC/m^2/day may be exceeded. In the west of the Arabian Sea, Ryther & Menzel (1965) obtained a production of 6.7 gC/m^2/day, estimated by ^{14}C, and Ryther et al. (1971) in the upwelling off Peru found, as already seen, a daily production 10 gC/m^2. These last authors obtained values of chlorophyll exceeding 600 mg/m^2, with a phytoplanktonic population composed mainly of diatoms.

7-12.—RELATION BETWEEN PRIMARY PRODUCTION AND BIOMASS

At the beginning of this chapter we justified the study of primary production by emphasizing its ecological interest, so as to know not only the phytoplanktonic biomass but also its growth, considering these two values were not constantly connected. We shall examine now the real phenomenon. Without doubt, there is in general, a parallelism between biomass and production as in the example of St. Margaret's Bay (Fig. 7-11); here there is the classic annual cycle of biomass at these latitudes with a very marked spring maximum and a smaller autumnal maximum. It is also in the regions of high biomass that in general, high production is found but, in detail, we note that the two are relatively independent. In Fig. 7-16 the curves A indicate the development of the biomass, expressed in organic C/m^2, from 0 to 50 m depth; they decrease from 30 to 10 g between the third and the fifth day but the curves B indicate that production is still high, as confirmed by the results using ^{14}C.

As another example of differences between biomass and production we show in (Fig. 7-17) the mean curves given by Parsons, LeBrasseur & Barraclough (1970) for the Strait of Georgia in different seasons. Generally, the measurable production is limited to the first 10 or 15 m while the biomass, determined by chlorophyll, may be traced to much greater depths. The distinction is particularly clear in the spring and summer. In other words, the ratio between production and the biomass expressed in terms of carbon gives the relative production which is a productivity and will show considerable variation as a function of depth and season.

The productivity so defined as a function of biomass, represents then a rate of relative growth (= rate of specific growth), and is not positively related to production. On the contrary, in the North Atlantic, Sutcliffe, Sheldon & Prakash (1970) showed an inverse relation between productivity and biomass (Fig. 7-18), the lowest biomasses having the highest productivities. The observed productivities vary considerably from 0.05 to 10.

Mullin (1969), using the data of Riley, estimated a mean primary production of 800 $mgC/m^2/day$ for Georges Bank, off Cape Cod (Mass. U.S.A.),

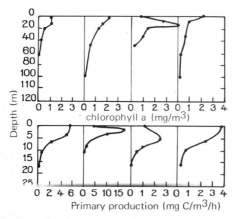

Fig. 7-17.—Seasonal variations during 1966 in Strait of Georgia (British Columbia, Canada) of the vertical distribution of chlorophyll *a* and primary production (note, depths scales different in the two diagrams): from left to right; winter, spring, summer, autumn (after Parsons, Le Brasseur & Barraclough, 1970).

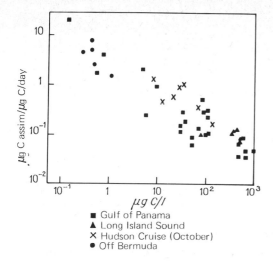

■ Gulf of Panama
▲ Long Island Sound
✕ Hudson Cruise (October)
● Off Bermuda

Fig. 7-18.—Phytoplankton productivity (μgC assimilated/μgC phytoplankton/day) as a function of biomass (μgC/l) in the North Atlantic (after Sutcliffe, Sheldon & Prakash, 1970).

with a mean biomass of 10 000 mgC/m², giving a productivity of 0.08, which agrees with the above data.

7-13.—ENERGY YIELD AND ENERGY FLOW

The idea of yield or energy efficiency, in primary production has given rise to much discussion (reviewed by Jacques, 1970). This is particularly useful for comparing the production of two given zones, after eliminating the differences in the light energy received.

Since the light energy is expressed in calories per unit surface per unit time, it will also be necessary to use this mode of expression for production which is usually given in mg C/m²/day. It will also be necessary to know the calorific value of the phytoplankton. Platt (1971) estimated it for St. Margaret's Bay (near Halifax) as 15.8 kcal/g of carbon. The mean value of the production per unit of radiation (kcal/m²/h) in this bay was $4.58 \cdot 10^{-4}$ (gC/m²/h)/(kcal/m²/h), and the primary production for 1 m² corresponds, therefore, to 0.72% of the light energy in the visible spectrum received at the surface (equal to half of the total energy). It varies during the year, attaining 0.92% in the summer, which is close to the value of 0.88% obtained by Patten (1961) near New-York.

Platt (1969), from theoretical considerations, compared the energy efficiency with the extinction coefficient and the total extinction coefficient then becomes the sum of two coefficients:

$K = K_b + K_p$

where K_b represents the biological contributions, i.e., the fraction of available light absorbed by photosynthesis and K_p the physical contribution which is much more important.

The coefficient K_b is calculated to a sufficient approximation from the following equation:

$$K_b = -\ln\left(1 - \frac{P_z}{I_z}\right)$$

where P_z is the production expressed in cal/m³ between the level z and the level $z+1$ and I_z is the irradiance at the level z. In two stations in St Margaret's Bay, Platt obtained values of K_b ranging from 0.0001 to 0.006, linearly related, for a given level, to the amount of chlorophyll; the total extinction coefficient is from 0.10 to 0.15 and most frequently K_b represents 0.5 to 0.1% of this coefficient.

Finally, to understand the ecosystem one should distinguish, according to Davis (1967), between the circulation of materials and that of energy flow. The gross production corresponds to the photosynthetic assimilation; the rate of energy storage is proportional to the rate of carbohydrate synthesis. The distinction begins at the net production level; Davis separated the net rate of the synthesis of organic matter (net organic primary production = P_{no} and the net rate of formation and storage of chemical energy (net energetic primary production = P_{ne}). On the one hand, we have, therefore,

$$P_{no} = P_{ho} - R_0 - L_0$$

where P_{ho} = gross photosynthetic production of carbohydrates;
R_0 = catabolic destruction by respiration;
L_0 = loss by diffusion out of the cells,
and on the other hand;

$$P_{ne} = P_{he} - R_e - L_e + S_e$$

where P_{he} = gross formation of potential energy by photosynthesis;
R_e = catabolic loss of energy;
L_e = loss of energy by diffusion of metabolites;
S_e = energy demand for cellular synthesis of metabolites other than carbohydrates.

P_{ho} is proportional to P_{he}, but it is not the same for P_{no} and P_{ne}; the ratio between R_0 and R_e varies according to the metabolites used for respiration; the oxidation of 1 g of carbohydrates gives 4 cal and of 1 g of lipids 9 cal. It is the same for the ratio between L_0 and L_e which varies according to the nature of the diffusing metabolites.

This distinction between energy flow and circulation of the matter, although only of theoretical importance, is nevertheless, important; photosynthesis, due to solar energy raises the energetic level of the matter which afterwards circulates in the alimentary chain by degradation of the energy received; while the matter will be able to circulate for a more or less long time in the ecosystem, the energy will run in it in an irreversible flow (McFadyen, 1963 in Jacques, 1970).

7-14.—THE CHEMICAL COMPOSITION OF PHYTOPLANKTON

The determination of primary production gives us the production of the plant organic matter in terms of carbon. This is only "primary" information, since this organic matter is distributed as proteins, carbohydrates, and lipids, and within each of these categories are numerous organic substances. These organic substances are far from being equivalent for the eventual consumers and a priori it is not certain that they are present uniformly in the different species of the phytoplankton. After a total estimation of the primary production, by the preceding methods, an attempt should be made to determine the different metabolic products; but we are still not far advanced in this subject.

Some data were given by Parsons, Stephens & Strickland (1961) after analysing the chemical composition many phytoplanktonic species grown in culture under as similar conditions as possible and their results, expressed in terms of dry weight, are given in Table 7-6. There are notable differences between the species; the diatoms *Coscinodiscus* and *Chaetoceros* have a small amount of carbohydrates: the peridinians *Amphidinium* and *Exuviella* are very rich in lipids. The ash weight is high in the diatoms because of their silicious frustules, in *Syracosphaera* which is covered by coccoliths, and also surprisingly in *Tetraselmis* because of the presence of chloride. The sum of the metabolites given in the last column of the Table, is sometimes clearly different from 100; it is higher when an inadequate factor of standard conversion was used (the proteins values were obtained by multiplying the nitrogen by 6.25); when the values are low it may be that there are other metabolites, not estimated by the methods used. Lewin, Lewin & Philpot (1958) showed that in *Phaeodactylum*, 12% could not be attributed to protein, lipids, carbohydrates, or to ash.

Among the carbohydrates, it is possible to distinguish a resistant fraction, largely constituting the cell walls, and a labile fraction, extracted in boiling water which on hydrolysis gives a very large proportion of glucose,

TABLE 7-6

Chemical composition of different species of unicellular algae grown in culture, expressed as a percentage of dry weight (after Parsons, Stephens & Strickland, 1961).

Species	Proteins	Carbohydrates	Lipids	Pigments	Ash	Total
Tetraselmis maculata	52	15.0	2.9	2.1	23.8	96
Dunaliella salina	57	31.6	6.4	3.0	7.6	106
Monochrysis lutheri	49	31.4	11.6	0.8	6.4	99
Syracosphaera carterae	56	17.8	4.6	1.1	36.5	116
Chaetoceros sp.	35	6.6	6.9	1.5	28.0	78
Skeletonema costatum	37	20.8	4.7	1.8	39.0	103
Coscinodiscus sp.	17	4.1	1.8	0.5	57.0	81
Phaeodactylum tricornutum	33	24.0	6.6	2.9	7.6	73
Amphidinium carteri	28	30.5	18.0	2.4	14.1	93
Exuviella sp.	31	37.0	15.0	1.1	8.3	92

representing the reserves of carbohydrates in the cell (Handa, 1969). Table 7-7 illustrates this differentiation between the two fractions in the diatom *Skeletonema costatum*. In this species, Handa studied the fate of these two fractions in a culture which, at the end of its rapid phase of growth, was placed in the dark (Fig. 7-19); while the resistant fraction hardly varied, the glucose and glucose polymers disappeared very quickly. This may be compared with the observations of Handa & Yanagi (1969) who observed, at a depth between 0 and 50 m, a decrease in the fraction extracted by boiling water of the particulate carbohydrates. They attribute this

TABLE 7-7

Monosaccharide composition (%) of carbohydrate from the diatom *Skeletonema costatum*: A, total carbohydrate; B, carbohydrate extracted by boiling water; C, resistant fraction; D, total carbohydrate after 8 days in the dark (after Handa & Yanagi, 1969; and Handa, 1969).

	Rhamnose	Fucose	Ribose	Arabinose	Xylose	Mannose	Galactose	Glucose
A	5.1	5.8	0.6	3.3	4.3	30.3	2.0	48.6
B	1.0	0.5	0.6	6.0	0.5	1.5	0.0	90.0
C	13.0	10.4	0.8	0.8	15.7	50.5	2.8	6.5
D	10.8	10.3	0.7	3.0	10.2	52.0	3.2	10.0

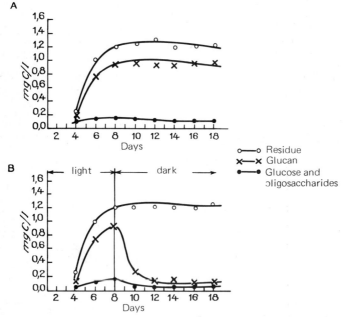

Fig. 7-19.—Development of cellular carbohydrate in a culture of *Skeletonema costatum* grown in continuous light (A) or in the light and then after 8 days placed in the dark: fraction extracted by boiling water separated into glucose and oligosaccharides, and into polysaccharides derived from glucose (glucan): the control shows that the cells remain alive in the dark until the 18th day: residue corresponds to the resistant fraction (after Handa, 1969).

decrease to the consumption of the carbohydrate reserves by the phytoplankton when the cells sink and do not receive enough light. This observation is of great ecological interest since, under these conditions, the cells which sink to the deeper layers and eventually to the bottom, bring with them organic carbon, but the utilization of this carbon is relatively difficult.

As regards the lipids, Table 7-8 gives the results of the analysis by Lee, Neuvenzel & Paffenhöfer (1971) on three diatoms, including *Skeletonema costatum*. Half of the lipids are represented by phospholipids and galactolipids. Hydrocarbons are present in diatoms and constitute about 1% of the dry weight. Blumer, Guillard & Chase (1971) analysed in detail the hydrocarbons produced by different species of the phytoplankton; the most abundant is an unsaturated hydrocarbon 3,6,9,12,15,18 hexeicosahexane or HEH ($C_{21}H_{32}$); traces of pristane were also identified. Since the hydrocarbons are particularly stable in the food web, they may be used as tracers in the studies of this web.

The above results were obtained with laboratory cultures and using concentrations of nutrient salts higher than those in the sea. A comparison with cells obtained by Strickland et al. (1969) in large tanks where conditions are nearer to natural shows, however, that the composition is not greatly affected by the differences in concentrations of the environment (Table 7-9).

There are clear differences dependent on the source of nitrogen; the diatom *Ditylum brightwellii* synthesizes more lipids when nitrogen is supplied in the form of nitrate rather than ammonium; in the peridinian, *Cachonina niei*, the difference is in the opposite sense—cells grown in nitrate are twice as rich in carbohydrates. In the same species, a deficiency in nutrient salts, leads to a considerable increase in carbohydrate. Prochazkova et al. (1970) showed on total phytoplankton, that at a reduced light intensity the production of proteins is increased.

In conclusion, these initial studies on the chemical composition of the phytoplankton show that, besides a general similarity, there are important differences as a function of the species and of the ecological conditions, which is a warning that the conclusions drawn from the total measurements of primary production could be in error.

TABLE 7-8

Lipid composition of three diatoms: the different categories of lipids are % of total lipids; total lipid as % dry weight (after Lee, Neuvenzel & Paffenhöfer, 1971).

	Lauderia borealis	*Skeletonema costatum*	*Chaetoceros curvisetus*
Triglycerides	16	14	12
Sterols	17	15	10
Free fatty acids	5	12	16
Phospholipids (+galactolipids)	51	57	50
Hydrocarbons	11	16	2
Total lipids	13.2	8.6	9.1

TABLE 7-9

Growth and composition of the diatom *Ditylum brightwelli* and the peridians *Cachonina niei* and *Gonyaulax polyedra* cultured under different conditions: DT, deep tank culture; L, laboratory culture; N, source of nitrogen; AC, age of culture (after Strickland, Holm-Hansen, Eppley & Linn, 1969).

	Ditylum brightwelli								*Cachonina niei*						*Gonyaulax*			
	DT, 14.5° N = NH_4-N AC = 3–8 days		DT 14.5° N = NO_3-N AC = 12 days		L 20°		DT, 20° N = NH_4-N AC = 1–5 days		DT, 20° Nutrient starved AC = 12 days		DT, 20° N = NO_3-N AC = 13 days		DT 22°		L 22°	
	µg	pg/cell	µg	pg/cell	µg	pg/cell	µg	pg/cell	µg	pg/cell	µg	pg/cell	µg		µg	
Carbon	100	1350	100	1550	100	680	100	150	100	263	100	230	100		100	
Nitrogen	22.5	300	20.5	320	18.0	120	16.2	245	8.2	21.5	14.5	37	13		14	
Phosphorus	4.8	65	5.1	80	3.4	23	4.4	6.7	1.9	5.0	3.1	7.2	1.8		0.75	
Proteins (N × 6.25)	140	1900	128	2000	112	760	101	155	51	135	91	210	79		87	
Lipids	23.5	320	38	590	—	—	57	86	42	110	35.5	83	72		63	
Carbohydrates	13	180	11.5	180	18.5	125	35	53	127	335	73.5	170	32		43	
Chlorophyll *a*	4.5	61	—	78	15.6	38	2.9	4.4	1.2	315	1.7	4.0	1.75		0.34	
Carotenoids	—	—	—	—	—	—	4.2	6.4	1.95	5.15	2.3	5.35	2.60		—	
DNA	2.3	31	1.95	30	—	—	4.0	6.1	2.9	7.6	3.2	7.5	2.4		1.7	
ATP	0.49	6.6	0.54	8.4	—	—	0.45	0.68	0.24	0.63	0.29	0.68	0.30		0.30	
Silicon	58	780	47	730	28	190	—	—	—	—	—	—	—		—	

	µg-at.		µg-at.		µg-at.		µg-at.		µg-at.		µg-at.		µg-at.		µg-at.	
Carbon	100		100		100		100		100		100		100		100	
Nitrogen	19.3		17.5		15.4		13.9		7.0		12.4		11.1		12.0	
Phosphorus	1.8		1.9		1.3		1.7		0.73		1.2		0.7		0.29	
Cells produced/l	0.74×10^5		0.645×10^5		1.78×10^5		0.66×10^6		0.38×10^6		0.43×10^6		1.0–10^4		1.0–10^3	
K_n (day^{-1})* cells	0.43		0.17		—		0.48		—		0.63		0.24		0.42	
K_n (day^{-1}) C	0.41		0.27		0.64		0.48		—		0.63		0.24		0.42	
K_n (day^{-1}) N	0.43		0.17		—		0.48		—		0.48		0.24		0.42	
K_n (day^{-1}) P	0.41		0.27		—		0.48		—		0.63		—		—	
K_n (day^{-1}) Chl *a*	0.41		0.27		—		0.48		—		0.63		0.24		0.42	
Average cell volume (µm³)	—		—		24000		—		—		—		3000		50000	

*For the determination of K_n see Appendix (Chapter 14-3) equation (8).

CHAPTER 8

THE ZOOPLANKTON: SYSTEMATIC COMPOSITION

The zooplankton consists of the whole of the heterotrophic planktonic organisms with animal nutrition. Since they cannot synthesize their organic needs they must obtain it from the external medium by ingestion of material, living or otherwise (phagotrophy).

The zooplankton (heterotrophic) contrasts, therefore, with the autotrophic phytoplankton. Most bacteria are also heterotrophic, but they are not included as zooplankton because they absorb organic nutrients directly from the external medium (osmotrophic).

Most groups of marine animals are represented in the zooplankton, either as holo- or meroplankton. In the latter category meroplanktonic larvae are the most abundant.

Most of the zooplankton belongs to the macro- or microplankton. In the nanoplankton (size $<50 \mu$m) the zooplankton includes only minute protozoa which are little known. It should be remembered that it is difficult to determine precisely a boundary between microplankton and small nekton or micronekton, and the difference is usually established arbitrarily as a function of the method of collection.

We shall now briefly review the different groups of zooplankton (for further detailed information, see the monograph of Tregouboff and Rose (1957). The most common forms are illustrated in Bougis (1967) and Hardy (1958), the latter giving excellent watercolour drawings. In the Appendix (Chapter 14) you will find a summary of a zoological classification which should help to refer the different groups or planktonic forms to their place in the various phyla.

8-1.—PROTOZOA

It should first be noted that some of the flagellates classified as phytoflagellates but which lack photosynthetic pigments, feed by predation and are therefore considered as zooplankton; these belong largely to the Peridinia, the best known of which is *Noctiluca miliaris* (= *scintillans*) (Fig. 8-1), with a diameter of 200–1200 μm, and characterized by a flagellum of similar length; it is bioluminescent.

Noctiluca was first kept in culture by feeding with unicellular flagellate algae, but subsequently it was found that in culture this species can also feed by osmotrophy (McGinn & Gold 1969; Gold 1970). Droop (1970) obtained similar results for another phagotrophic peridinian, *Oxyrrhis marina*.

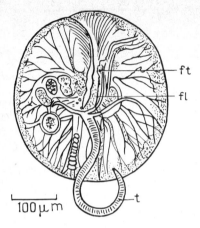

Fig. 8-1.—*Noctiluca miliaris*, Peridiales, ventral posterior view: t, tentacle; fl, longitudinal flagellum; ft, rudiment of transverse flagellum (after Robin in Grassé, 1952).

A large part of the zooplanktonic protozoa belongs to the Rhizopoda; their protoplasm is not limited to the interior of a rigid membrane, but can change its form and give rise to ramified expansions, the pseudopodia, which capture prey. Among the rhizopods we find the Foraminifera, Acantharia and Radiolaria.

In the Foraminifera a small shell, the so-called test, protects the protoplasm. Most of the many species of the Foraminifera are benthic. The planktonic forms belong to the family Globigerinidae (Fig. 8-2) within which there is little variety but the species are very abundant. Their shells fall to

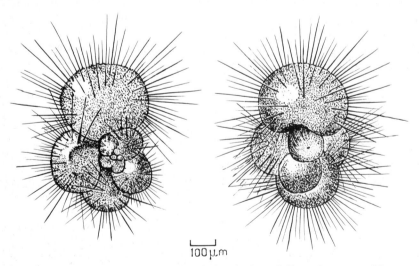

Fig. 8-2.—*Globigerinella aequilateralis*, planktonic foraminiferan from two different angles (after Bé, 1967).

the bottom of the ocean in large quantities forming the mud known as Globigerina ooze and this covers wide areas of the ocean bottom.

The Actinatia, formerly considered as belonging to the Radiolaria have a skeleton which consists of a fixed number of spicules with definite directions; the number of these radial spicules is twenty while in other species they are joined to give ten diametrically opposite spicules. In the most simple species the spicules are identical; in other species there is spicule differentiation, some being markedly larger than others (*Acanthochiasma*, Fig. 8-3). In *Lithoptera*, four of the spicules are distally ramified to form a small "grid" at their apex. The chemical constitutions of the spicules is unusual in that they contain strontium sulphate. The Radiolaria are characterized by a shell, the central capsule, situated within protoplasm; the capsule contains the nucleus and has one or many openings. In *Aulacantha*, a very common form, the radial spicules are numerous, and small tangential spicules form a very simple skeleton. This skeleton, made up of silica may, however, in some species be extremely complicated. In *Hexacontium*, for example, it is made of three perforated spheres attached to each other by rods. Colonial forms with reduced skeletons or without skeletons have many cells and are sometimes found within a gelatinous cylinder which may reach a size of 1–2 cm (*Collozoum*).

The Ciliata are the last group of the Protozoa to be considered which are represented in the zooplankton. The external covering of the cells possesses cilia which are used for locomotion, and these may be considerably modified. In the Tintinnidae the cilia join to form flexible lamellae located in spirals at one end of the animal and form a test in the shape of a bell or a bowl into which the animal can contract and so be protected. One important genus is *Tintinnopsis* (Fig. 8-4); many of its species have been maintained in culture (Gold, 1970). *Favella* and *Dictyocysta* have a test in the shape of a vase, while *Eutintinnus* has a cylindrical test open at its two edges. Besides the Tintinnidae, which form a very characteristic group, there are other planktonic ciliates. Beers & Stewart (1969) have shown that 50–80% of the total number of zooplankton organisms have a size below 30 μm and that an important fraction of the microzooplankton is made up of ciliates. Unfortunately our knowledge of these forms is extremely poor.

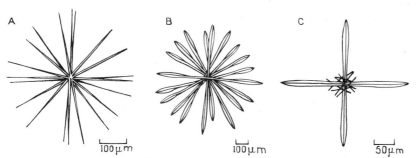

Fig. 8-3.—Skeletal spicules of three species of *Acanthochiasma* (Acantharia): A, *A. fusiformis*; B, *A. quadrangulum*; C, *A. hertwigi* (after Bottazzi, Massera & Nencini, 1969).

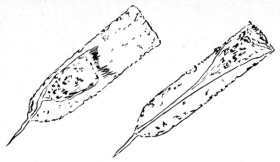

Fig. 8-4.—*Tintinnopsis platensis*; left, contracted in its theca; right, extended (length of the theca is of the order of 100 μm) (after a photograph by Gold, 1970).

8-2.—CNIDARIA

Although their morphology is sometimes very complicated, the Cnidaria have a simple structure. They have two layers of cells an external (ectoderm) and an internal (endoderm) layer, which are separated by a gelatinous non-cellular layer, the mesoglea. A fundamental character of the Cnidaria is the possession of stinging cells (nematocysts) which inject venum that paralyses their prey; they are carnivores. Two types of medusae are mainly found in the plankton, namely Hydromedusae and Scyphomedusae, as well as colonial Cnidaria, the Siphonophora. The shape of the body or umbrella of Hydromedusa is like a bell, parachute or a disc; the margins of the body carry tentacles. From the middle of the under surface hangs a tube-like appendage, the manubrium, at the free end of which is found the mouth; it also contains the stomach. The medusae swims by repeated contractions of the umbrella. Many of the Hydromedusae are produced from a hydroid, a colonial polyp living on the sea floor, which is only a stage in the life history of a given species namely, the reproductive and dispersal stages; the sexes are separated. Beside these meroplanktonic forms other Hydromedusae, complete their whole life cycle in the plankton. Among the first we find the Anthomedusae (e.g., *Sarsia*, *Pandea*), the Leptomedusae (e.g., *Phialidium*, *Obelia*, *Aequorea*) and the Limnomedusae (e.g., *Olindias*). Among the holoplanktonic Hydromedusae are the Narcomedusae (e.g., *Solmaris*, *Cunina*). Most of the Hydromedusae are small—a few millimeters or centimeters—but there are exceptions such as *Aequorea* with a size of 20–25 cm (Fig. 8-5).

The Scyphomedusae, or jellyfish, are generally larger (usually about 10 cm) but sometimes they reach a size of even 1 m. There is variability in the number and the length of the tentacles and sometimes they are absent but the mouth present on the under surface of the umbrella is accompanied by mouth lobes which may be much developed and have a very complicated structure. The development is sometimes direct as in *Pelagia noctiluca*, a luminescent species, in which the egg develops into a small medusa. Usually

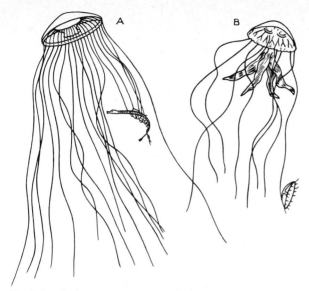

Fig. 8-5.—A, *Aequorea* sp., a young hydromedusa, diameter 32 mm, catching a young syngnatid; B, *Chrysaora hysoscella*, a young scyphomedusa, diameter 25 mm, catching a hydromedusa from the genus *Phialidium*; note the difference between the marginal tentacles and the buccal lobes (drawn from living animals, after Lebour, 1923).

the larva attaches itself to the same bottom material and forms a polyp (scyphistoma). After some months it gives the pelagic medusa by segmentation (strobilisation). In addition to the common species, *Aurelia aurita* and *Chrysaora hysoscella* (Fig. 8-5) the cold water species *Cyanea capillata* should be mentioned; this jellyfish may attain a diameter of 2 m and the length of the tentacles may be tens of meters. In *Rhizostoma pulmo* the single mouth is replaced by a number of openings on the mouth lobes. According to recent observations they are not used to attach to a large prey and sucking it but are true mouths which capture small crustacean and pelagic larvae.

The Siphonophora, colonial organisms, consists of different individuals which are not similar to each other; they are connected by a common cord, the stolon. In the Physonectes (*Forskalia, Nanomia* (Fig. 8-6)) the stolon begins as a small vesicle full of gas secreted by its walls to form a float; underneath the float are the swimming bells which propel the colony. The cormidia are groups of polyps found on the stolon. The cormidia consist of gastrozoids—tubular feeding polyps—each of which carries on its base a long tentacle armed with nematocysts and forming the so-called fishing filament, a few dactylozoids which are simplified polyps used for defence, the gonozoids which contain the male and female reproduction organs, and finally the bracts—leaflike polyps which give mechanical protection. In the Cystonectes there is no swimming bell and in the most common species, *Physalia physalis*, the Portuguese man-of-war, the violet blue float is much developed and attains a size of 10–30 cm; it has a membranous sail over a

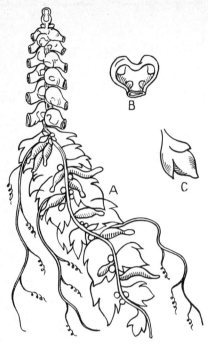

Fig. 8-6. — *Nanomia cara*, Siphonophora, Physonecta: A, whole colony, below the float, a series of swimming bells and stolon carrying the gastrozoid armed with long ramified fishing filaments: gonozoids surrounded by bracts; B, details of a swimming bell; C, details of a bract; length of zone of swimming bell 1.5 cm (after Hardy 1958).

part of its length. The gastrozoids, dactylozoids, and gonozoids are found on the under-side of the float without a stolon. On the other hand, *Calycophora* has no float and the colony begins by one, two or three swimming bells (*Sulculeolaria* (Fig. 8-7 and 8-8)). *Chelophyes, Muggiaea*); the cormidia are often free and named euxodia.

Velella velella (Fig. 8-9) was for a long time classified among the Siphonophora but has recently been considered by Brinckmann-Voss (1970) as a pelagic hydrozoan. It has the shape of an oval disc from which arises a blue-violet triangular sail. *Velella* contains an internal float made of transparent and resistant matter, with numerous concentric walls forming the skeleton. Around its margins *Velella* has tentacles (dactylozoids) and on the underside it has one large gastrozoid encircled by the gastrogonozoids, polyps which give rise to small medusae. Like the Portuguese man-of-war, *Velella* floats on the surface of the sea (pleuston) and, driven by the wind, is frequently found stranded on the shores in large quantities.

8-3. — CTENOPHORA

The Ctenophora are simple in their structure; like the Cnidaria they also have only two layers of cells. For a long time these two phyla were

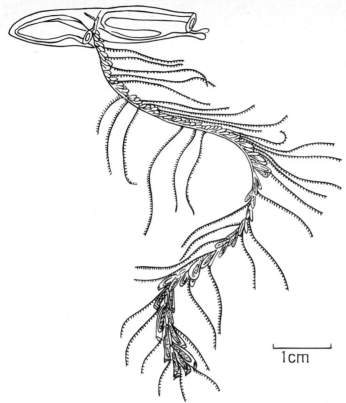

Fig. 8-7.—*Sulculeolaria biloba*, Siphonophora Calycophora, swimming: upper, two swimming bells; on the stolon the gonozoids with fishing filaments; towards the lower part developed gonozoids (after a drawing of Delap in Totton, 1965).

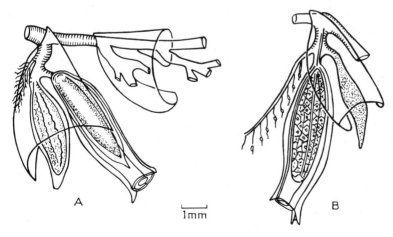

Fig. 8-8.—*Sulculeolaria biloba*, details of cromidia: A, bract, male gonozoid (right), gastrozoid (centre), a contracted fishing filament (left); B, bract, female gonozoid (centre), gastrozoid (right), a part of an extended fishing filament with tentillia (left) (after Delap in Totton, 1965).

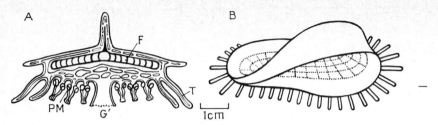

Fig. 8-9.—*Velella velella*, pelagic hydrozoan (Pseudosiphonophora): A, diagrammatic section perpendicular to the sail through the centre; F, float; T, tentacle; G, central gastrozoid; PM, peripheric polyps budding medusae: B, a sketch of a colony swimming at the surface (A after Hardy, 1958; B after Brinckmann-Voss, 1970).

considered as one, the Coelenterata. The Ctenophora differ from the Cnidaria in the absence of nematocysts, and in having adhesive cells, the coloblasts, which cling to the prey. Their shape is usually globular but their most marked character is the possession of comb-like plates arranged in eight meridial rows, each comb plate consisting of fused cilia. A mouth is present at one pole of the animal, while at the other is a statocyst.

In the Tentaculata there are two lateral ramifying tentacles covered with coloblasts, termed tentillia. In *Pleurobrachia* (Fig. 8-10) the tentillia extended to a very considerable length and contract completely into the tentacle sheaths. The tentillia collect food, of which copepods are an important part. This simple morphological type may be considerably modified. In *Bolina* the mouth is encircled by large lobes, in *Cestus veneris*, Venus' girdle, the body is flattened, elongated, transparent, and undulates; this species can attain the length of some tens of centimeters.

The naked ctenophores (*Beroe*) lack tentacles, but the large mouth

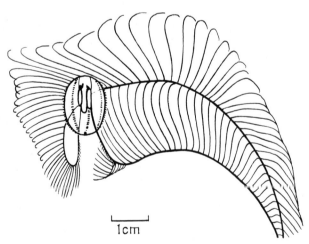

Fig. 8-10.—*Pleurobrachia pileus*, Ctenophora Tentaculata: showing extended tentacles with their tentillia forming a "cobweb" (after a photograph by Greve, 1970).

enables them easily to capture their prey, among which tentaculated Ctenophora are frequently captured. Like many other Ctenophora *Beroe* is luminescent.

8-4.—THE WORMS—ANNELIDA AND CHAETOGNATHA

It will be most convenient to deal together with all the animals having soft bodies, bilateral symmetry, and without articulated appendages, and with all worms, although they belong to different phyla.

The worms are represented in the plankton by some Polychaeta, Annelida, and by the Chaetognatha, a very peculiar group whose position in the zoological classification is not precisely clear.

The Polychaeta, segmented worms covered with bristles, or chaetae, are essentially benthic animals and although some genera are exclusively planktonic, they are usually carnivorous. On each segment of *Tomopteris* (Fig. 8-11), length 1–2 cm, there is a pair of appendages, the parapodia, ending in a bilobed "paddles" and between the two lobes there is a small luminescent organ. The head carries two long antennae. *Alciope* has two large globular eyes. Besides the holoplanktonic Annelida there are also meroplanktonic species, which are pelagic only during the reproductive phase. When the gonads of Nereis are ripe the animal is considerably changed and is even given a special name—heteronereis. In the Samoan Islands during the breeding season of the Pallalo worm, *Eunice*, it is found on the surface in enormous quantities.

The Chaetognatha, or arrow worms, are transparent, elongated spindle-like animals not more than 2 cm long; the head bears dark spots, the eyes, and on both sides of the head there are strong curved bristles which act as jaws. It is these bristles which give the name Chaetognatha to the group. On the trunk there are two pairs of horizontal lateral fins and the tail also ends within a horizontal fin. In order to catch its prey, mainly fish larvae and small crustaceans, the animal is propelled by up and down movements of this tail and the animal darts like an arrow. The most important genus is *Sagitta*

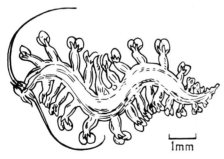

Fig. 8-11.—*Tomopteris helgolandica*, Annelida: sketch of a swimming animal (after Hardy, 1958).

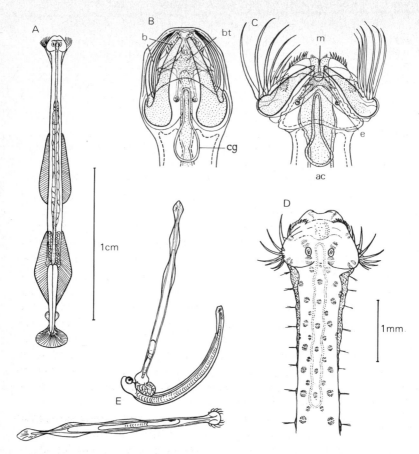

Fig. 8-12.—Chaetognatha: A, *Sagitta megalophthalma*, dorsal view; B, *Sagitta* enlarged dorsal view of the head showing closed bristle-jaws; C, bristle jaws extended; D, *Sagitta megalophthelma*, anterior region, E, *Sagitta bipunctata* feeding on a young herring 6 to 7 mm long: ac, alimentary canal; b, bristles; br, brain; cg, cilio-glandular wreath; e, eyes; m, mouth (A and D after Dallot & Ducret, 1969; C, after Beauchamp, 1960; E, after Lebour, 1923).

(Fig. 8-12) with its many species, some of which are very common. It is a very important group of the zooplankton (Alvarino, 1965).

8-5.—CRUSTACEA

In their adult stage many crustaceans live in the plankton; they are members of the following groups: Cladocera, Ostracoda, Amphipoda, Mysidacea, Euphausiacea, Decapoda. We shall deal with these first and subsequently with the planktonic larvae.

The Cladocera are well represented in fresh water plankton mainly by *Daphnia*. The body, a few millimeters in size, is enclosed in a carapace with only the head free. Few species are found in the sea, but in certain condition they may be quite abundant. Among the marine Cladocera we find the genera *Podon* and *Evadne* and also *Penilia ovirostris*.

The Ostracoda are mainly deep-water animals. There body is entirely enclosed in a bivalve carapace which looks like a small shell. In the plankton we find the genus *Conchoecia* (Fig. 8-13) 1–2 m long and in deep water *Gigantocypris* up to 1 cm long and with very well-developed eyes.

The Copepoda is one of the most dominant groups of the plankton. Our type for this group will be the extremely common genus *Calanus* (Fig. 8-14, 8-16) and in particular two related species *Calanus finmarchicus* and *C. helgolandicus*. The body, a few millimeters long, has an oval forepart, the metasome, which is formed by the fusion of head and first thoracic segment. This is followed by five, free thoracic segments. Behind the thoracic segments is a short and slender urosome, composed of four to five segments. In the middle of the head there is a single eye. The head bears five pairs of appendages, namely first pair of antennae as long as the total length of the animal, second pair of antennae, mandibles, maxillules and maxillae. (Fig. 8-15). The thoracic segment, fused to the head, carries maxillipeds and the five free thoracic segments have pairs of swimming limbs; the urosome has no appendages, except for a pair of tail extensions, forming the telson. The genital pores open on the first urosome segment and the anus on the last. The animal swims by moving its large first antennae. These morphological features are found with many but slight variations, in all the planktonic copepods. The very common genera are: *Calanus, Euchaeta, Temora, Centropages, Acartia* (Fig. 8-16) *Gaussia*, from the Calanoidea, *Euterpina* (Fig. 8-16) from the Harpacticoidea, mainly a benthic group, and from the Cyclopoidea *Oithona*. Their size varies in length from a millimeter to a few tens of a millimeter.

The Amphipoda, have a segmented and arched trunk without a distinct carapace, as typically seen in *Talitrus*. The planktonic species are numbers of the group Hyperidea, often characterized by two well-developed eyes.

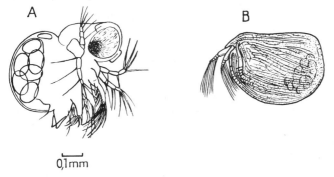

0,1mm

Fig. 8-13.—A, *Podon polyphemoides* Cladocera, a female; note the one large compound eye, the developed second antenna used for swimming, and the eggs filling the incubation cavity: B, *Conchoecia curta*, Ostracoda, a female; the animal is completely enclosed in a bivalve carapace only the swimming antennae protrude (length of the order of mm) (A, after Massuti & Margalef, 1958; B, after Rose & Tregouboff, 1957).

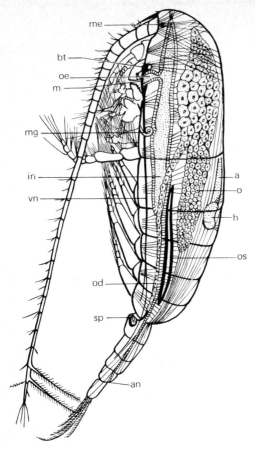

Fig. 8-14.—Morphology and anatomy of a female *Calanus*, Copepoda: side view; a, aorta; an, anus; br, brain; h, heart; in, intestine; m, mouth; me, medium eye; mg, maxillary gland; o, ovary; od, oviduct; oe, oesophagus; sp, spermatheca; vn, ventral nerve (after Marshall & Orr, 1955).

The best known genus *Phronima*, 1–2 cm long, lives in a small gelatinous, transparent barrel, cut out by the animal from the substance of other planktonic organisms such as *Siphonophora* or *Pyrosoma* (Fig. 8-17). Other hyperids are associated with jelly-fish and salps (*Hyperia* (Fig. 8-17); *Vibilia*) and some species may often form large swarms (*Euthemisto*).

In the Mysidacea the head and the thorax are fused giving the cephalothorax, which is covered by a carapace; in the primitive members of the group the last thoracic segment is still free. Most of the Mysidacea are benthic but many regularly leave the sea bottom and ascend into the water column. Some of them are found in very deep water (*Eucopia*). Their size range is few centimeters.

The Euphausiacea (Fig. 8-18), are very close to the Mysidacea; they have a complete carapace which covers their cephalothorax, and are disting-

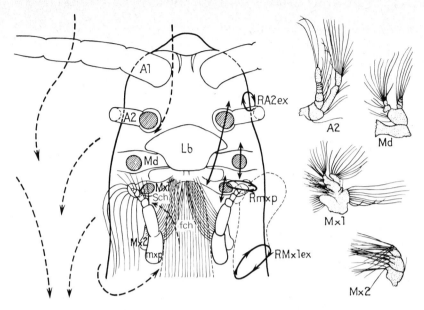

Fig. 8-15.—Anterior region of *Calanus finmarchicus*: a general diagram showing position of different appendages, their movements (heavy arrows), the induced water current (broken arrows), and the structure of the appendages: A1, first antenna; A2, second antenna; md, mandible; Mx 1, maxillule; Mx 2, maxilla; Mxp, maxilliped; Lb, labrum; Fch, filtration chamber; Sch, suction chamber; RA 2 ex, rotation path of tip of exopodite of antenna 2; RMax, rotation path of tip of maxillipede; RMxL ex, rotation path of tips of setae of maxillulary exopodite (after Marshall & Orr, 1955).

uished from the Mysidacea by the presence of dense and well-developed gills at the base of the thoracic appendages, as well as by a complex luminescent organ build of lens and reflector. They sometimes reach a size of 5 cm. Their habitat is deep water—sometimes very deep even up to thousands of meters. Some species show a diurnal migration. Among the most common species are *Meganyctiphanes norvegica* and *Euphausia pacifica*. The diet of the whales is comprised of Euphausiacea and their stomachs contain large quantities; the whalers call them krill.

The Decapoda are the most advanced crustacea; most of them are benthic but *Sergestes* is a pelagic shrimp. They may be considered to be on the border between plankton and necton.

8-6.—MOLLUSCA

Besides the larvae of molluscs, which are very common in the plankton, this phylum is also represented in the plankton by three specialized groups; the Heteropoda and two members of the order Pteropoda, the Thecosoma and Gymnosoma.

The morphology of some of the heteropods is very "classical" to other

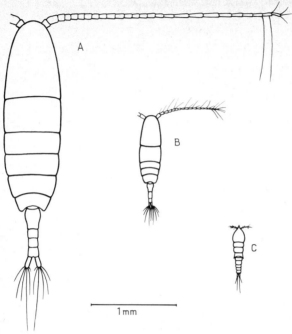

Fig. 8-16.—Three planktonic copepods drawn to the same scale, dorsal view: A, *Calanus finmarchicus*, Calanoidea, ♀; B, *Acartia clausi* Calanoidea; C, *Euterpina acutifrons*, Harpacticoidea (after Fiches d'indentification du zooplankton du conseil international pour l'exploration de la mer, Nos. 4, 12, 32).

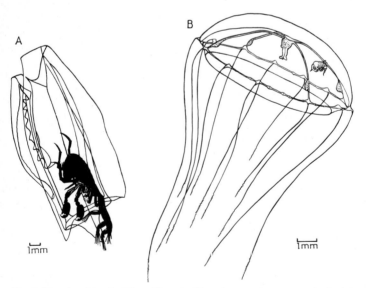

Fig. 8-17.—Amphipoda, Hyperidea: A, *Phronima curvipes* grown in the laboratory, building its living barrel out of the upper bell of an entire individual of *Abylopsis tetragona* (Siphonophora): B, *Hyperia schizogeneios*, a juvenile individual (stage IV, ♂) on its host, a leptomedusa of the genus *Phialidium* (after Laval, 1968 (A) and 1972 (B)).

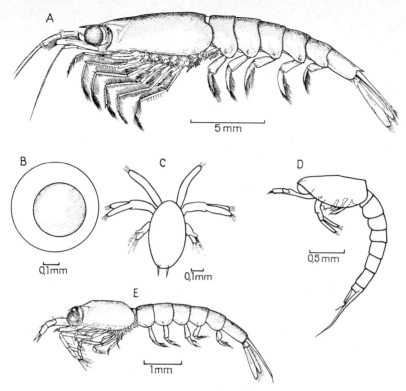

Fig. 8-18.— *Thysanoessa inermis* Euphausiacea; A, an adult female; B, an egg; C, second stage of nauplius with its three pairs of appendages; D, third stage calyptopis with the eyes covered by the carapace but still without pigmentation; E, seventh stage furcilia (after Einarsson, 1945).

gastropods, with a spiral shell the principal difference being the presence of a flattened fin rather than a foot. In *Carinaria* the shell is reduced and becomes conical while in *Pterotrachea* there is no shell but a transparent elongated animal 10 to 20 cm long with a large round fin. At the head extremity there are two compound eyes and an elongated snout ending in a mouth. At the posterior end there is a spindle shaped body corresponding to the viscera.

The Thecosoma are characterized by the presence of two symmetrical wing-like fins and a shell; their common name is sea butterflies. In the primitive forms the shell is spiral (*Limacina*, Fig. 8-19) but in more advanced forms straight (*Creseis*). The mouth is found between the fins and a ciliary system streams unicellular algae stuck together by mucus towards the mouth. Some species of Thecosoma are very abundant and sometimes form swarms. Their shells, with a maximal length of few centimeters, sink to the sea floor and form Pteropod ooze.

The Gymnosoma are very different in their morphology and anatomy, the only common character with the Thecosoma being the presence of two fins; this is the only justification for putting them together with the Thecosoma in

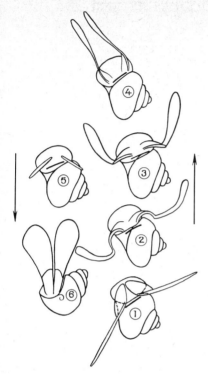

Fig. 8-19.—*Limacina retroversa* Pteropoda Thecosoma; outline sketch showing successive positions of wings in swimming upwards (1-4), and in descending (5); 6 shows the wing held motionless and vertical in a different view from 5; shell diameter 2 to 3 mm (after Morton, 1954).

the order Pteropoda. They have no shell and are not herbivores, but carnivores, often feeding on Thecosoma.

For the capture of prey there are complex tentacles possessing suckers with hooks; the tentacles can contract into the mouth cavity when the animal is resting. The Gymnosoma are of a very small size, about one centimeter (*Pneumodermopsis, Clione* (Fig. 8-20)).

Another planktonic mollusc is adapted to pelagic life in an unusual way. *Janthina* has a typical molluscan shell, lavender blue in colour. Its foot secretes mucus which forms a bubble and hardens forming a "bubble" float which enables the animal to hang upside down under the surface of the water. *Janthina* is a predator of *Velella* (3–5 cm).

Phylliroe is another pelagic mollusc, transparent and leaf-like.

8-7.—TUNICATA

The structure of the Tunicata will be described briefly. The Tunicata are characterized by possessing a notochord which generally disappears in the adult. They have a gill sack and are enclosed in a thick "jacket", the tunica, secreted by the integument. Such a structure, more or less modified, is found

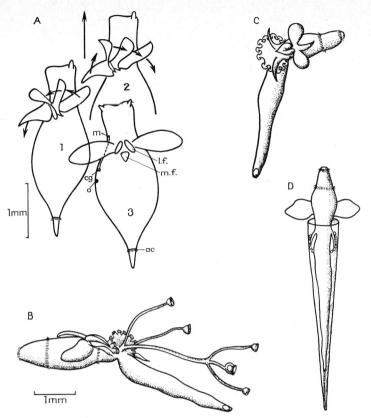

Fig. 8-20.—Pteropoda, Gymnosoma: A, *Clione limacina*, outline sketches showing successive positions of the wings, downward (1) and upward (2) strokes left side view, and (3) ventral view: a, anus; cg, common genital aperture; lf, lateral lobe of vestigial foot; m, male aperture; m.f; median lobe of foot: B, C, D, *Pneumodermopsis paucidens*; individual in extension showing its trunk, the wreath of suckers and 5 suckers on peduncules (B), an individual searching for prey (C) an individual penetrated into a thecosome pteropod of the genus *Creseis* whose retractor muscle was cut (D) (A, after Morton, 1953; B, C, D, after Sentz-Braconnot, 1965).

in all four groups of planktonic tunicates, *Pyrosoma, Doliolum, Salpa* and the Appendicularia.

The Pyrosoma are cylindrical, hard and strong (*Pyrosoma atlanticum*) or gelatinous (*P. spinosum*), from several cm up to tenths of a centimeter long. The cylinder is hollow, closed at one end and open at the other. The walls of the cylinder are made of many juxtaposed individuals. Each animal has one gill-sack which opens to the exterior by the mouth and to inside by a cloaca. Water is filtered and passes through the wall of the colony. The entire colony can contract and so force the water through the opening of the cylinder, propelling the colony by the exhaled jet of water. Each individual of the colony has small luminescent glands and the whole cylinder may be seen shining; this gives the group its name (pur = fire, soma = body).

The most common form of *Doliolum* has a small transparent barrel-like body (several mm to 1 cm long) open at its two edges and traversed by the gills with two rows of openings; the life history is very complex. The eggs develop into a characteristic larvae, tadpole-like with a notochord in its tail; it loses its tail and develops into a small doliolid-like creature. On its dorsal side an appendage grows backwards while on the underside a ventral stolon develops. This young individual (oozoid) develops into a "nurse". The stolon buds off to give fragments which are carried by ameboid cells on the barrel towards the dorsal appendage. These fragments multiply to give three types of buds; one, namely, the gastrozoids, filter water, respire and feed the whole colony; a second series of buds, the phorozoids, present on the dorsal appendage develop into a normal small doliolum but one carried on a stalk; the third group of buds are present on the stalk of the phorozoid. The phorozoid is liberated from the dorsal appendage of the oozoid. The buds on the stalk of the phorozoids divide and give secondary buds which develop into normal doliolids; the gonozoids, which carry the gonads and which give rise to the eggs which begin this life cycle. In some cases shorter cycles have been observed and this facilitates multiplication (Braconnot, 1967).

The morphology of the salps is very close to that of *Doliolum*, except that their simple gills with two large slits separated by a bar, are present across the large cavity between the mouth and the cloaca. Like *Doliolum*, a ciliary system covered by mucus collects food, mainly phytoplanktonic particles. Each species has two slightly different forms. The asexual form hatched from an egg (oozoid), has a ventral stolon which elongates and is segmented into a series of individuals, the blastozoids, attached to each other in a chain, several tens of a centimeter long and lasting more or less for a long period. The blastozoids are sexual individuals and give eggs. Due to the asexual reproduction of the stolon, *Salpa* are able to multiply extremely rapidly.

The Appendicularia (*Oikopleura, Fritillaria*), the last group of planktonic Tunicata, are very different from the preceding groups. They pass all their life in a larval structured tunicate, with its fundamental characteristic, namely, a tail supported by a notochord. Their ectoderm secretes a delicate transparent gelatinous envelope, the logette, which is inflated so that the animal swims in an interval cavity of the envelope (Fig. 8-21); this causes a stream of water which passes through fine filters present inside the logette and so nanoplankton is collected. After a certain time the appendicularian leaves its logette and rapidly secretes another one. They are small creatures, several mm long, but very abundant (Fenaux, 1967).

8-8.—PLANKTONIC LARVAE

Besides adults, the zooplankton contain numerous larvae of benthic and nectonic organisms.

The Echinodermata have numbers of larval forms, characterized by bilateral symmetry, while the adult itself shows a five-fold radial symmetry. Sea urchins have a pluteus larva, which has the shape of a pyramid at the

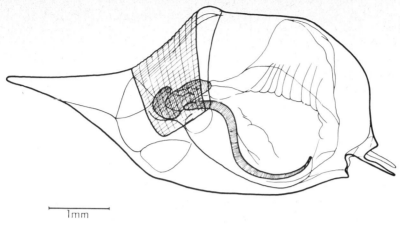

Fig. 8-21.—*Oikopleura* Appendicularia, in its house (original drawing by R. Fenaux).

base of which are prolonged ciliated arms, supported by skeletal, calcareous bars (Fig. 8-22). The ophiopluteus, very similar in its general shape to a pluteus, is the larva of brittle-stars. The starfishes, Asteroidea and sea cucumbers (Holothuridea) have larvae called, respectively, bipinnaria and auricularia, with a flattened oval shape, bordered laterally with more-or-less developed lobes.

In the Crustacea there is a very characteristic primitive larvae, the nauplius, a compact triangular or oval larva with three pairs of appendages

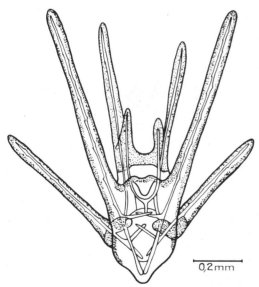

Fig. 8-22.—Echinopluteus of the sea urchin *Paracentrotus lividus*, at the stage of 8 arms, ventral view: note the rods supporting the arms, the ciliated bands on the arms and body, and the transversal ciliated shoulders (after Fenaux, 1968).

(first and second antennae, mandibles) and a median eye—the nauplius eye. In the Copepoda after several moults the nauplius changes to a copepodid which resembles the adult, but has to pass through more moults to reach the adult shape. In the Cirripedia the nauplius gives rise to a very peculiar larva with a bivalve carapace, the cypris. In the higher crustaceans the nauplius, or its successor, the metanauplius, changes by metamorphosis into a zoea (Fig. 8-23). Often the last form hatch directly from the egg but a nauplius stage is found during its embryonic development. The zoea has a segmented abdomen and several appendages on the thorax, besides the head appendages; it has a dorsal carapace, two large eyes, and very often long spines. By successive moults the zoea of crabs gives rise to the metazoea and later to the megalopa with large stalked eyes; in the megalopa attributes of the adult form may already be seen, namely, a voluminous cephalothorax and a reduced abdomen. The zoea of shrimps gives rise to the mysis which superficially resembles a mysidacean. Finally, the lobster larva is very peculiar, much flattened with long appendages—phylosoma (phyllon = leaf; soma = body). *Sergestes* has a very spiny larva the elaphocaris.

In the Polychaeta (Annelida) the first larva, called a trochophore, is similar to a spinning top and is encircled by a band of cilia (Fig. 8-24); it elongates, becomes segmented, develops bristles, and becomes a nectocheta, which exist in numerous variations, one of them the mitraria, looking rather like a small hat whose margins are more or less gadrooned and carry a large bundle of bristles. The Nemertea, unsegmented benthic worms without bristles have a larva known as pilidium (Fig. 8-24), which has a ciliated belt, marking two

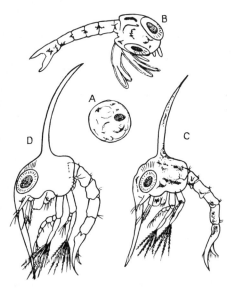

Fig. 8-23.—*Portunus latipes*, Crustacea Decapoda, larval stages: A, an egg (0.32 mm diameter); B, pre-zoea stage, hatching of the egg (12 mm length); C, zoea, first stage (1.28 mm between the extremities of the two spines); D, zoea second stage (1.8 mm between the spines) (after Lebour, 1944).

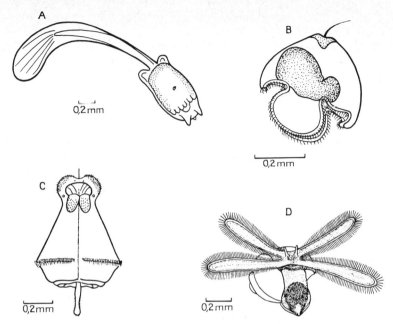

Fig. 8-24.—Planktonic larvae: A, tadpole of *Distomus variolosus* (ascidian); B: pilidium of nemertean C, trochophora, a very young larvae with one circle of cilia, of the polychaete *Mesochaetopterus sagittarius*; D, veliger of the gastropod *Simnia patula* (Cypraeidae) (A, after Berrill, 1948; B, after Sentz, 1962; C, after Bhaud, 1971; D, after Fretter & Pilkington, 1971).

large lobes, and a bundle of cilia at one end. *Phoronis* (Phoronidae) whose adults live in tubes attached to the sea floor have a very beautiful and characteristic larva the actinotrocha with numerous ciliated tentacles. Finally, the cyphonautus larvae belong to certain Bryozoa, which are very common colonial benthic organisms.

The Mollusca, Gastropoda, and Bivalvia has a swimming larva the veliger (Fig. 8-24), characterized by a strong ciliary band, the velum,

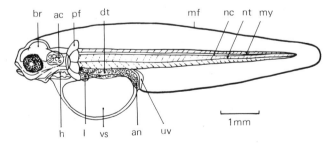

Fig. 8-25.—Scheme of a new hatched larva of plaice (*Pleuronectes platessa*): ac, auditiary capsule; an, anus; br, brain; dt, digestive tube; h, heart; l, liver; mf, marginal fin; my, myotomes; nc, nerve cord; nt, notochord; pf, pectoral fin; uv, urinary vesicle; vs, vitelline sac (after Shelbourne, 1956).

multilobed. The larva builds the primitive shell of the adult, bivalve or spiral, according to the group.

During their development the Enteropneusta, (*Balanoglosus*), have a very curious larva, the tornaria, in the shape of a spinning top, with a strong ciliary band which spins the larva. The benthic ascidian, or tunicate larvae when hatched are tadpole-like (Fig. 8-24) and generally fix themselves to the substratum a short time after the hatching; they appear in the plankton for only a short period.

Most of the bony fish (Teleostei) have planktonic eggs and larvae (Fig. 8-25). The eggs are generally spherical measuring some tenths of a millimeter up to several in diameter and often containing an oil droplet. The larvae or "fry" after hatching have relicts of the egg in an antero-ventral position. Initially they are very delicate and transparent but slowly become pigmented as their fins differentiate.

CHAPTER 9

QUANTITATIVE STUDIES AND THE DISTRIBUTION OF ZOOPLANKTON

9-1.—COLLECTION METHODS

Apart from large macroplankton taken from a boat or by diving, the collection of zooplankton is made essentially by filtration of water by nets in situ or on the deck, by collecting the water in bottles, or by pumping. Nets for qualitative work—large cones made of material of different mesh size (Tregouboff & Ross, 1957) were developed into quantitative nets by ensuring uniformity in the mesh size in order to define more accurately the sizes collected and by adding equipment to measure the volume of water filtered. For ecological work it is necessary to know the volume of water sampled and this has lead to the development of closing nets. In vertical sampling the equipment used is quite simple (Fig. 9-1) but in horizontal collections they are more complicated. Besides the Clarke-Bumpus net we may mention the equipment used by Bary (Bary et al., 1968) with remote closing and opening devices. For small zooplankton, pumps and bottles are used. Rapid sampling methods have been developed to obtain data on the distribution of plankton over large distances in a quite short time. Various torpedo shaped samplers have been designed; the Hardy plankton recorder which has been used by the Edinburgh Oceanographic Laboratory for a long time and on a large scale for investigations covering the North Sea and North Atlantic (Fig. 9-2) may be specially mentioned. Finally, we may mention the method originally introduced by Maddux & Kanwisher (1965), in which the zooplankton is concentrated by a silk cone and counted in situ by a system of Coulter counters (for more details on sampling and collection methods see Bourdillon, 1971).

9-2.—TREATMENT OF SAMPLES

After collection the zooplankton must usually be fixed, pending sorting and counting. The fixation is usually done with neutralized formalin but some animals may need special treatment such as anaesthesia or delicate fixation in order subsequently to be able to identify them. The problem of fixation, especially in the field in tropical regions, is more difficult than might be expected, and an international working group is trying now to issue a directive on this subject. The systematic sorting of the sample often takes a long time and this puts certain limits on the ecological study of zooplankton. In order to simplify this procedure several techniques relying on density

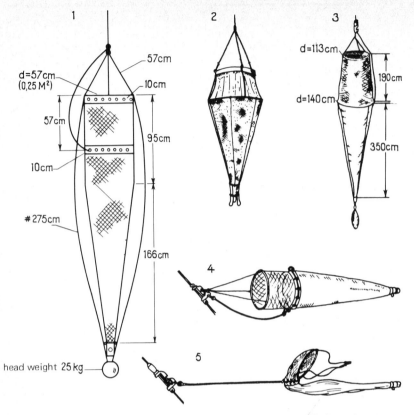

Fig. 9-1.—Plankton nets: 1, International WP-2 closing net, modified version, 200 μm-mesh aperture, filtration ratio, 6–1; 2, Hensen net; 3, Bogorov-Rass vertical closing net, mouth area 1 m² (reference BR 113/14); 4 and 5, closing net, open and closed (1 and 2 after Anonymous, 1968; 3 after Vinogradov, 1970; 4 and 5 after Hardy, 1958).

differences have been tried. Animals with calcified shells may be separated using a saturated salt solution (McGowan & Faundorf, 1964), gelatinous organisms (Siphonophora, jelly-fish, *Salpa* and *Doliolum*) rise in a stream of rising water (Michel & Grandperrin, 1971) and eggs and fish larvae have been separated by centrifugation in sucrose or silicate solutions (Bowen et al., 1972). In other techniques the zooplankton may be separated by size groups using special graded sieves and this often facilitates subsequent identification (Michel & Grandperrin, 1971). Bé (1968) has proposed a scheme using a theoretical system with sieves, which can estimate correctly the total plankton using different equipment for collection and elimination of any overlap (Fig. 9-3).

Usually sorting is facilitated by counting only sub-samples for the abundant taxa and the whole sample for rare species. Different procedures such as the Motoda box, a cylinder with partitions, a stempel pipette, or rubber-pear are used to obtain aliquots. All these procedures may give rise

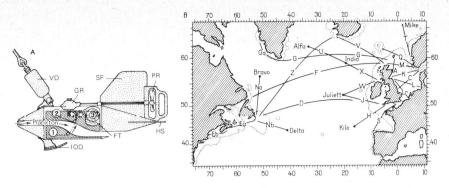

Fig. 9-2.—A, diagrammatic longitudinal section of Hardy's Planktonrecorder: the propeller (PR) is turned by the passage of water while being towed, driven by inside mechanism by gears (GR). There are two rolls of silk (27 meshes/cm) of about 15 cm width; gauze from the lower band (1) pass across a tunnel, where the sea water enters and filters off the plankton; gauze from the upper band (2) joins to the first one and keeps the plankton on it; the two are wound on the storage spool (3) in a formalin tank (FT). The recorder has a vibration damper (VD), a vertical stabilizing fin (SF) which serves as a direction guide, a horizontal stabilizing fin (HS), and a diving plane (DP). When towing at 12–18 kt with 18 to 24 m of wire out the recorder stays at a constant depth of 10 m. B, the routes of Hardy's plankton recorder survey along which the recorder was used once each month. The routes are identified by their code letters, the names of the meteorological stations are shown, the dotted line indicates the position of the 100 fathom (180 m) depth contour (after Glover, 1967).

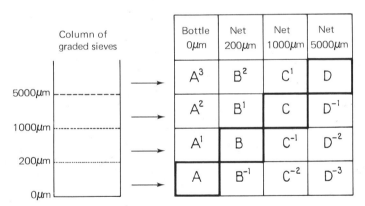

Fig. 9-3.—Scheme of theoretical method for the estimation of the total biomass of the zooplankton: the material obtained by different samplers, bottle or nets, is filtered through a set of sieves, and divided into many fractions, numbered at the left; fractions A, B, C, D, contain the plankton organisms which are ideally collected by each sampler; their sum, after correcting for the volumes of water filtered by each sampler, should give the best estimate of the total biomass without overlap (after Bé, 1968).

to additional problems regarding the reliability of the sub-samples; this problem will be discussed later.

It is possible to use, as for phytoplankton, a particle counter in order to get a numerical estimate of the organisms, but in this case they are not identified. With further developments such automatic techniques may enable forms to be recognized but the price of such equipment will not allow its use by many laboratories. A more economic way is a rapid analysis in which total counting is replaced by abundance classes. Frontier (1969) used abundance classes based on a geometrical progression with a basis of 4.3; 1–3 individuals is abundance class 1; 3–17, class 2; 18–80 class 3; 80–350 class 4, etc. As regards the expression of the results we are faced by the same problem already met with in the case of phytoplankton. The total counts of associated organisms which differ considerably from each other by volume and shape, are not significant unless the precise size categories are also given. Counting by taxa is more satisfactory but does not always allow easy comparisons.

Since the volume of zooplankton collected in any one sample may be relatively large, an alternative method of expressing the results is to determine its wet volume; in view of the large differences in the quantities of water in different organisms, the expression of the results on a dry weight basis is often preferred. Drying the sample has the disadvantage that it is, thereby, destroyed; an alternative is to calculate the dry weight from previously prepared tables. The best way may often be to determine the dry weight of an aliquot.

When the total dry weight of the material collected is used for the estimation of the biomass of the zooplankton, the technique of analysis is rapid but needs care. The biomass of the zooplankton may be expressed in terms of organic carbon, organic nitrogen, or sometimes organic phosphorus and it must be remembered that fixation may considerably effect the results, as a result of diffusion of organic matter into the fixation liquid; such diffusion causes a distinct error in dry weight estimation.

9-3.—VALIDITY OF THE RESULTS

We shall examine to what extent the accuracy of the results obtained using the available collecting equipment, mainly nets, may be considered to reflect the real distribution of the zooplankton, supposing that the net was built and used in accordance with the remarks mentioned in the Appendix (Chapter 14).

The first important cause of errors is the avoidance of the net by organisms. The organisms are rarely completely passive and many of them can move as the net is approaching. The net and its cable may initiate an acceleration of the water in front of itself (Fig. 9-4); such an acceleration or change in pressure may be detected by the organisms. If the organism reacts by avoiding the turbulent water some will avoid the net. Net perception by animals resulting from "sound" caused by the net has not been proved (Clutter & Anraku, 1968) but visual perception of the net appears to take place. LeBrasseur & McAllister (in Clutter & Anraku, 1968) observed that dark-coloured nets capture more Euphausiacea than light-coloured ones, either in day-time or at night. The bioluminescence of organisms touched by the net may also interfere in the perception of the latter.

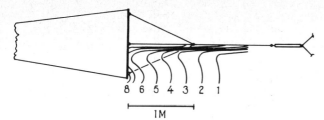

Fig. 9-4.—Velocity profile of the water in front of an approaching plankton net of a diameter of 1 m, towed at a velocity of 2.8 kt (144 cm/sec); numbers indicate the water velocity in cm/sec. (after Smith et al. in Clutter & Anraku, 1968).

It must, then, be accepted that not all the organisms of the zooplankton which are in the way of the net will be captured, even if the water filtration coefficient of the net (F) is 100%. It is, therefore, necessary to define a coefficient of capture (C) which is the quantity of plankton captured to that in the volume of water filtered, by the passage of the net. There are few estimations of this coefficient. Gilfillan (in Clutter & Anraku, 1968) gave the following equation

$$C = \left(\frac{R - (K/V)}{R}\right)^2$$

Where V is the velocity of towing, R, the radius of the net, and K the product of the mean velocity of the animal and the distance at which it can detect the approach of the net and react to it. (The velocity of escape is the perpendicular component of the animals velocity to the line of tow, the other component being parallel to the line of tow). It is possible to determine the value of K by towing at different speeds and experience shows that this factor is more or less constant with changing speed. By the analysis of collections of *Calanus* sp., *Euchaeta japonica* and Euphausiacea, made at 0.7, 2, and 4 kt with different nets Gilfillan concluded that the larger animals had more ability to avoid the net and that the capture coefficient dropped with speed, especially below 2 kt. Although this conclusion seems to be logical and has been found in several investigations it should be mentioned that some authors did not obtain the same results. Barnes & Tranter (1965) compared the Indian Ocean standard net (113 cm opening) Juday net (80 cm) and Clarke-Bumpus (12 cm), and did not find that the larger zooplankton avoided the small nets more than the larger nets. The problem of net avoidance has not been resolved (see also the work of Laval, 1974, on this subject); the perception of the approaching net as well as the reaction of the organisms to it probably varies considerably. Until there is more information it seems best to take into account the general recommendations of Clutter & Anraku (1968), namely, to use as large a net as possible at a constant speed, which is as high as possible, to maintain a filtration coefficient of at least 85%, to keep the opening as free as possible, and to use dark coloured, non-shiny materials.

Besides the avoidance of the net another source of error in the estimation of zooplankton is loss of organisms through the mesh of the net itself

(Vannuci, 1968) (we neglect the eventual loss of contents by the opening and closing, passive or active, since this may be eliminated by correct technique). This loss must certainly be considered since the filtering materials do select only particles whose diameter is higher than the mesh size; however, this "mesh selection" does give rise to some problems. With few exceptions the elements of the plankton are not spherical but have various diameters along different axes. Saville (1948) has shown that for the same mesh size the curve of selection of *Oithona* is not the same as that for *Calanus* (Fig. 9-5A,B). This may be explained by the fact that *Oithona* holds its antennae at right angles to the body. Even with spherical organisms it seems that a certain plasticity of an organism may interfere; even fish eggs

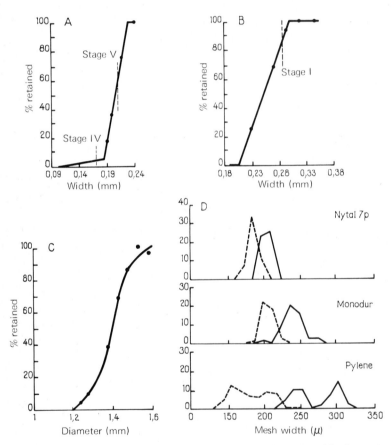

Fig. 9-5.—Selection of 60 mesh silk for *Oithona*: A, maximum breadth of organisms used; the dashed lines indicates mean sizes of copepodid stages IV and V: B, selection curve for the same gauze for *Calanus*: C, selection curve of 16 in. mesh silk, mean size 1.188 ± 0.010, for fish eggs: D, mesh width measurements for three plankton gauzes Nytal 7P, nylon monofilament mean mesh width 197 μm S.D. ± 32; Monodur perlon monofilament, mean mesh width 224 μm S.D. $= \pm 69$; Pylene, polythene monofilament (220 μm, S.D. $= \pm 169$; solid line warp to warp, broken line weft to weft (A, B, C after Saville, 1958; D, after Heron, 1968).

may pass a mesh which theoretically should catch them; with a mesh of 1.188 ± 0.01 mm 50% of eggs of the 1.403 mm diameter escape (Fig. 9-5C). It should also be remembered that the dimensions of the filtration tissue itself shows a more or less large variability (Fig. 9-5D).

The results of mesh selection may interfere and bias the qualitative and quantitative results of any collection and should be kept as constant as possible. The clogging and ageing of the net should be taken into account. The velocity of the water in the mesh or "mesh velocity" (p. 322) should be maintained as low and constant as possible since any increase favours the escape of organisms.

If then we take into account the uncertainties and variations of the net avoidance, variations in escaping from it, and the other hazards it is evident that there will be more or less large errors in the estimation of the zooplankton population of a given zone. In order to use the data obtained as efficiently as possible it is necessary to know the error or at least an approximation to it. The estimation of the error gives rise to difficult problems because any observed variability is due not only to the net errors but also to the natural variability of the plankton, as we shall see later, when discussing the microdistribution. It is not easy, a priori to separate these two components, i.e., instrumental error and variation in distribution. Cassie (1968) considered that under good working conditions, (especially without clogging) most of the errors, or better the variations in capture, may be attributed to variations in the plankton itself.

Finally, we are left with the problem of the validity of sub-samples. Venrick (1971) has studied the statistical implications of this procedure, and by examining the variance of the different steps of the procedure he came to the conclusion that an increase in the level of sub-sampling diminishes considerably that part of the variance of the original sample in the resulting variance. This operation is important if we wish to reduce the variance of the original; large samples should be collected and relatively small aliquots examined. On the contrary, if we wish to study the variance of the natural population ideally the total sample rather than subsamples should be examined.

9-4.—SAMPLING AND MICRODISTRIBUTION

Experience shows that even when the collections are made under optimal technical conditions, there are very important variations in the results (numbers, dry weight ....) variations which must be due to the plankton itself, in particular, to its type of statistical distribution. Cassie (1963), examining the results of many authors, found that the coefficient of variation ($V = s/m$) often lies between 22–44% but that higher coefficients of variation are not rare. The calculation of the standard deviation s, supposes that the samples of one lot are normally distributed; however, with high coefficients of variation there are some absurdities. For example, with a coefficient variation of 110% (or 1.10) and mean 100 two thirds of the samples, if the distribution is normal, will be found between

$100 \pm (1.10 \cdot 100) = -10$ or 210.

The negative value is not acceptable, and in fact, the normality of the results obtained is generally an illusion. On the contrary, in many cases normality can be obtained by taking the logarithm of the value; we have then, a logarithmic variation coefficient which is (on the basis of $\log_{10}$):

$V' = 10^{s'} - 1$

when s' is the standard deviation calculated on the logarithms of the raw data. A coefficient of variation of 110% is then significant for a mean of 100, since the interval where two thirds of the count is found is entirely positive

$100 \times$ or: $2.10 = 48$ or 210

($V' = 10^{s'} - 1$ gives $s' = \log 2.10$ and the interval $m' \pm s'$ ($m' = \log m$ becomes $\log 100 \pm \log 2.10$). By using the logarithmic transformation of the results the range in which most values of the coefficient variation becomes 23–53%. The logarithmic transformation has another characteristic as was shown by Winsor & Clark (1940) (Table 9-1): the coefficient of variation V ($V = s/m$) is then independent of the mean, so that large samples are as variable as small ones; originally the standard deviation is proportioned to the mean s ($s = Vm$) and the logarithmic transformation eliminates this proportionally by stabilizing the variance and hence the standard deviation. This enables an analysis of variance to be correctly used.

Theoretically this logarithmic transformation normalizes the counts of plankton only with large numbers—over 50; in practice, however, it may be used for smaller numbers without significant loss of precision (Cassie, 1962).

The important variations met with in the quantitative estimations of zooplankton abundance makes the comparison of samples difficult, and it is necessary to be cautious as regards the possible significance of any observed differences. Silliman (1946) working on sardine eggs came to the conclusion that the number of eggs in one tow may only be considered significantly different from that in other tows when it is less than half or more than double the number. Although according to Cassie (1968) such a rigid criteria of significance may not be necessary for organisms other than fish eggs that are found permanently in the plankton (holoplankton) this does not appear to be so from the table given by Wiebe & Holland (1968) (Table 9-2). This table gives the limits of confidence, with a probability of 95%, of an isolated observation and these limits are very variable.

The coefficient of variation ($V = s/m$) is, as we have seen, largely independent of the mean, which means that s is proportional to m, or s^2 the variance to m^2.

$$s^2 = V^2 m^2 \text{ or } s^2 = cm^2 \tag{1}$$

where $V = c$. In reality, this may only hold for numbers greater than 50; in general the variance of the quantitative measurement of plankton is expressed;

$$s^2 = m + cm^2 \qquad (2)$$

The variance is equal to the mean, m, with the addition by a correction term m^2. Since c is small it is only for large numbers that the term in m^2 becomes large with respect to m. (With $V = 32\%$, $c = 0.1$ the difference between the values of s^2 given by the expressions (1) and (2) is 20% for $m = 50$, 10% for $m = 100$ but 50% for $m = 10$). The constant, c, varies generally between 0.05 and 0.1 but for the coefficients V and V' higher values are not rare.

The most frequently observed variance in the quantitative studies of plankton differs only slightly from that expected on a Poisson distribution, in which the variance is equal to the mean:

$$s^2 = m$$

When the variance is greater than the mean it indicates the existence of aggregation; such a distribution could also be overdispersed or contagious as

TABLE 9-1

Results of enumeration of four groups of organisms in a series of 12 successive hauls; net, 75 cm aperture diameter, 10 pores/cm: oblique tows from 30 m to the surface, the net being raised at 3 m/min; 24th August 1936, at the entrance to Vineyard Sound, Atlantic coast of the United States, depth of bottom 38 m. The Arabic numbers are the total number estimated in the catch, the decimal logarithms of the estimated catch is given in italics: m, mean; r, range of variation: the standard deviation is not given but the range of variation (r) which approximates to the S.D. varies considerably for the total numbers and is proportional to the mean (m); with the logarithmic values it is stabilized (after Winsor & Clarke, 1940).

	Calanus finmarchicus				Centropages typicus			
	Copepodid IV		Copepodid V		Female		Male	
	895	2.95	1520	3.18	43300	4.64	11000	4.04
	540	2.73	1610	3.21	32800	4.52	8600	3.93
	1020	3.01	1900	3.28	28800	4.46	8260	3.92
	470	2.67	1350	3.13	34600	4.54	9830	3.99
	428	2.63	980	2.99	27800	4.44	7600	3.88
	620	2.79	1710	3.23	32800	4.52	9650	3.98
	760	2.88	1930	3.29	28100	4.45	8900	3.95
	537	2.73	1960	3.29	18900	4.28	6060	3.78
	845	2.93	1840	3.26	31400	4.50	10200	4.01
	1050	3.02	2410	3.38	39500	4.60	15500	4.19
	387	2.59	1520	3.18	29000	4.46	9250	3.97
	497	2.70	1685	3.23	22300	4.35	7900	3.90
m	671	2.80	1701	3.22	30775	4.48	9396	3.96
z	663	0.43	1480	0.39	24400	0.36	9440	0.41

if the animals were attracted to each other. Such an aggregation may sometimes be very strong, visible even to a naked eye on the surface of the water, and the term swarm is then used. A distribution as a swarm is an extreme expression of the more generally aggregated distribution or patchiness.

After the elimination of all the errors of technique and instrumentation and in spite of the mixing of the pelagic environment and the ability of the plankton according to its definition to follow the movements of the water, a random distribution of plankton is extremely rare, as is the distribution of animals in most other biotopes. Aggregation is the most general form of distribution; however, sometimes another type of distribution is found which is more regular than random distribution, and is called "under dispersed" or negative association (contagion); statistically this is characterized by variance less than the mean.

The terms underdispersed and over dispersed are used and accepted; but the terms aggregated and regular which seem more expressive are preferred. Taylor (1961) has studied aggregated distribution and suggests using the following equation for the variance,

$$s^2 = am^b$$

where a and b are constants. He shows that much data from different biotopes is satisfactorily expressed by such an equation. The constant, a, depends largely on the sample characteristics while b represents a real "degree or index of aggregation", less than 1 for a regular distribution, equal to one for a random distribution, and greater than 1 for an aggregated distribution. Frontier (1972) from 32 taxa counted on 3 subsamples of 28 collections of zooplankton and obtained an index of aggregation of 1.23 ($s^2 = 0.61 \, m^{1.23}$). In samples with a particularly low mean (especially when $m < 1$) m is the preponderant term in the equation $s^2 = m + cm^2$, unless c is unusually high, and so the variance becomes equal to the mean; the rare species will appear randomly distributed. With large samples giving high numbers this random distribution should disappear (Cassie, 1968). Wiebe (1970) in a study of the distribution on a small scale found such a relation between the density of the organisms and their distribution (Fig. 9-6). From such a hypothesis of aggregation and more particularly from patchy distributions Wiebe & Holland (1968) attempted by a simulation technique to estimate the effect of this distribution on the result of sampling. Fig. 9-7 shows the variation index of accuracy obtained by dividing the number of the total individuals estimated in the simulated sample in the theoretical volume of water by the total number actually found. It appears that the precision of the mean obtained for this index increases with the diameter of the patch; this means that there is a decrease in limits of confidence at 95%, when the dimensions increase; the precision is also better for big nets. In general, however, and mainly for small nets this precision is of the order that was obtained under natural conditions (Table 9-2), which confirms the suggestion that the variations in sampling in the natural milieu are essentially

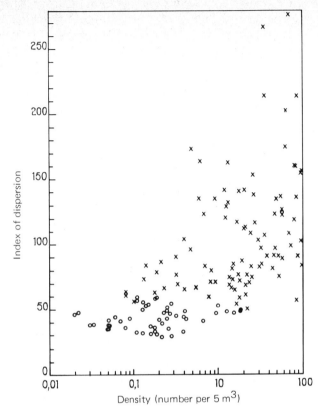

Fig. 9-6.—Plot of density of common species against the index of dispersion (calculated according to Fisher): the high density associated with departure from randomness of distribution (aggregate distribution): crosses, significant departure from random; circles, non-significant departure from random (after Wiebe, 1970).

due to the aggregated distribution of the plankton. This Fig. (9-7) also shows the extreme bias that the sampling may produce for one of the random distributions of the centre of the patch (RD-2).

Wiebe (1970) tried to verify in situ the existence of aggregated distribution during an investigation made in the south east of Guadalup Island (28°54′N; 118°16′W) using a plankton recorder towed horizontally at 90 m and running a grid of approximately 600 m each side; each sample corresponds to a tow of 11.7 to 15.9 m. The entire collections were made at day time from 10.00 to 16.33 h, a period during which vertical migration may be neglected. Fig. 9-8 shows the abundance of the 6 most abundant species as a function of distance. This distribution is extremely variable and illustrates the existence of successive patches of great abundance. The mean radius of such patches for the principle species is not very variable—of the order of 15 m. Finally, statistical analysis indicates that high or low concentration of individuals of these 6 species have a tendency to coincide more often than expected on a random distribution. The patches, therefore, tend to be of

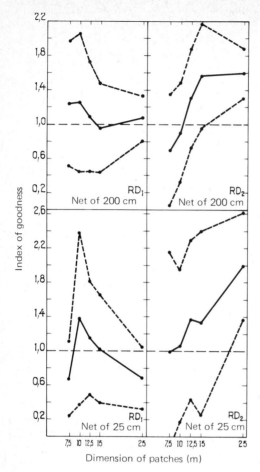

Fig. 9-7.—Comparison of mean exactitude index (continuous line) and confidence limits of the mean (broken lines) with $P = 95\%$ based on nine replicate tows with changing size of patch (abscissa), two random distributions (RD_1 and RD_2) and for 2 different size of nets opening (200 m and 25 cm) (After Wiebe & Holland, 1968).

many species; it is extremely important to confirm this last point: it agrees with the fact that, even in measurements of biomass without specific distinction, aggregated distributions are observed, and this suggests that the reason for patchiness is of a physical rather than a biological nature.

Stommel (1949) emphasizes the possible role of Langmuir cells, movements induced by the friction of the wind on the surface of the water and creating turbulence with the axes of rotation parallel to the direction of the wind. Perpendicular to this direction a series of successive small convergences and divergences are established. Particles with a tendency to sink should accumulate in the zone of "upwelling" and those which have a tendency to float in the zones of "down welling". Only particles with neutral buoyancy should be dispersed throughout the water mass. The tendency to

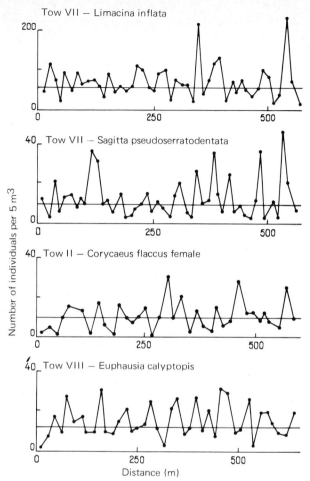

Fig. 9-8.—Plot of abundance against distance used to estimate size of patches of zooplankton: these show some of the most extreme changes in abundance observed over a short distance on the grid; median value also plotted (after Wiebe, 1970).

rise may be true not only for particles with positive buoyancy but also for animals showing vertical migration towards the surface (Fig. 9-9, upper).

Stavn (1971) has examined the problem experimentally using *Daphnia* and gives a model of distribution slightly different, in which the zone of accumulation of organisms should be a function of the velocity of the current in the cell, and the capability of the animal to orientate actively to this current in relation to light (Fig. 9-9 lower). Whatever the details of the theory it is interesting to mention an observation by Neess (1949), according to which the samples of plankton collected up-wind in a parallel direction show greater variability than those collected perpendicular to this direction. We may recall the fact that it is well known that phytoplankton are also found in patches (see, particularly, the work of Platt).

TABLE 9-2

Estimation of the validity of quantitative observations of zooplankton (after Wiebe & Holland, 1968).

Investigator	Limits of confidence with 95% confidence coefficient	Organisms used in calculation	Details of collection Sn, station; T, tow; OD, opening diameter of the net; M, mesh.
Hopkins	42.3–236%	*Acartia tonsa*	16 Sns (4–6 T/Sn); 5 min oblique T; Clarke–Bumpus net; No 10 nylon netting
	41.7–239%	*Oithona brevicornis*	
	44.7–226%	*Paracalanus crassirostris*	
Brinton (Wiebe analysis)	38.1–263%	*Euphausia eximia*	25 Sns (2T/Sn, 24 h apart); 14 min oblique T., 0–140 m; OD, 100 cm; M, 505 μm
	14.9–671%	*Euphausia gibboides*	
	39.1–256%	*Nyctiphanes simplex*	
	31.9–314%	*Thysanoessa gregarin*	
	14.3–698%	*Thysanoessa spinifera*	
Silliman	44.0–225%	*Sardinops caerulea* eggs	24 Sns (2 T/Sn); 12–20 min oblique T; OD, 160 cm
Taft	6.33–158%	*Sardinops caerulea* eggs	25 Sns (5T/Sn, replicate T 24 h apart); 14 min oblique T; 0–140 m; OD, 100 cm; M, 505 μm
Winsor & Clarke	59.0–169%	*Calanus finmarchicus* CIV	Series of 12, 10 min oblique T; 0–30 m; OD, 75 cm; 10 M/cm
		Calanus finmarchicus CV	
		Centropages typicus ♀	
		Centropages typicus ♂	
Barnes	47.0–213%	*Pseudocalanus minutus*	Series of 4, 2 min vertical T; 0–60 m; OD, 45 cm; 333 μm
		Microcalanus pygmaeus	
		Centropages hamatus	
		Temora longicornis	
		Acartia clausi	
		Oithona similis	
Cushing	24.4–485%	*Paracalanus parvus*	59 Sns (3T/Sn) ⎫ Vertical T
	17.9–560%	*Pseudocalanus elongatus*	59 Sns (3T/Sn) ⎬ OD:73 cm
	23.9–418%	*Temora sp.*	37 Sns (3T/Sn) ⎭ Hensen egg net
	38.0–263%	*Calanus finmarchicus*	18 Sns (3T/Sn)
Gardiner (Winsor & Clarke analysis)	44.9–200%	*Calanus finmarchicus*	Series of 18 vertical; T, 0–80 m; OD, 45 cm; M, 333 μm
		Paracalanus parvus	
		Temora longicornis	
		Pseudocalanus elongatus	

Künne (Winsor & Clarke analysis)	47.1–213%	Hydromedusae Terebellid larvae *Temopteris* sp. *Sagitta* sp. *Calanus* sp.	8 series of 4 vertical T, Helgoland larval net; OD, 143 cm and Hensen egg net; OD, 73 cm
Motoda & Anraku	47.0–213%	*Paracalanus* sp. *Clausocalanus* sp. *Oithona* sp. *Oncaea* sp. *Corycaeus* sp. Amphipods *Sagitta* sp. *Oikopleura* sp.	Series of 5 vertical T, 0–±50 m; OD, 45 cm; M, 327 μm
Winsor & Clarke	43.4–321%	*Limacina retroversa* *Centropages typicus* ♂ *Centropages typicus* ♀ *Centropages hamatus* *Pseudocalanus minutus* *Tortanus discaudatus*	Series of 8 vertical T; 0–31 m, alternatively with 2 nets; OD, 75 cm and OD, 30.5 cm; M, 10/cm
Barnes	49.3–203%	Copepod copepodite stages *Sagitta* sp.	Series of 4, 20 min, horizontal T at 30 m with 2 nets; OD, 100 cm and OD, 200 cm; stramin
Barnes	53.0–189%	*Meganyctiphanes norvegica* *Euchaeta norvegica*	Series of 4, 20 min, horizontal T at 30 m with 2 nets; OD, 100 cm and OD, 200 cm; stramin
Winsor & Clarke	26.8–373%	*Calanus finmarchicus* ♀ *Calanus finmarchicus* CV *Metridia lucens* ♀ *Centropages typicus* ♂ *Centropages typicus* ♀ *Euthemisto* juvenile *Sagitta elegans*	Series of 18, 10 to 15 min horizontal T with 2 nets of OD, 12.5 (at 29 and 31 m level) and 1 net OD, 75 cm (at 30 m level); M, 10/cm.

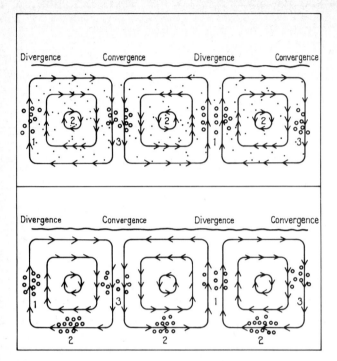

Fig. 9-9.—Langmuir spirals and the distribution of zooplankton. The sections are perpendicular to the direction of the wind, the arrows show the direction of the currents. Upper, Stommel model with particles of different densities; 1, particles with tendency to sink and to be aggregated in upwellings; 2, particles with neutral buoyancy and a tendency to be scattered randomly; 3, particles with tendency to float and be aggregated in downwellings. Lower, Stavn model with zooplankton actively oriented into a current in relation to light; 1, 2 and 3 represent aggregation zones when the current velocity is high, medium and low, respectively (after Stavn, 1971).

We shall end this discussion of sampling and microdistribution by three comments on the size of samples, the strategy of sampling, and the transformation of raw data.

Size of the sample: For small samples where the term m^2 in equation (2) is negligible, $s = \sqrt{m}$ and so $V = 1/\sqrt{m}$; increasing of the size of the sample will decrease the coefficient of variation and will give greater precision. For larger samples where $s = m + cm^2$ the calculation of V as a function of different values of c give the values shown in Table 9-3, according to which above 100 larger samples do not improve the precision; with the most variable species ($c = 100$) even above 10 the precision does not substantially improve. Large samples are only justified when taken by large nets in order to reduce net avoidance. This conclusion is somewhat paradoxical and needs further experimental examination (Cassie, 1968).

Strategy of sampling: In order that classical statistical theory may be used to estimate errors, sampling must be done at random in space and time;

TABLE 9-3

Values of the coefficient variation (V in %) for different values of the mean, m, and of the constant c (after Cassie, 1968).

m	c		
	0.05	0.2	1.0
0.1	317	319	332
1.0	102	110	141
10	39	55	105
100	24.5	45.8	100.5
1000	22.5	44.8	100.0
∞	22.3	44.7	100.0

this is rarely done and most samples are collected according to a systematic scheme fixed in advance; however, an improved strategy would be to carry out stratified sampling, the stratum, in the statistical meaning of the term, should be a zone where the physical, chemical, and biological conditions are relatively homogeneous. At each stratum two or more samples should be taken at random. The major difficulty is to define the stratum in advance (Laevastu, 1962). With continuous sampling equipment, such as the Hardy recorder, the count of a certain number of spaced samples can lead to the definition of the appropriate strata and to decide on a new series of samples to be counted (Cassie, 1968).

Transformation of the raw data: The logarithmic transformation is often used. In practice the transformation used is $\log x + k$, where x is the raw number and k is a constant, generally 1. This overcomes the difficulty of zero counts in a logarithmic transformation and gives a better normalization (Cassie, 1968). The transformation necessary to make s^2 independent of m and to be able to use the technique of variance analysis is only logarithmic when Taylor's index of aggregation is equal to 2. The last index enables one to calculate the necessary transformation (Taylor, 1961). With an index lower than 2 the best transformation will be to a fractional power, between 0.5 and 0. For instance, Frontier (1971, 1972) with a coefficient b of 1.23 used the cube root (transformation $x^{0.33}$) with good results. For a very large aggregation, b higher than 2, the transformation will be a negative fraction power (for $5 = 2.35$ the transformation will be $1/x^{0.18}$).

9-5.—QUANTITATIVE VARIATION WITH TIME

In this paragraph we shall discuss successive seasonal variation (from one season to another) and annual variations (from one year to another). As regards daily variations (from one day to another) there are only few data.

The observed variations may be due to microdistribution or to movements of water masses renewing the sampled stratum (in its statistical sense). Diurnal variations (between day and night), will be discussed in a separate chapter dealing with vertical migration.

Seasonal variations

In a given place and at a fixed level (the diurnal variation not being taken into account, and considering only day or night collections), the zooplankton is far from constant in its composition over a year; on the contrary, there are considerable monthly variations which do, however, in general show considerable similarity from year to year, so that it is possible to establish for a well studied station a kind of "plankton calendar" which gives the probable composition of the zooplankton at different periods of the year. In examining these variations the holoplankton and meroplankton should be dealt with separately.

Seasonal variation of the meroplankton: Numerous animals whose larvae belong to the meroplankton have a limited period of reproduction during the year; only few species reproduce continuously. Their eggs and larvae will, therefore, appear in the plankton for a more or less long period according to the species.

The pluteus of the sea-urchin *Arbacia lixula* is found in Villefranche (France) only in the hot season (Fig. 9-10A), since reproduction only takes place at temperatures higher than 20°C (Fenaux, 1962). The mitraria (larvae of the Oweniidea, Polychaeta) appear in Villefranche only from November to March (Fig. 9–10B) (Sentz, 1962).

In northern waters similar phenomenon but much more pronounced, are evident. In the east of the Murmansk Sea, according to Rzepishevsky (1962) the nauplii of the barnacle *Balanus balanoides* appear as scattered individuals in the plankton in January, and with time their density increases slowly until at the end of March it is about 100 nauplii/m^3. Suddenly, in two or three days the population is increased by about hundred-fold and sometimes the density may reach 30000 individuals/m^3 and constitute more than 99% of the organisms in the plankton at this time (Fig. 9-10C).

Many other examples, concerning different elements of the meroplankton may be given, e.g., fish eggs and medusae released by Hydrozoa, the breeding forms of certain polychaetes etc. The factors affecting the reproduction of species with planktonic larvae or planktonic forms will, therefore, influence the composition of the zooplankton.

Seasonal variations of the holoplankton: These organisms which spend their entire life-cycle in the water sometimes show important quantitative variations over the year. At Villefranche the two main families of Appendicularia alternate with each other in abundance; the Fritillariidea are predominant during the cold season and the Oikopleuridea during the rest of

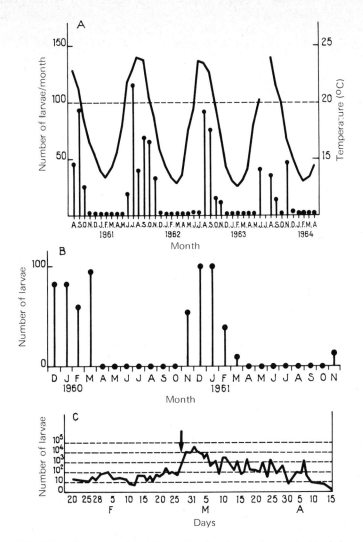

Fig. 9-10.—Examples of variation in the abundance of meroplanktonic larvae: A, changes in the number of pluteii of the sea urchin *Arbacia lixula* per sampling (at 5 m depth, in similar conditions), and temperature changes at Villefranche; B, variation in the number of mitraria larvae (Polychaeta, Oweniidae) collected per month per 10 net tows at Villefranche; C, variations in number of *Balanus balanoides* nauplii (Cirripedia) per m^3 in the bay of Dalne-Zelenetskaya (east of Murmansk Sea) in 1958; arrows indicate the beginning of a massive appearance of nauplii (logarithmic ordinate) (A, after Fenaux, 1962; B, after Sentz, 1962; after Rzepishevsky, 1962).

the year (Fig. 9-11) (Fenaux, 1963). The tunicate, *Salpa fusiformis*, is found in the bay of Villefranche only from the end of December to May (Fig. 9-12), and it is abundant largely when the temperature is between 13°C and 15°C (Braconnot, 1968). Among the microzooplankton the Tintinnida show considerable variations related to those of temperature. *Proplectella claparedei*

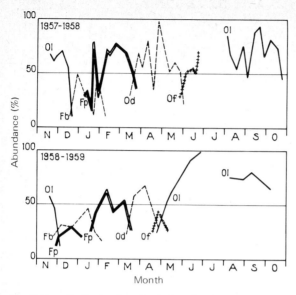

Fig. 9-11.—Cyclical order of relative abundance, at point A at the entrance of Villefranche roadstead for 5 principal species of appendicularians: Ol, *Oikopleura longicauda*; Fb, *Fritillaria borealis*; Fb, *Fritillaria pellucida*; Od, *Oikopleura dioica*; Of, *Oikopleura fusiformis* (after Fenaux, 1963).

appears in the plankton at temperatures below 15°–16°C, while *Tintinnopsis* sp. is found in considerable amounts only above 20°C (Fig. 9-12A) (Posta, 1963).

In the holoplanktonic Crustacea similar phenomena are found. The annual development of *Calanus finmarchicus* has been thoroughly studied in Loch Striven in Scotland not only as regards the adults but also with respect to the eggs and different stages of development of both nauplius and copepodid I to V (Fig. 9-13) (Marshall & Orr, 1955). The eggs appear in February but are numerous mainly from April to August, which corresponds to the time when nearly all the different nauplius stages and adults, except copepodid V are present; stage V is, however, abundant and dominant in autumn and winter and represents a form of hibernation. From such quantitative research it is possible to identify the successive generations and recognize the annual cycle of the species (Fig. 9-14). These changes are frequently accompanied by variations in the size of the adults; the largest are developed at the lowest temperature (Fig. 9-15).

Total seasonal variations: Considering the whole plankton and its numerical distribution by groups, it is not surprising that monthly changes in composition are found as shown in Fig. 9-16. Each group is represented by more or less numerous species which differ in their appearance "calendar". The figure only gives a general picture of the seasonal changes but nevertheless the important part played by the Copepoda is very clear, and

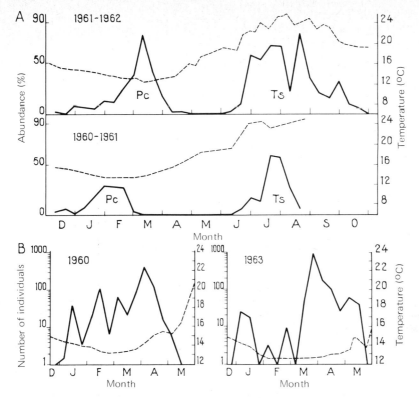

Fig. 9-12.—A, changes in relative abundance (solid lines) of two tintinnids and the changes of temperature (broken lines) in the shallow water of Villefranche roadstead: Pc, *Proplectella claparedei*; Ts, *Tintinnopsis* sp: percentages established on 200 individuals: B, changes in the mean number of individuals (solid curves) of the tunicate *Salpa fusiformis* caught each ten days (only blastozoids) and changes of the temperature (means from 0 to 50 m) (broken lines) in Villefranche roadstead (A, after Posta, 1963; B, after Braconnot, 1968).

this is in general the case. Amongst the copepods themselves there is also much seasonal variation in the different genera or species; this is illustrated in Fig. 9-17. *Pseudocalanus* is well represented throughout much of the year while, in contrast, *Temora* is abundant only during the first half of the year and *Oithona* increases only during the second half.

Using the data obtained by a plankton recorder from 1948 to 1962 in the waters around the British Isles, Colebrook (1965) analysed statistically the seasonal fluctuations after eliminating annual variations and those due to geographical differences. After the seasonal variations had been standardized he calculated the correlation coefficient between seasonal cycles of different species and organized the results in a matrix so that the species with high correlation coefficients are found together (Fig. 9-18). It is then possible to recognize, among the eighteen species studied, three types of seasonal cycles: A, the earliest with maximal abundance in May, June; C the latest with maximum abundance in September, October.

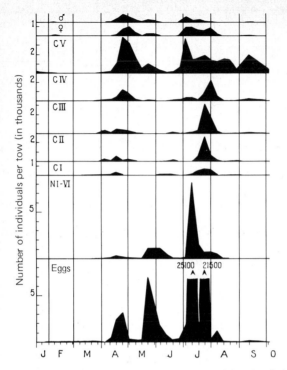

Fig. 9-13.—Variation in the number of individuals of the copepod *Calanus finmarchicus*, appearing at each stage from January to October 1933 in Loch Striven (West of Scotland); weekly observations combined 5 vertical tows from 70 m depth to the surface with a quantitative net of 77 meshes/cm: N, nauplius; C, copepodid (after Marshall & Orr, 1955).

As already indicated, the estimation of numbers does not take into account important differences in the size of the organisms; from this point of view the measurement of volume of zooplankton/m^3 is more satisfactory; Fig. 9-19 shows the changes in volume corresponding to the series of collection shown in Fig. 9-16, from which it is clear that the volume maxima are somewhat different from the numerical maxima, except for the poor months (under 10 mm^3/m^3) October to February when the two curves are similar. Fig. 9-19A also gives the biomass of the zooplankton expressed as organic phosphorus and measured on aliquots of the sample; the results are very similar to those obtained by volume. Finally, Fig. 9-19B shows the biomass under a surface of 1 m^2, taking into account the depth of the station (50 m) first in organic phosphorus and then as the weight of organic matter (by subtracting the ash weight from the dry weight). From these graphs the following changes appear clear; there is an increase in the biomass at the end of March and in April with a large maximum in May followed by a decrease in June–July and a slight increase in September; there is a decrease in autumn. Such a clear seasonal cycle is found in the waters of temperate zone and in the cold northern waters.

The question, already posed for phytoplankton, as to whether there are

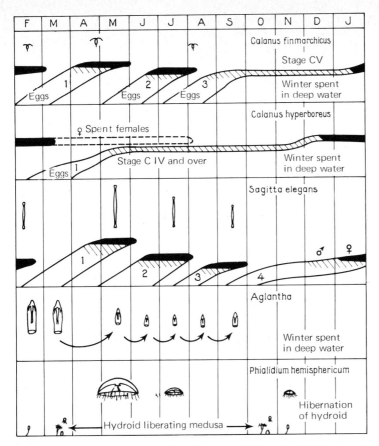

Fig. 9-14.—Schematic representation of an annual life cycle of 5 elements of the zooplankton: relative dimensions of the individuals are retained: CV, copepodid V; 1, first generation, etc. . . . (after Russell, 1935).

seasonal quantitative variations in tropical water where the ecological conditions are more uniform also arises in the case of the zooplankton. In the Sargasso Sea, Deevey (1971) observed notable oscillations (a factor greater than 2) and she considered that there was an annual cycle with a maximum abundance (volume and dry weight) in March–April even though it is undoubtedly less pronounced than that in cold waters.

Annual variations

Besides seasonal variations, annual variations in the abundance of the organisms of the zooplankton have been observed. Fraser (1970) showed that in Scottish water of the North Sea, *Pleurobrachia pileus* (Ctenophora), had a considerably higher density in 1965 than the mean density, with an earlier annual maximum (Fig. 9-20). Hardy's plankton recorder is particularly useful in examining important variations from one year to another; Fig.

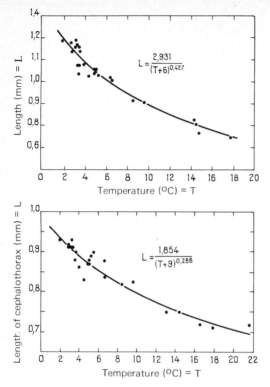

Fig. 9-15.—Length–temperature relations: A, relation between the mean length of adults (stage III) of the chaetognath *Sagitta elegans* and the mean temperature at 25 m depth at Plymouth during the development period: B, relation between the cephalothorax length of the copepod *Acartia clausi* females and the temperature at the time of collection in Long Island Sound (Atlantic Coast of U.S.A.) (after McLaren, 1963).

9-21 shows for the pteropoda, *Spiratella retroversa*, that to the seasonal cycle, annual variations may be added; the species was very rare in the zone C5 from 1957 to 1962 (Glover, 1967). A statistical study of data from the recorder made it possible to show that hidden by the variations from one year to another, there are tendencies in the development of annual mean abundance of numerous species. Fig. 9-22B shows that in the waters around the British Isles the three species of copepods, *Pleuromamma borealis*, *Euchaeta norvegica* and *Acartia clausi*, have over about fifteen years shown an increase in abundance, and yet *Temora longicornis*, *Clione limacina* (Pteropoda) remained stable over the same period; on the contrary, 7 other species of copepods along with *Spiratella retroversa*, have declined. The analysis of such tendencies also showed (Fig. 9-22C) that the number of copepods per sample or the index of biomass characterizing the zooplankton, continuously decreased both in the North Sea and the Atlantic around the British Isles.

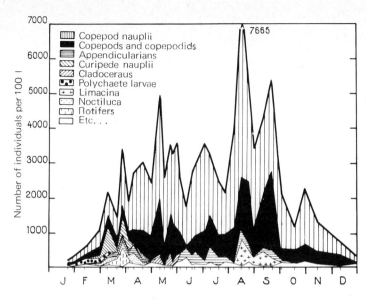

Fig. 9-16. — Changes in the total number and composition of zooplankton organisms/100 l water, between the surface and 45 m during 1934, in the English Channel 5 miles off Plymouth; the distribution between the principal groups is represented; depth of the station 50 m; fine silk net width of mesh not indicated (after Harvey et al., 1935).

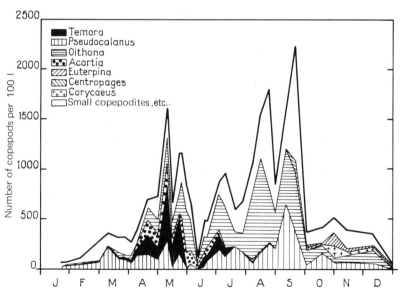

Fig. 9-17. — Changes in the total number of copepods caught/100 l of water in the sampling corresponding to Fig. 9-16 and the main generic composition of the copepod population (after Harvey et al., 1935).

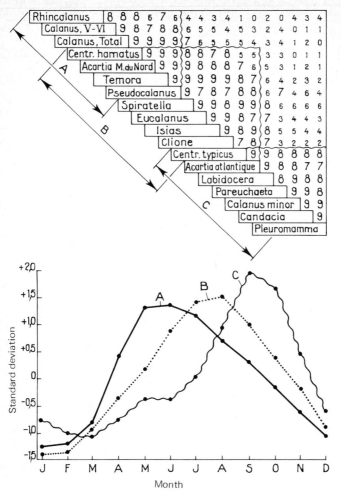

Fig. 9-18.—Top, a matrix of correlation coefficients (multiplied by 10) between 18 seasonal cycles of abundance of 18 species of zooplankton based on the results of the plankton recorder for the period 1948–1962 for the water surrounding the British Islands (North Sea, English Channel, the near Atlantic at the following boundaries, the parallel of the north of Spain, the median meridian of Iceland and the parallel of the south of Iceland (see charts in Fig. 9-26)): specific names are omitted, where possible, in the interest of clarity of the diagram. It is possible to distinguish three groups A, B, C, of species with high correlation between them as regards their seasonal abundance. Bottom, graphs representing the mean seasonal cycles of abundance standardized for each of the three groups of species shown in the above matrix (after Colebrook, 1965).

9-6.—DISTRIBUTION IN SPACE

The continuity of the marine environment and its mixture by the currents may suggest that there should be numerous cosmopolitan species in the zooplankton; these are, however, rare. The marine environment is very

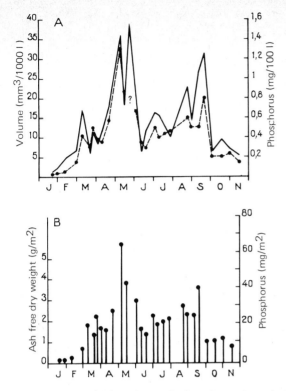

Fig. 9-19.—A, solid line, changes in the volume of zooplankton caught/100 l of water between the surface and 45 m during 1934 in the English Channel 5 miles off Plymouth: volume calculated from the numbers represented in Fig. 9-16; broken line, changes in the organic phosphorus/m^3 measured on aliquot of the catch: B, organic phosphorus/m^3 from the preceding data (depth, 50 m); the left ordinate expresses the values of the biomass in ash free dry weight (organic matter dry weight) by multiplication of the values of phosphorus by a factor of 87 (A, after Harvey et al., 1935; B, after Harvey 1950).

diverse as regards its salinity, temperature, light etc., and it is only in the depths that there are relatively large homogeneous areas. It is not surprising, therefore, that it is among deep water species that many may be considered as cosmopolitan. A good example is the large, red mysid *Gnathophausia gigas*. More numerous are species whose area of distribution forms a belt which surrounds the globe and is limited to a specific zone. The small copepod, *Paracalanus parvus*, exists in all tropical and temperate waters but is absent in arctic and antarctic waters (Davis, 1955); the Hydromedusa, *Sarsia tubulosa*, with a circumboreal distribution is found in all the seas of the north temperate zone.

Organisms of bipolar distribution are those which are found on both sides of the tropical zone. *Thysanopoda acutifrons* (Euphausiacea) is known from subarctic and subantarctic waters (Einarsson, 1945). The distribution of a cold water species may also be continued by submergence into deeper, colder water levels near the tropics: a classical example of this phenomenon

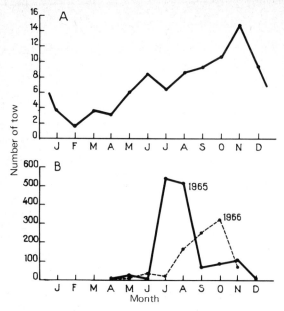

Fig. 9-20.—Changes in the annual abundance of the ctenophore *Pleurobrachia pileus*, in the Scottish water of the North Sea: A, mean monthly variation based on the years 1925–1939 and 1946–1954; B, monthly variations in 1965 and 1966: filtered water per haul is of the order of 450 m^3; since the richest tows had 1000 and 3000 individuals the density/m^3 can attain a maximum of 2 to 7 (after Fraser, 1970).

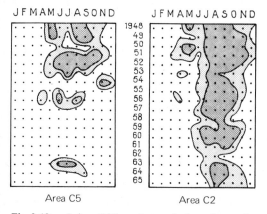

Fig. 9-19.—A, in solid line, changes in the volume of zooplankton caught/100 l of water between sampled with plankton recorder: dots show month in which samples were available; contours drawn at abundance intervals of approximately threefold based on the average number per sample of the organisms in each month; because the averages are calculated from logarithmic transformation of the original counts it is not possible to express them precisely as number per sample; observations outside the stippled areas do not represent absence of the organisms but only the lowest category of abundance: area C_2 is in the North-Sea, area C_5 in the Atlantic (see Fig. 9-22 A) (after Glover, 1967).

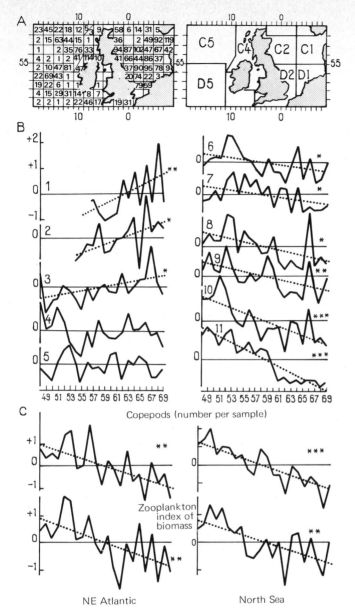

Fig. 9-22.—Fluctuation of zooplankton abundance in time. A, left, total number of samples analysed (divided by 10) in the standard rectangles in 1948–1969; right, boundaries of the seven areas studied below. B, fluctuation in abundance in the 7 areas of the following species; 1, *Pleuromamma borealis*; 2, *Euchaeta norvegica*; 3, *Acartia clausi*; 4, *Temora longicornis*; 5, *Clione limacina*; 6, *Calanus helgolandicus* and *C. finmarchicus* (copepodids V and adults); 7, *Metridia lucens*; 8, *Candacia armata*; 9, *Centropages typicus*; 10, *Spiratella retroversa*; 11, *Pseudocalanus*, combined. C, fluctuation in the plankton in the north-east Atlantic (areas C_4, C_5 and D_5 combined) and in the North Sea (areas C_1, C_2, D_1 and D_2 combined), the total number of copepods per sample and the biomass index of the zooplankton. The graphs in B and C show the average annual abundance of each species or each parameter standardized so that the general mean will be equal to zero (the scale of standard deviation units is at top left). Calculated trend lines are drawn for those graphs which give a significant fit to the straight line indicated by one two or three asterisks for $P = <0.05, <0.01 <0.001$, respectively. Biomass index is calculated by multiplying the number of the organisms and their mean wet weight (after Glover, Robinson & Colebrook 1970).

known as tropical submergence is *Eukrohnia hamata* (Chaetognatha) whose upper limit of distribution at the surface is 60°N and 60°S but at depth of 800 m between 0° and 20°N (Alvariño, 1964). Species which live in deep (bathypelagic) regions most often show a bipolar distribution. In the upper levels in the Arctic and Antarctic the same groups are most often found represented not by the same species but by closely related species which replace each other (vicarious species) so that they have the same function in the ecosystem (ecological niche).

Most planktonic species have a relatively limited area of distribution. Fig. 9-23B, of data from Hardy's plankton recorder results for the North Atlantic, illustrates this for two copepods, *Calanus helgolandicus* and

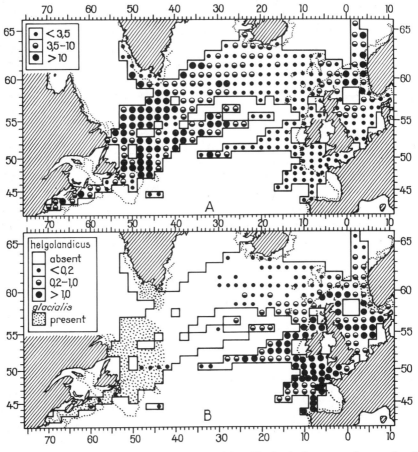

Fig. 9-23.—Distribution of three copepods based on Hardy plankton recorder samples during the period 1958–1964: results (positive or zero) are given for those statistical rectangles (1° latitude, 2° longitude) sampled five or more times: a sample corresponds to about 3 m^3 taken a 10-mile tow. A, average numbers of *Calanus finmarchicus*, per sample; B, average number of *Calanus helgolandicus* per sample, the presence of *Calanus glacialis* is indicated by the dotted areas (after Glover 1967).

Calanus glacialis; the latter is found west to 40° longitude while the former is abundant mainly in the east towards Europe. The very closely related species, *Calanus finmarchicus* (Fig. 9-23A) is common throughout that area but is more abundant in the west and a comparison of the two distribution maps allow one to establish a vicarious relation—if not total at least important—between the two species. Fig. 9-24 also obtained from plankton recorder data, shows the limited distribution of another copepod, *Labidocera wollastoni*, in a smaller area, the North Sea.

Although the areas of distribution of zooplankton species is very variable it is possible, nevertheless to distinguish two general types of distribution, namely, neritic and oceanic. In the first of these the species are abundant in coastal water, where the ecological conditions in particular, temperature and salinity show large changes. In the oceanic area, the ecological conditions are much more stable as regards temperature and in general the salinity. The difference between the two biotopes, neritic and oceanic, also arises from other physical and chemical factors such as the amount of particles in suspension and the dissolved elements. Among the neritic elements of the zooplankton the following may be noted, *Pleurobrachia pileus* (Ctenophora) *Sagitta setosa* (Chaetognatha) the copepods *Labidocera wollastoni* (whose typical distribution in the North Sea has already been mentioned), *Temora longicornis*, *Acartia tonsa*, and *Centropages hamatus*. The neritic plankton is also characterized by an abundance of larvae of benthic species; in contrast in the oceanic plankton the meroplankton is rarer and consists largely of the larvae of nectonic organisms. Among the oceanic elements we find the Heteropoda (Mollusca), most of the Siphonophora, the copepods *Pareuchaeta norvigica*, *Pleuromamma gracilis*, and *Centropages bradyi*. Some organisms may have "demands" lying between the conditions of the neritic and oceanic areas and, therefore, have "intermediate" distributions.

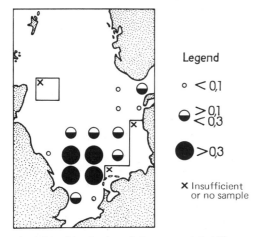

Fig. 9-24.—Distribution of the copepod *Labidocera wollastoni* in the North Sea based on the results of the plankton recorder: abundance indicated by average number of animals per sample, which is about 3 m³ on a 10-mile tow (after Colebrook 1969).

With the knowledge of these kinds of distributions, Colebrook (1964) has made an interesting study using data from the plankton recorder surveys and applying a technique known as principle components analysis to the

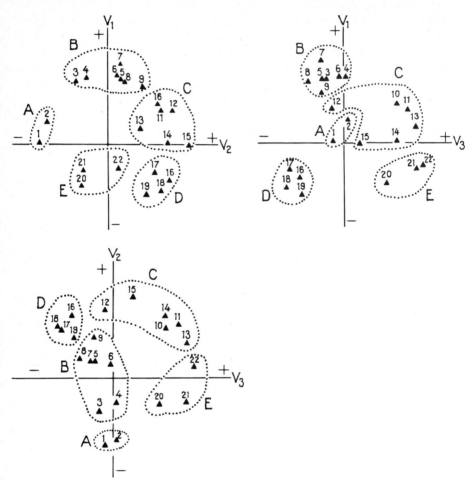

Fig. 9-25.—Results of the principal component analysis for the geographic distribution of 22 taxa collected with Hardy plankton recorder around the British Islands in the period 1948–1960. A scatter diagram for the first three vectors V_1, V_2, and V_3. Each point in the diagram represents a taxon and the identity of it is indicated by a number given to each point. A key to this number is given in the form of a list in each of the groups. The distribution of these points enables one to distinguish 5 groups of taxa A, B, C, D, and E. The groups are indicated in the diagrams by a broken line and are labelled A to E. A, northern oceanic group; 1, Pareuchaeta norvegica 2, Acartia sp. (Atlantic): B, southern oceanic group; 3, Pleuromamma robusta; 4, Rhinocalanus nasutus; 5, Calanus gracilis, 6, Calanus minor, 7, Pleuromamma gracilis; 8, Pleuromamma borealis; 9, Centropages bradyi: C, southern intermediate group; 10, Centropages typicus; 11, Candacia armata; 12, Corycaeus sp.; 13, Euchaeta hebes; 14, Calanus f. helgolandicus; 15, Paracalanus sp. and Pseudocalanus sp.: D, neritic group; 16, Acartia sp. (North Sea); 17, Labidocera wollastoni; 18, Centropages hamatus; 19, Temora longicornis: E, northern intermediate group; 20, Calanus f. finmarchicus; 21, Clione limacina; 22, Spiratella retroversa (after Colebrook, 1967).

geographical distribution of the zooplankton. Figs 9-25 and 9-26 show the results obtained for 22 species divided into five groups, namely, one neritic, two oceanic, and two intermediate. The three first components of the analysis account for 65% of the total variance; the first was identified as the surface salinity; the second as a complex function of the summer surface temperature and the range of the seasonal variation of temperature and is related to the vertical stability of the water while the third was provisionally identified with the distribution of mixed oceanic and coastal water.

Under certain conditions the neritic—oceanic contrast may sometimes appear in an extremely small area; at Villefranche where the bottom reaches a depth of about 1000 m just in front of the bay it is possible to characterize the plankton at the entrance of the bay as oceanic and that at the end of the bay separated by a distance of only 2 nautical miles as neritic (Sentz, 1963).

The presence of definite geographical areas of distribution is true for the plankton but such areas are often labile and vary in their outlines over the

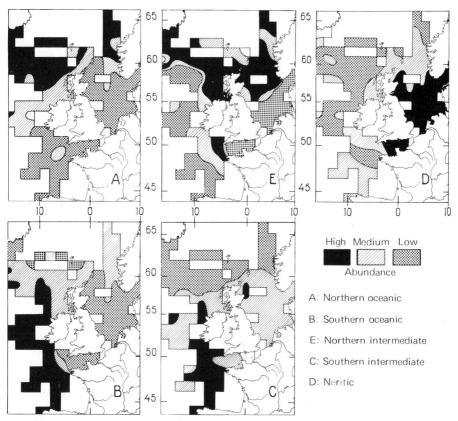

Fig. 9-26.—Charts of geographical distributions of the five taxa groups, A, B, C, D, and E, determined in Fig. 9-25; distributions shown by mean of contours on an arbitrary scale; clear zones correspond to water not sampled (after Colebrook, 1964).

year and from one year to another, largely as a function of the variations in the movements of the water due to currents. An excellent example has been given by Fraser (1969) for *Salpa fusiformis* whose distribution, limited to the autumn in the north of Scotland is shown in Fig. 9-27.

Since the water movements may cause changes in the planktonic fauna one may consider that faunistic changes indicate the arrival of a different water mass. Russell (1939), for example, distinguished at the entrance to the Channel in the area off Plymouth, two types of water, namely, that carrying *Sagitta setosa* and with neritic plankton and that with *Sagitta elegans* a mixture of coastal and oceanic water with an intermediate plankton population. Before 1930 the *Sagitta elegans* water occupied the area surrounding Plymouth, but since 1930 *Sagitta setosa* water has been predominant and *Sagitta elegans* has been found further away from the entrance to the English Channel. In the North Sea, Fraser (1939) again found the same contrast between water with *Sagitta elegans* and that with *Sagitta setosa*; the latter occupied the meridional part of this sea.

As a consequence of the existence of planktonic species whose ecological demands characterize waters of a particular type such species may serve as indicators of water movement or even of the presence of a given water in a mixture of water masses. Where they are generally abundant such planktonic species may be called "indicator species". Although the use of

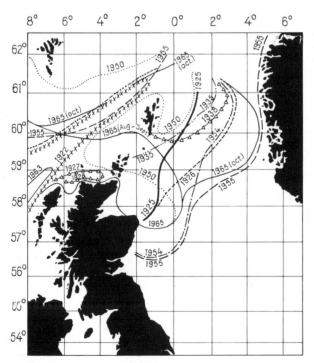

Fig. 9-27.—Variations in the distribution of the tunicate *Salpa fusiformis* during the autumn of several years in the northern part of the North-Sea (after Fraser, 1969).

"indicator species" has been very successful it should be kept in mind that the characteristics of an indicator species must be very carefully defined by a detailed ecological study and that the value of this term is relative rather than absolute. Finally, to be used effectively as an indicator species it must be easily recognized. Fig. 9-28 gives the excellent example of the distribution of water containing *Sagitta elegans* and *Sagitta setosa* around the British Isles.

The interest in these differences is more than simply biogeographical; waters containing *Sagitta elegans* are generally richer in plankton than those with *Sagitta setosa*. The replacement of one by the other near Plymouth corresponded to a decrease in the winter maximum of phosphate (Fig. 4-17). In the North Sea, Fraser (1969) showed that fish larvae can survive better in *Sagitta elegans* water, Wilson (1951) observed in in vitro experiments that

Fig. 9-28.—Generalized view of the distribution of water characterized by different species of *Sagitta*: the conditions shown are such as might be found in the autumn of a year in which the influx of Atlantic water into the North Sea is strong. SETOSA (or SET), *Sagitta setosa*; ELEGANS, *Sagitta elegans*; SER, *Sagitta serratodentata*: arrows indicate the general water circulation (after Russell, 1939).

Sagitta elegans water is favourable for the development of sea urchin larvae used as a biological test for subtle differences, and such exist between these waters and are not related to temperature, to salinity and/or the concentration of phosphates (Bougis, 1964).

At the end of this chapter we must mention the distribution of swarms*. Large swarms of the Hydromedusae *Tiaropsis multicirrata* and *Rathkea octopunctata* have been described in the Jarnyshnaya Fjord in the Barents Sea (Zelickman, Gelfand & Shifrin, 1969). These may vary in dimensions from several square kilometers to several hundreds of square meters and are related to the conditions in the sea and are established in July and August at the depth of several meters. Another example is that of the Euphausiacea which are able to form a shoal in day time in the superficial waters with the individuals swimming at a distance of 1 to 2 cm from one another all in the same direction. The shoal has the appearance of a cloud without a definite form often changing its outline. In Japan, Komaki (1967) has studied the frequency and abundance of swarms, collected by fishermen with fine pyramidal nets (Figs 13-4 and 13-15) in the water of Kinkazan, 38°20′ N; 141°35′ E). The fishery during the favourable season reach several tons per day and some time up to 100 tons. Finally, we may note the existence of shoals of the sea butterflies (Pteropoda: Thecosoma) and particularly those of *Creseis*.

---

*The author distinguishes between two types of swarms; both are included under the same term, "swarm" in English literature. The term swarm (in French "essaim") is used to describe small-sized swarms which are maxima in the microdistribution; the other term, shoal (in French "banc") is used when describing aggregations of a larger dimension.

CHAPTER 10

VERTICAL DISTRIBUTION AND DIURNAL MIGRATION OF ZOOPLANKTON

10-1.—THE VERTICAL DISTRIBUTION

In the previous chapter we examined the distribution of the zooplankton in space without taking into account the depth of the water. The limits of any distribution is not only horizontal but also vertical. The distribution of the Euphausiacea in the North Atlantic gives an excellent example of this (Fig. 10-1). Six different species are present, three of them living at a relatively shallow depth, notably *Meganyctiphanes norvigica* which lives at about 200 m and so is found on the borders of the continental shelf. Three others are found at greater depths and two species of the genus *Thysanopoda* are found only below 1500 m and so may be regarded as bathypelagic; on this figure the submergence of *Thysanoessa longicaudata* towards the south should be noticed.

The depth limitation may be very much reduced, as is the case with elements which belong to the pleuston and the hyponeuston. Fig. 10-2 shows the remarkable distribution of the copepod *Anomalocera patersoni*, which is found in the first meter under the surface (Champalbert, 1969). The distribution in depth may also show some seasonal variations, for example, the hydromedusan *Persa incolorata* is absent at Villefranche, except in winter from the layer between 0–50 m (Goy, 1964).

Besides these qualitative variations there is the problem of how the zooplankton is quantitatively distributed as a function of the depth. The general conclusions from a considerable amount of work on this subject by Russian authors has been given by Vinogradov (1970). As we go deeper the biomass decreases, this decrease being towards the greater depths. As a first approximation this decrease may be expressed by a logarithmic relation.

$$\log y = a - kz$$

when y is the biomass (in wet weight) z is the depth, a is a constant and k is a coefficient of decrease. On a natural basis this gives ($a = \log a'$)

$$y = a'e^{-kz}$$

and correspond to an exponential law. In the region of the abyssal of the Kurile and Kamchatka Trenches the biomass is several hundreds of milligrams/m^3 at the surface, but only tenths (1/10) of a mg/m^3 at a depth of 7000 to 8000 m (Fig. 10-3), and the changes are represented by the following equation

$$y = 56.2e^{-6.5 \times 10^{-4}z}$$

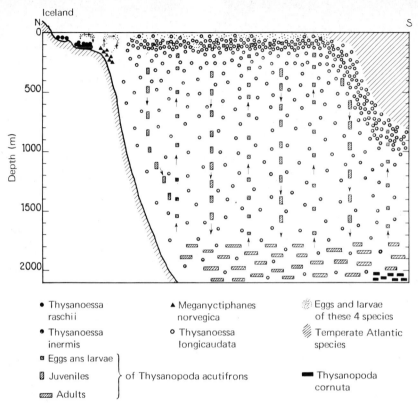

Fig. 10-1.—Schematic representation of the vertical and horizontal distribution of a North-Atlantic euphausiid, on a north-south section passing by Iceland: all the stages of *Thysanopoda cornuta* are limited to deep water (after Einarsson, 1945).

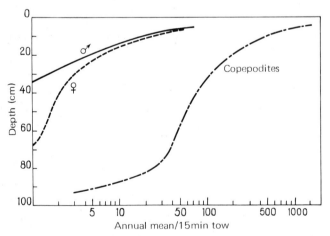

Fig. 10-2.—Vertical distribution of the copepod *Anomalocera patersoni* in the Gulf of Marseille; annual average 1967–1969 (after Champalbert, 1969).

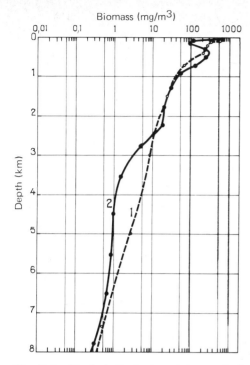

Fig. 10-3.—Vertical distribution of the zooplankton biomass (in wet weight) in the Kurile–Kamchatka Trench: 1, spring 1953, average of 6 station of which the layers sampled were thick; 2, summer 1966, average of 9 stations the layers sampled were thinner (after Vinogradov, 1970).

Fig. 10-4 shows such a vertical decrease for the Mediterranean where it is more accentuated than in near Atlantic waters with the same superficial concentrations.

From the numerous observations on the research vessels *Vitiaz* and *Ob*, Vinogradov (1970) was able to establish a general vertical change in the plankton biomass from north to south of the Pacific Ocean, with poverty in the tropical water at all levels as distinct from the richness beyond 40°N and 40°S (Fig. 10-5).

In addition to a distinct vertical distribution of plankton at a given level, very short period variations in the distribution are observed and these variations are related to the alternation of day and night and correspond to the migration of individual animals; this migration is termed diurnal migration. We shall discuss successively the meaning and determination of this remarkable phenomenon.

10-2.—THE CHARACTERISTICS OF DIURNAL MIGRATION

Regular observations show that samples of night plankton at the surface are notably different from those taken during the day, not only in their

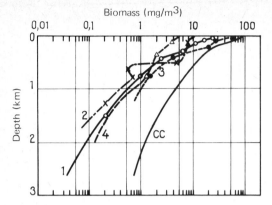

Fig. 10-4.—Vertical distribution of the zooplankton biomass (in wet weight) in different regions of the Mediterranean: 1, Ionian Sea (37°42′ N: 19°07′ E) in winter; 2, Ionian Sea (36°07′ N: 22°09′ E) in winter; 3, Ionian Sea in summer; 4, Western Basin; CC, for comparison, the distribution of the biomass in the Canary Current in the Atlantic Ocean (after Vinogradov, 1970).

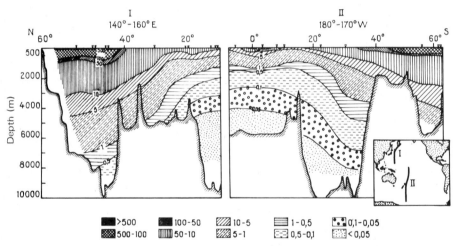

Fig. 10-5.—Scheme of the zooplankton biomass distribution (mg wet weight/m³) along two south-north transects in the west of the Pacific: triangles at the top margins of the scheme indicate the position of the station; map shows the mean position of the transects (after Vinogradov, 1970).

faunistic composition but also quantitatively. Fig. 10-6 shows the abundance of adults of the copepod *Calanus finmarchicus*; the number of individuals collected in the surface layers become ten times higher between 16.00 and 20.00 h. Since these observations were made over several weeks these changes cannot be attributed to the interference of currents regularly replacing "poor" water; the only possible explanation of these variations is vertical migration, representing a rhythm of 24 h; similar phenomena have been frequently observed. Russell (1931a) found important differences in the

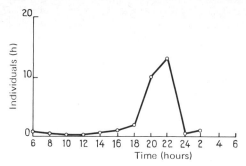

Fig. 10-6.—Changes in the abundance of the copepod *Calanus finmarchicus* in the surface, between 06.00 and 14.00 h during 6 weeks from 21st June to 11th August, 1911, off Californian coast: abscissa, time in hours; ordinate, the mean number of individuals/1 hour net tow (after Esterly, 1914 in Cushing, 1951).

distribution of the arrow worm *Sagitta elegans* (Chaetognatha) (Fig. 10-7); during the day, the percentage of individuals near the surface is almost zero and, at sunset and during the night the numbers increase and then decrease at sunrise. The pelagic polychaete *Tomopteris helgolandicus* also accumulates at the surface during the night (Fig. 10-8).

It is mainly in numerous species of planktonic crustacea that this phenomenon is most marked and has been most studied. Nicholls (1933) gives an excellent series of observations and much information on the migration of the copepod *Calanus finmarchicus*; Fig. 10-9 concerns the adult females in the summer and winter. Absent from the surface during the day the females appear there after sunset on sampling at 22.00 h in July and

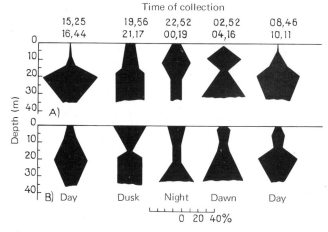

Fig. 10-7.—Percentage of vertical distribution of the chaetognath *Sagitta elegans* from the 15th to the 16th of July 1924, at the times shown at a station near Plymouth: A, individuals, length 12 to 13.5 mm; B, individuals length 5 to 7.5 mm: for each level the number of individuals is expressed as percentage of the total number of individuals caught in a given series of catches (after Russell, 1931).

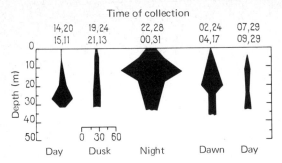

Fig. 10-8.—Changes in the vertical distribution of the annelid *Tomopteris helgolandicus* (Polychaeta) from the 3rd to the 4th June 1926 at the given times at a station near Plymouth; number of animals collected at each level (after Russell, 1931).

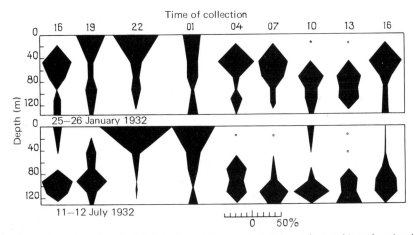

Fig. 10-9.—Diurnal vertical distribution of the copepod *Calanus finmarchicus*, females, in Loch Fyne, Clyde Sea Area, Scotland: sunset at 16.27 h on the 25th of January, at 20.27 h on the 11th of July; sunrise at 08.27 h on the 26th January, at 03.58 h on the 12th July: at each level the number of animals as a percentage of the total animals caught in the given hour is indicated. The points correspond to no capture (after Nicholls, 1933).

at 19.00 h in January and the population accumulates in the superficial waters. At about midnight there is a tendency for the population to become uniform in its vertical distribution and then in the morning it sinks, to rise again in the evening. The total amplitude of the movement is of the order of 100 m. Higher amplitudes of several hundred meters have been observed in the migration of some species of eupausiids (Fig. 10-10) and in the large copepod *Gaussia princeps* living at great depths (Fig. 10-11). Some crustaceans living during the day near the bottom (Mysidacea) or on the bottom (Cumacea) rise to the surface during the night. They constitute, therefore, a particular category of meroplankton with a diurnal rhythm (Fig. 10-12). The pelagic crab of Lower California, *Pleurocondes planipes*, belongs to this nocturnal plankton. The juveniles, up to the length of several centimeters,

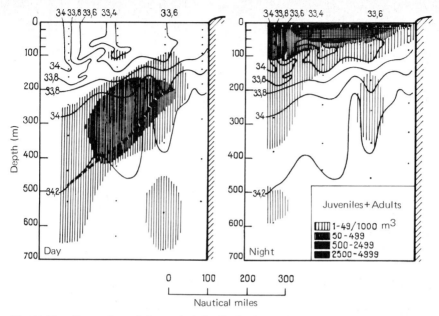

Fig. 10-10.—Comparison of the vertical distribution in daytime and at night of the euphausiid, *Euphausia recurva*, along a line of stations south west of Punta Banda (north of the Californian Peninsula 31°30′N; 116°40′W): the interrupted heavy line represents the depth where the maximal number of adults are found: isohalines indicated (after Brinton, 1967).

are present in the plankton with a markedly daily rhythm, being more abundant in the surface at night than during daytime (Boyd, 1967).

Although the occurrence of vertical migration has been amply verified, more detailed investigations indicate important variations in the behaviour of migratory species and even in the same species the amplitude of the migration may change during its development. Juvenile *Sagitta elegans* (Fig. 10-7B) are closer to the surface in day time than are the adults. In crustaceans, certain stages may not show any migration; Nicholls found that in *Calanus finmarchicus* in the Clyde Area the males differed from the females in not showing a very distinct migration and that the copepodid V (the last stage in the development before sexual maturity) remain at a constant depth day and night (Fig. 10-13). It also appears that the migration in the same species may differ considerably from one generation to another. Russell (1937) observed that in the neighbourhood of Plymouth the maximal day level of *Calanus* density gradually changes from 10 m in April to 20 m in June, which is in accordance with the increase in light intensity penetrating the sea which as will be shown later is a very important factor. Yet in July, August, and September the optimal day level rose to the surface which cannot be related to a decreasing in light intensity. Russell considers that it is due to the presence of a new generation which developed under different conditions and so modified the physiological requirements. Clarke (1934) also attributes the considerable difference that he found at the same period

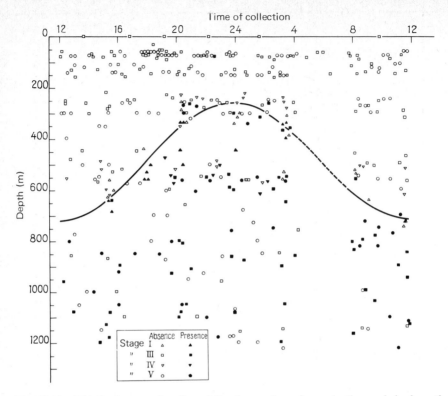

Fig. 10-11.—Distribution as a function of the time and maximum depth sampled where the copepod *Gaussia princeps* is found: solid line, regression curve $z = z_0 + a \cos \omega t = 487 - 230 \cos \omega t$, giving the depth, in meters, of the high limit of the species (t in h; $\omega = 2\pi/24$ h): collections made during the *Caride* I, III, IV and V cruises in the equatorial central Pacific using 10 ft Isaacs–Kidd midwater trawl, 9 m² opening, and mesh aperture of 0.9–1 mm in the cod end (after Gueredrat & Friess, 1971).

between two zones of the Gulf of Maine to a divergence in behaviour of different generations or populations: above a bottom of 200 m depth *Calanus* shows a typical migration from 120 m to the surface (Fig. 10-14A), but at the edge of Georges Bank, it is found mainly in the upper 30 m and shows only a small amplitude of movement (Fig. 10-14b). Clarke suggested that the two populations corresponded to different generations.

There are some observations which seem to be completely anomalous; it is possible to observe an accumulation of *Calanus* close to the surface in day time in a region where at this time of day the individuals would be expected to be at the bottom (Marshall & Orr, 1955). The migratory Euphausiacea at certain times during the day form superficial swarms colouring the water red or brownish-red (Komaki, 1967).

In spite of these anomalies diurnal vertical migration is a very common phenomenon whose meaning has been variously explained. Cushing (1951) analysed all the work on crustacea and considered that it is possible to

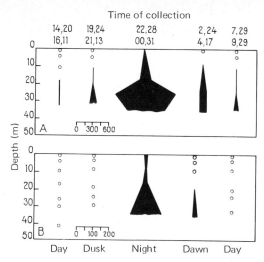

Fig. 10-12.—Changes in vertical distribution at a station in the region of Plymouth from the 3rd to the 4th June, 1926: A, mysids, B, cumaceans: the isolated point corresponds to no capture: results expressed in number of individuals collected at each level (after Russell, 1931).

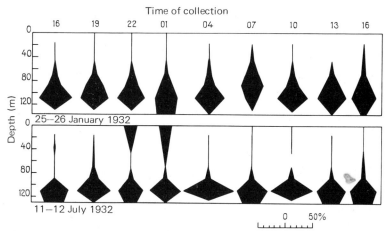

Fig. 10-13.—Diurnal vertical distribution of stage V copepodids of *Calanus finmarchicus* in Loch Fyne, the Clyde Sea Area, Scotland: sunset at 16.27 h on the 25th January, at 20.27 h on the 11th July; sunrise at 08.37 h on the 26th January, at 03.58 h on the 12th July: at each level number of animals as a percentage of the total animals caught in the given hour is indicated (after Nicholls, 1933).

arrange their patterns of migration in the following general scheme:

1. Rising in the evening from the "day depth" towards the surface.
2. Departure from the surface at about midnight or before it.
3. Rising to the surface just before dawn—"dawn rise".
4. Rapid sinking at dawn to the "day depth".
5. Remaining during the day at a "day depth" more or less variable.

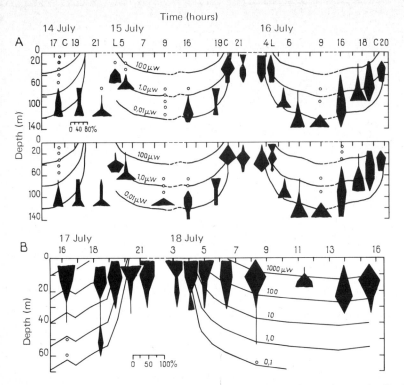

Fig. 10-14.—Vertical distribution of the copepod *Calanus finmarchicus*: A, distribution of females (top) and of copepodid V (bottom) from the 14th to the 16th July, 1933 at a deep water station in the Gulf of Maine: B, distribution of females on the 17th and the 18th July 1933 in a station in Georges Bank: isolumes in $\mu W/cm^2$; R, sunrise; S, sunset: isolated points indicate zero capture: at the different levels the catch is expressed as percentage of the total catch for the given hour (after Clarke, 1934).

As a general rule, when there is a vertical migration the animals are present in the surface layers during the night; however, there may be exceptions for Bhaud (1969) has described a nightly sinking of the larvae of *Mesochaetopterus sagittarius* (Polychaeta) at Nosy-Bé (Madagascar) (Fig. 8-24C).

Some authors had earlier considered this vertical migration to be an artefact due to the animals being able to see and avoid the net during the day while at night any avoidance is more difficult. The use of large nets will eliminate this possible source of error. By calculating the number of each species of euphausiids in a water column under the surface of 1 square meter, Brinton (1967) show that in California some species were sampled with more efficiency at night than during the day and he could distinguish the species which were able to avoid the net from those which are unable to do so (Fig. 10-15).

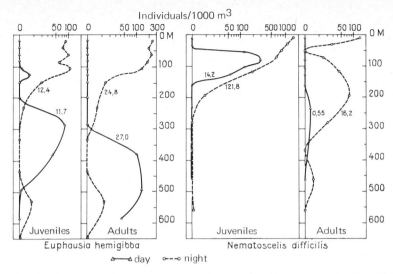

Fig. 10-15.—Comparison of day and night captures for two euphausiids off the coast of California: number of individuals under a surface of 1 m^2 indicated at the side of the curves; *Euphausia hemigibba* caught in equal quantities during the day and the night; *Nematoscelis difficilis* caught in much larger quantities during the night, during the day it avoids the net (after Brinton, 1967).

10-3.—CAUSES OF DIURNAL MIGRATION

The fact that the vertical migrations are related to the alternation of day and night suggests that light intensity is of primary importance.

The observations on migrations at high latitudes where during part of the year there is no alternation between days and nights gives a kind of "natural" experiment. According to Bogorov (1946) in continuous light *Calanus* and other principal species of the zooplankton remain in the superficial layer of water with no migration; this also appears again in winter with the return of the change of day to night. During the winter night, Bogorov observed that *Calanus* stays in the depths at a constant level. Some differences may exist according to the particular region or species (Raymont, 1963), but it appears that continuous light masks or cancels migratory phenomenon.

The simplest interpretation would be that the migrants show negative phototaxis and so will retreat during the day to the bottom; however, an experiment of Rose (1925) lead long ago to a modification of this conclusion. In his experiment when the planktonic copepods were held in a long tube (Fig. 10-16) covered by a black screen most of the animals did not remain in the darkest zone, but in an intermediate zone; they seem to stay at an "optimal" light intensity and show, then, not an absolute negative phototaxis but a negative phototaxis above certain levels and positive one below

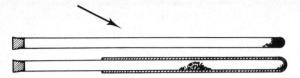

Fig. 10-16.—Distribution of photonegative copepods exposed to strong diffused light; light direction indicated by the arrow: upper, uncovered tube; lower, tube covered with a black sheet at the end of 30 min (after Rose, 1925).

that level. Russell (1927) adopted this hypothesis and from it deduced a theoretical distribution for a "eurylume" species, i.e., one which is capable of withstanding rather large variations around the optimal light intensity and for a "stenolume" species which is much more restricted (Fig. 10-17) shows these two kinds of distribution and the coincidence between the scheme for eurylume species and the observed distribution of females of *Calanus finmarchicus*. Relying on this, Russell then calculated a theoretical distribution based on the hypothesis of an optimal light intensity, shown for females of *Calanus finmarchicus* in Fig. 10-18. The changes during 24 h very much resemble the observations of Nicholls (Fig. 10-9) and agree with the general idea of Cushing, notably the night distribution and a rise to the surface at dawn.

The hypothesis of an optimal light intensity must, therefore, be considered seriously and a better way to demonstrate it would consist in measuring the light level and the distribution of the animals simultaneously. This was done by Ullyott (1939) for the fresh-water copepod *Cyclops strenuus*, Fig. 10-19 shows clearly the parallelism in time between the

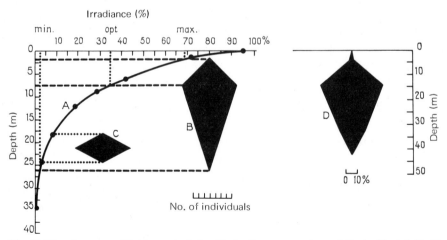

Fig. 10-17.—A, curve of light penetration with depth, as percentage of the incident light: B, theoretical vertical distribution of an eurylume species: C, theoretical vertical distribution of a stenolime species: D, vertical distribution of *Calanus finmarchicus* females adults off Plymouth on the 13th April, 1925, expressed as percentage of the total catch; notice the similarity of B and D (after Russell, 1927).

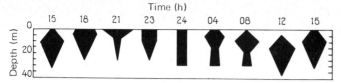

Fig. 10-18.—Changes in the theoretical vertical distribution of an eurylume species behaving like the copepod *Calanus finmarchicus* Sunset at 20.00 h (after Russell, 1927).

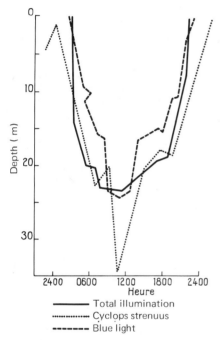

——— Total illumination
·········· Cyclops strenuus
------- Blue light

Fig. 10-19.—Changes in irradiance with depth during the day, in June, in Lake Windermere (England) and changes in the level with the largest number of the copepod *Cyclops strenuus*: total illumination corresponds to an irradiance of 10 800 $\mu W/cm^2$ and blue light to an irradiance of 30.5 $\mu W/cm^2$ (after Ullyott, 1939).

maximum density of the copepods and the depth at which a given light intensity was found. During their migration the organism adjusts so that it is at a light intensity which represents the optimal value. Recently this was also shown for *Daphnia* experimentally by Lincoln (1970). Lincoln added Indian ink to the water so increasing the extinction coefficient and enabling him to use experimental vessels of some tens of centimeters in length yet equivalent to a large range of natural depths (Fig. 10-20). He showed that the vertical migration of *Daphnia* is controlled by light, the animals sink with an increase in light intensity and rise when it decreases; the optimal light intensity was about 200 lux, which corresponds to 84 $\mu W/cm^2$. One of the consequences of the existence of an optimum is clearly seen; at the

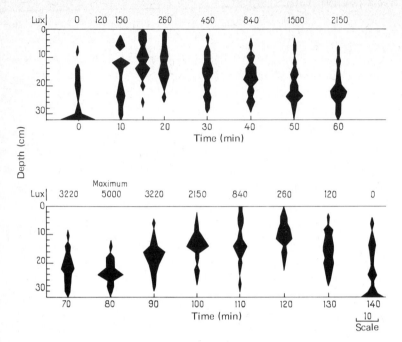

Fig. 10-20.—Distribution of 35 adult females of *Daphnia magna* during vertical migration in the laboratory; Indian Ink was added to the water to give an extinction coefficient, k, of about 15.3: variations in light at the surface are indicated at the top part of the diagram; time in min: the polygons show the distribution of the animals, for each centimeter, following the scale drawn on the base of the diagram (after Lincoln, 1970).

beginning of the experiment the animals were attracted to the surface so long as the light intensity remained below the optimum but in the absence of light they were dispersed in the medium and even had a tendency to accumulate at the bottom.

Postulating an optimal light intensity enables us to explain satisfactorily the most common characteristics of diurnal migration but it should be kept in mind that the optimum intensity is far from being an absolute value: Fig. 10-14 for females of *Calanus finmarchicus* in the Gulf of Maine indicates an optimum lower than 0.01 μW while the optimum for animals from George's Bank is in the order of several hundreds of microwatts. The difference in migratory behaviour of the population from the two regions, attributed by Clarke to differences in generations or populations, ought to be explained more precisely in terms of physiological differences. The work of Beeton (1960) on *Mysis relicta* (Mysidacea) in the large American Lakes is pertinent. This species remains at the bottom during the day, rises to the surface in the evening, but does not show the "dawn rising" which accompanies an optimal light intensity. Beeton has shown experimentally that after several hours of darkness the phototaxis of this species is inverted and at the end of the night they become photonegative and therefore go down even under low light intensity. This modification may be related to a biochemical substrate responsible for the photonegative characteristics, such a substrate probably needing a long enough period of darkness to be synthesised (Beeton, 1959).

That light is the principal factor in diurnal migration cannot be doubted but some authors have insisted on the role which is played by other factors such as temperature, pressure or gravity. A good example of the role of temperature is given by Beeton (1960) for sexually undifferentiated *Mysis relicta* (Fig. 10-21; when the thermocline is clearly established the migrating individuals rise only to it and remain there, but when only a weak thermocline is present they reach the surface waters. Cushing (1951) also stressed the relation between temperature and vertical migration and concluded that an homogeneous group of zooplanktonic organisms (homogeneous concerning sex, stage, or generation) will avoid temperatures outside a certain range; a thermocline, situated near the limits of this range is a much more efficient barrier to migration than a thermocline in the middle of that temperature range. It should be added that with regard to the observations of Beeton shown in Fig. 10-21 the efficiency of the thermocline also depends on its gradient.

During their vertical migration the organisms of the zooplankton are exposed to considerable variations in pressure, from several atmospheres to some tenths of an atmosphere. Pressure cannot, therefore, be neglected particularly since it has been shown that numerous marine species are barosensitive, notably so in *Calanus finmarchicus* (Rice, 1962). Moore & Corwin (1956) introduced into a mathematical equation, besides the light, temperature and the depth, (i.e., the pressure) in order to explain the variation of distribution of some siphonophores. Lincoln (1970), however, in his experiments on *Daphnia* did not find distinct variations of pressure to give any permanent change in the vertical migration pattern of *Daphnia*. *Daphnia* is not, however, barosensitive and this type of experiment should be repeated with other organisms before any definite conclusions are drawn regarding the role of pressure in migration.

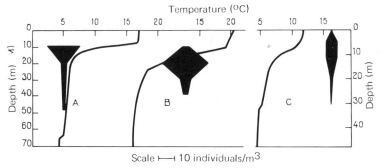

Fig. 10-21.—The influence of temperature on the diurnal migration in Lake Michigan for sexually undifferentiated *Mysis relicta* (Mysidacea): continuous line represent the temperature as a function of the depth: A, 27th June 1954, 24 h, Station 13; B, 27th August, 1954, 23 h, Station 13; C, 24th July 1955, 22.45 h, Station 76e (after Beeton, 1960).

With respect to gravity, organisms react negatively or positively, by negative or positive geotaxis. Some authors particularly Clarke (in Raymont, 1963) regard it as a complementary factor; the geotactic reaction is induced by light, a decrease of light intensity causing negative geotaxis and, therefore, a rising of the animals, and vice versa. This theory enables one to explain deviations from the optimal light intensity theory, shown in Fig. 10-14A. Only by more precise work and if possible, experimental studies will the exact role of geotaxis be elucidated.

This is a field where speculation might become rife; for instance it is very difficult under natural conditions to separate negative geotaxis and positive phototaxis. The nature of the reaction to light has given rise to much discussion and certain authors have examined whether photokineses, i.e., undirected movements related to light intensity rather than phototactic movements are involved. According to Raymont (1963), however, it is the latter type of reaction which is important.

Recently, Rudjakov (1970) has given a new theory, according to which only the rise is active, sinking being passive. The migration should, therefore, be a diurnal alternation of periods of activity and inactivity. The principal argument of Rudjakov is that the speed of sinking of anaesthetized organisms is of the same order as the velocity of sinking observed in vertical migration. The question as to whether the amplitude of migration is not sometimes greater than that "attainable" by the animal was posed long ago. The measurements of Hardy & Bainbridge using a tube in the shape of a large vertical wheel answered this question, as may be seen in Table 10-1. Copepods, larvae, zoea or nauplii can reach a climbing speed of 10–30 m/h, which could account for the amplitude of their migration from 20 to 100 m. The Euphausiacea which have a very marked migration attain higher speeds of the order of 100 m/h. The sinking speed is higher than the climbing speed and this may help to explain the asymmetry, sometimes observed between sinking and climbing migrations, for example, in *Calanus* in the Gulf of Maine (Fig. 10-14A), where the sinking speed is of the order of those measured and much higher than that of passive falling which is in the order of 15 m/h.

The notion of passive sinking hardly agrees with the results of Lincoln in which the depth range was artificially contracted. The *Daphnia* stayed at well defined levels which were a function of light intensity. On the contrary, in these experiments, in periods of darkness passive falling seems able to explain the rapid accumulation towards the bottom.

In summary, from all these studies on diurnal migration it must be concluded that light is without any doubt the fundamental factor, temperature is an important accessory factor, and pressure and gravity are auxiliary factors which may probably not be neglected. Although we have no precise data on the subject we may note the possibility of the existence of secondary vertical migration, namely, migration of predatory species which follow the migratory organisms on which they are feeding. We must also consider the possibility of the existence of an internal biological rhythm, which may

TABLE 10-1

Distance travelled and speed of vertical swimming of some crustaceans measured in the plankton wheel (after Hardy & Bainbridge in Bainbridge, 1961).

Organism	Up			Down		
	Duration (min)	Distance (m)	Speed (m/h)	Duration (min)	Distance (m)	Speed (m/h)
Calanus finmarchicus	2	2.2	66	2	3.57	107
	30	17.8	35.6	30	29.9	59.8
	60	15	15	60	47	47
Acartia clausi	2	1.12	33.6	2	0.57	17.1
	30	4.43	8.86	8	1.61	12.1
	60	8.81	8.81	—	—	—
Balanus nauplius	2	0.75	22.5	—	—	—
	30	7.63	15.26	—	—	—
	60	14.9	14.9	—	—	—
Zoea of Brachyura	1	1.17	70.2	—	—	—
	2	1.95	58.5	—	—	—
	10	4.89	29.34	—	—	—
Meganyctiphauus norvegica	2	5.77	173.1	2	7.17	215.1
	30	62.21	124.4	30	68.2	136.4
	60	92.8	92.8	60	128.8	128.8

maintain the vertical migration when the external conditions have become constant. Esterly (in Raymont, 1963), has shown the existence of such a rhythm in *Acartia*. This rhythm explains some abnormalities.

10-4.—ECOLOGICAL INCIDENCE OF DIURNAL MIGRATION

After examining successively the characteristics and what is known of the causes of diurnal migration one should consider what are the responses in the pelagic ecosystem.

Although these migrations are very common, they do not effect the whole zooplankton, and migratory and non-migratory species are found in the same groups. Baker (1970), studying the vertical distribution of Euphausiacea in the water of the Canary Islands, found that more than half of the species did not migrate (Fig. 10-22). Of population of Euphausiacea (except for larvae) migratory species made up of 46% of the individuals present and 63% of the biomass. The entire copepods from the order Cyclopoidea (Oithonidea, Oncaeidae, Corycaeidae) were studied by Zalkina (1970) in the tropical region of the Pacific; out of 17 species only 5 showed an effective migration, 3 a slight migration and 2 were of an intermediate nature (*Oncaea mediterranea, Oncaea conifera*). It should be remembered that the copepodid stage of *Calanus finmarchicus* does not show vertical migration.

The migrating species do not represent the entire population and different

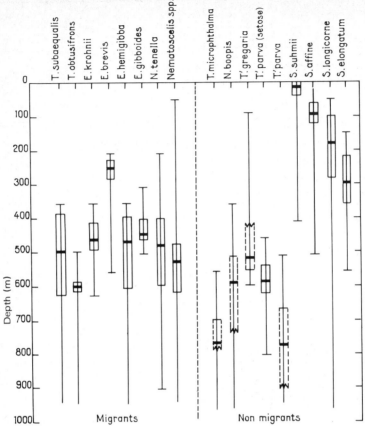

Fig. 10-22.—Daytime distribution of adult euphausiids in the Canaries zone during the SOND cruise (September to December 1965): depths above which 25 and 75% occurred are indicated by the top and bottom of the rectangle, the 50% by the solid black bars: vertical centre line indicates the overall range for each species: where the lower limit of the range was not reached or sampling was thought to be inadequate the rectangle is pecked: T, *Thysanopoda*; E, *Euphausia*; N, *Nematoscelis*; T', *Thysanoessa*; S, *Stylocheiron* (after de Baker, 1970).

amplitudes of migration exist in migrating species. Any given species is, therefore, during any 24 h period surrounded by several types of population, differing in the predators from which they have to escape. The differences may be accentuated when the current regimes at depth and at the surface are different. According to the conception of Hardy (1935), the migratory species "should try" continuously different biological conditions. Miller (1970) tried to verify this phenomenon; even if the results are far from definite, they do show that the composition of the sub-shallow-water (10 m) nightly zooplankton develops only slowly, although as a result of vertical migration several more general ecological consequences must follow (Fig. 10-23). Herbivorous organisms of the zooplankton will feed during the night on phytoplankton present in the superficial layers of the sea and then transfer organic matter to the depth during daytime (Raymont, 1970); in this

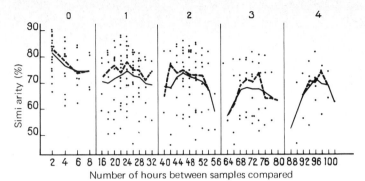

Fig. 10-23.—Changes in percentage similarity indices for night samples (at 10 m) plotted against the intervals between the pair of samples compared by the indices: Section 0 corresponds to samples collected in the same night; Section 1 night interval, etc: sampling with a Bongo net with 505 mean mesh: index of similarity expressed as a percentage is calculated from the equation

$$PS = 100 - 50 \left(\sum_{i=1}^{s} |a_i - a_b| \right)$$

in which a_i and b_i are percent of total individuals that belong to i^{th} category (taxon or stage) in the samples A and B respectively; S is the number of categories: points indicate individual values, the solid line the average for each interval, and the broken lines show the medians (after Miller, 1970).

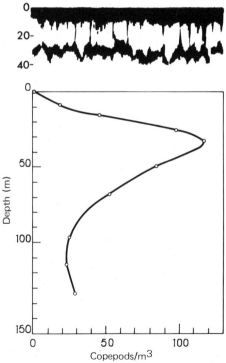

Fig. 10-24.—Shallow scattering layer in the North Pacific, recorded between 20 and 40 m with a 200 kHz Furuno recorder: upper, echogram; bottom, distribution of copepods *Calanus cristatus* caught with nets (after Barraclough, LeBrasseur & Kennedy in Parsons, 1972).

way the production of the limited photic zone will be distributed down to several hundreds of meters. This is also done by the sinking of organic particles but is accelerated by diurnal migration. Besides, as Roger (1971) has shown for Euphausiacea in the tropical pacific, the daily sinking enables the migratory species to avoid predation by shallow water fishes feeding only at daytime, and this is the case for certain tuna fish of this region.

Vertical migration allows the existence of relatively independent trophic chains which it is necessary to understand in any rational exploitation of the sea. Finally, vertical migration, because of the consequent exchange between deep and shallow water, accelerates the rise to the surface of any polluting elements which might be present at that depth.

McLaren (1964) relying on the fact that size is related inversely to temperature (Fig. 9-15) and that large sizes correspond to greater fecundity has suggested that vertical migration might be an advantage for some species, when the shallow waters are warmer than deep waters. According

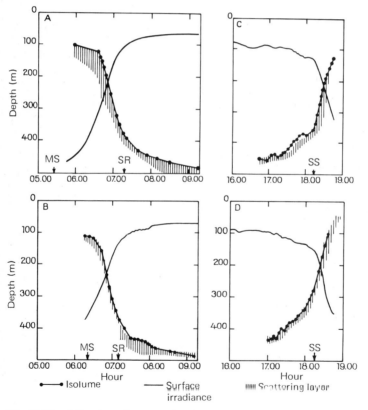

Fig. 10-25.—Migration of a scattering layer: changes in the depth of the $5 \times 10^{-4}\ \mu W/cm^2$, 474 nm isolume and the surface irradiance: A, dawn on 7th November, 1965; B, dawn 8th November, 1965; C, dusk on 3rd November, 1965; D, dusk on 4th November, 1965: Canary Islands: echosounder frequency, 10 KHz: MS, moonset; SR, sunrise; SS, sunset (after Boden & Kampa, 1967).

to his analysis an organism feeding at night near the surface in warm water and "resting" during the day in colder water at depth, is at an energetic advantage which can be invested in the production of eggs and an increase in the population. Only little work on the energetics of vertical migration has been done and much of it is contradictory. According to Petipa (1967) the migration of *Calanus finmarchicus*, with an amplitude of 50–100 m should increase the animals metabolism 31–35 times; this does not agree with the theory of McLaren. According to Petipa in *Acartia clausi* migrating 10–15 m, the metabolism does not increase. We shall see later (Chapter 11-5) that the calculations of Vlymen (1970) support this point of view.

10-5.—VERTICAL MIGRATION AND SCATTERING LAYERS

Among the interpretations given for the presence of scattering layers recorded by echosounders are those which attributed it to plankton. Gas bubbles are good reflectors of sound and the floats of Siphonophora Physonectes or bubbles escaping have been considered as playing a role in certain scattering layers. A direct relation may sometimes be shown. Fig. 10-24 shows a shallow scattering layer recorded with a frequency at 200 kHz and the numerical distribution of the copepod *Calanus cristatus*; the scattering layers coincide with the peak in the abundance of the copepods. Hansen & Dunbar (1970) made an experiment "in situ", injecting a suspension

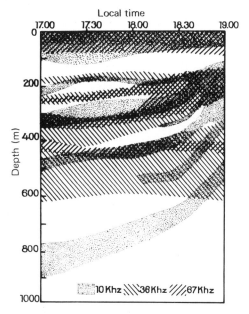

Fig. 10-26.—Upward movement of scattering layers at dusk recorded at different frequencies; Canary Islands, 9th October, 1965 (after Blaxter & Currie, 1967).

of Pteropoda and causing a scattering layer at this level. The plankton may, therefore, be regarded as one of the possible reasons for diffuse layers.

Scattering layers are known to show vertical migrations of a diurnal nature, rising to the surface at night and sinking in the morning. Kampa (1967) has shown (Fig. 10-25) that the movements may correspond very closely to that of a given light intensity, so that it is very tempting to attribute these scattering layers to living organisms undergoing vertical migration and including planktonic forms such as Euphausiacea; but, micronecton and notably pelagic fishes may also be concerned. The verification of this kind of hypothesis by the collection of the responsible organisms has so far, however, not been very satisfactory and the relation between plankton and scattering layers is not yet well understood. It should be kept in mind that the observed scattering layers are a function of the method of observations and very particularly the wavelength used. Fig. 10-26 clearly shows the complexity of these layers when they are detected with a frequency of 10, 36 or 67 kHz. Their behaviour varies, some migrate, others do not and this recalls the diversity already met within plankton groups*.

---

*Since the preparation of this chapter the work of Cushing (1973) has appeared and this contains an excellent review of scattering layers.

CHAPTER 11

NUTRITION, METABOLISM AND ENERGETIC BUDGET OF ZOOPLANKTON

Our knowledge of the nutrition and metabolism is largely restricted to the Crustacea and in particular to the copepods, which play a considerable quantitative role in the zooplankton. We shall discuss here the excellent pioneer work done by Marshall and Orr on *Calanus finmarchicus*.

11-1.—MODE OF NUTRITION AND FEEDING

Copepoda

The copepods are, in general, filter feeders, largely using the maxillae whose setae have fine setules forming a kind of filtering net (Fig. 11-1). In *Calanus finmarchicus* the ingoing water passes the maxillae; in *Acartia clausi*, on the contrary, the maxillae are actively used as a scoop. The distance between the setae changes according to the species and stage of development, ranging from several microns up to some tens of microns. In *Calanus finmarchicus* the surface of the maxillae is of the order of 0.3 to 0.4 mm^2 and the volume of filtered water attains 100 cm^3 per day.

Filtration is not, however, the only method of feeding used by copepods, but is often associated with capture of prey or predation, and this last method of feeding may sometimes be dominant as in *Pareuchaeta norvegica*; sometimes, the prey is not ingested but only the contents sucked out, e.g., in the case of large diatoms and *Noctiluca*.

The contents of its alimentary canal enables one to discern whether a copepod has fed. After digestion, material is expelled as a faecal pellet, elongate in form, enveloped by a pellicule and in which sometimes tests of ingested cells can be distinguished. A partial determination of the food of copepods may be deduced by an examination of the contents of the alimentary canal and the composition of the faecal pellets. It is, however, difficult to recognize organisms which do not have tests and sometimes the tests are ground up and hardly recognizable so that such observations must be supplemented by a study of cultures and trials of different kinds of food. Our present knowledge for most of the copepods is still not precise; only for a few species have we adequate information.

Calanus finmarchicus, is essentially an herbivore, taking very little animal food. It feeds on a great variety of diatoms from small species such as *Skeletonema costatum* (4–15 μm) up to large ones of the order of 50–100 μm, such as *Biddulphia*, *Cosinodiscus* or *Ditylum*. It ingests peridians, *Pro-*

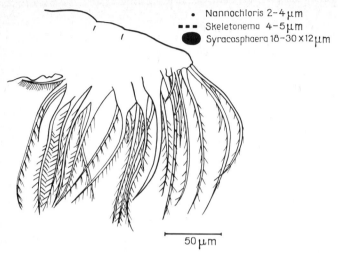

Fig. 11-1.—Right maxilla of the copepod *Temora Longicornis* female, and dimensions of some organisms of the phytoplankton (after Beruer, 1962).

rocentrum, *Peridinium*, *Gymnodinium*, but curiously it seems to avoid *Ceratium*, a fact which cannot be attributed to the dimensions or to the spines of *Ceratium*. The silicoflagellates and coccolithophorids are also taken and, as has been shown in cultures also small flagellates like *Dicrateria*, *Pyramimonas*, *Chlamydomonas*. The lower limit of filtration seems to be at about 3–4 μm; the very small flagellates such as *Nannochloris oculata* are not retained very efficiently. The role of bacteria in nutrition appears to be very uncertain (Marshall & Orr, 1955). In the animal part of the diet small crustaceans, tintinnidae and radiolarians are found.

It is not always easy to determine, even experimentally, the dimensional range of particles retained by a given organism, as was shown for *Euphausia pacifica* studied by Parsons, LeBrasseur & Fulton (1967). An initial experiment suggested that 12 μm was the lower limit, a second experiment, made a month later indicated a lower limit of 5 μm (Fig. 11-2).

Marshall & Orr determined, over one year at Millport the recognizable food in *Calanus finmarchicus* (20 to 70%; mean 52%) and the percentage of individuals containing different groups of organisms with identifiable material including diatoms, peridians, silicoflagellates, cocolithophorids, crustaceans, and radiolarians. The diatoms were clearly dominant and this corresponds to their importance in the nutrition. On the whole, during this investigation, the food of *Calanus* reflected the composition of the microplankton (sensu lato) at the moment of their collection.

The question that should be asked now is whether *Calanus finmarchicus* feeds automatically and without discrimination in the environment in which it is found. It would seem that there is often automatic feeding for more or less long periods but that is not permanent; it was observed that in cultures of phytoplankton copepods will cease filtration and delay the ingestion of

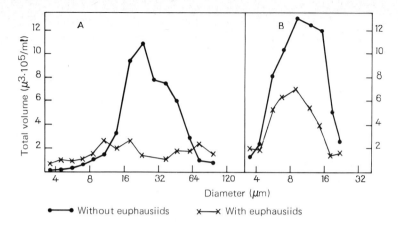

Fig. 11-2.—Determination of the dimensional range of the particles ingested by *Euphausia pacifica* using a particle counter: A, on a mixture of diatoms of the genus *Chaetoceros*; B, on a population of small flagellates (after Parsons, LeBrasseur & Fulton, 1967).

food. The experiments of Harvey (1937) had earlier shown the capability of *Calanus* to some extent to select food and to take in more *Ditylum* than *Lauderia* from a mixture of the two and, in a mixture of *Lauderia* to *Chaetoceros* to prefer the former. This has recently been confirmed for *Calanus finmarchicus* and other of copepods and it is generally accepted that automatic filtration can be replaced by selection (Conover, 1966c) or there may be completely selection (Richman & Rogers, 1969). Besides predation by capture of a chosen cell, selection may be made after a particle has contacted the mouth parts (Conover, 1966c) which is called encounter feeding and which we shall call passive predation in contrast to active predation. Whatever is the mode of predation it is accepted by many that different species of *Calanus* select (Fig. 11-3) the largest cells or organisms. Table 11-1 gives a good example of this tendency in a very closely related species, *Rhinocalanus nasatus*.

In a culture of *Ditylum brightwellii* containing paired cells, after division but not yet separated (Fig. 1-1g) and larger than others (136×26 μm against 100×24 μm) *Calanus helgolandicus* showed a rather lower consumption of *Ditylum* (estimated by the amount of water filtered and cleared of cells) as long as the quantity of paired cells did not exceed 20% of the whole; presumably, by a coincident interference of selection by predation the consumption greatly increased (Fig. 11-3).

Selection is not, however, always for the largest forms as shown by the data of Martin (1970) on mixed plankton with *Acartia clausi* (53%) dominated together by *Balanus* (20%) nauplii. In a natural phytoplankton population with *Rhizoselenia delicatula* (40–60 μm long and 16–22 μm diameter) and *Skeletonema costatum* (4–12 μm long, 8.15 μm diameter) the former diatom, although larger is used only when the density of *Skeletonema* decreases to several hundreds of cells per 1 cm^3 (Fig. 11-4).

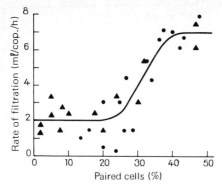

Fig. 11-3.—Filtration rate of *Calanus helgolandicus* females in a culture of *Ditylum brightwellii* at a constant density as a function of the paired cells: triangles, results obtained by ^{14}C method; circles, results obtained by counting of cells (after Richman & Rogers, 1969).

TABLE 11-1

Food selection of the copepod *Rhinocalanus nasutus* from a mixture of organisms available for consumption (after Mullin & Brooks, 1967).

Organism	Artemia salina nauplius	Coscinodiscus wailessi	Ditylum brightwellii	Thalassiosira fluviatilis	Cyclotella nana
Initial concentration	0.13/ml	0.95/ml	120/ml	1081/ml	7090/ml
Volume per individual (μm^3)	2.9×10^7	1.6×10^7	5×10^4	4.3×10^3	313
pgC/organism	5×10^5	1.5×10^5	1.9×10^3	3.0×10^2	41
Proportion in the mixture (in C)	6%	14%	32%	30%	28%

Stage	Total C ingested in 18 h (μg)	C(μg) ingested in 18 h/μg C of copepod	Contribution of each organism in total C ingested per each stage (%)				
			A.s.	C.w.	D.b.	T.f.	C.n.
Nauplius II and III	0.2	0.6	—	0	100	0	0
Nauplius IV	1.9	1.7	—	0	46	54	0
Nauplius V and VI	1.3	0.7	—	62	37	0	0
Copepodid I	2.2	1.2	0	14	49	37	0
Copepodid II	4.4	1.2	0	13	51	32	4
Copepodid III	5.7	0.4	0	0	28	40	31
Copepodid IV	8.2	0.2	0	95	3	2	0
Copepodid V	21.8	0.4	4	54	23	19	0
Adult males	4.7	—	25	75	0	0	0
Adult females	16.5	0.1	37	45	18	0	0

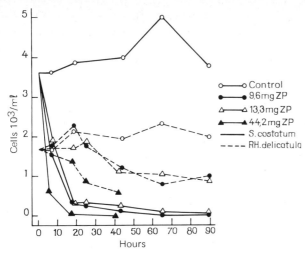

Fig. 11-4.—Grazing experiment on a mixture of phytoplankton initially composed of *Skeletonema costatum* (3600 cell/ml) and of *Rhizosolenia delicatula* (1700 cells/ml) for different concentrations of zooplankton (/160 μm): 53% of *Acartia clausi*; 20% of *Balanus* (nauplii) expressed in mg dry weight/ml (9.6; 13.3 and 44.2 mg zooplankton, ZP) (after Martin, 1970).

The observations reported above on the food of *Calanus finmarchicus* refer to the adults or the copepodid stage V. The copepodid stages III and IV take similar food; for copepodid stages I and II the data are still inadequate but their food seems not to differ from that of the adults and their mouth parts have the same form. This is not the case for the nauplii whose mode of feeding is probably different. An examination of the faecal pellets of the nauplii stages III and IV has shown that they take small-sized cells, smaller than 25 μm (*Prorocentrum triestinum*, *Peridinium trochoideum*) or even smaller than 10 μm (*Skeletonema costatum* of 5 μm). It is very probable that they ingest small flagellates of several microns. Certain differences do, however, exist between the food of the adults and that of the young stages. It seems reasonable to accept that nauplii stages I and II do not feed but live on their reserves so problems associated with feeding first arise for stage III nauplii. Mullin & Brooks (1967, 1970) showed that in cultures of *Calanus helgolandicus* (= *pacificus*), very close to *Calanus finmarchicus*, in contrast to the copepodids, the first nauplii stages do not feed and that the subsequent stages could not feed on *Ditylum* of which the cells are too large. This differs from species to species; *Rhinocalanus nasutus* begins to feed at stage II, and even at the nauplius stage selects relatively large cells (Table 11-1).

Calanus finmarchicus is an example of an essentially herbivorous species, animal food being taken only as a minor component. To the same category belongs *Pseudocalanus* and *Paracalanus* (Raymont, 1963). Contrary to this *Centropages*, *Temora*, *Acartia*, *Metridia* and also *Rhinocalanus* (Table 11-1) show an equilibrium between the vegetable and animal food. Finally, *Pareuchaeta norvegica*, appears to be a carnivorous species which in experiments feeds easily on small copepods such as *Acartia* and *Centrophages* (Conover, 1960). If this species is presented with mixed food its faecal pellets will contain only animal debris. Carnivorous feeding is also shown in copepods such as *Pareuchaeta norvegica* which lives at depths where phytoplankton is rare or absent, but it can also pertain to shallow water species. Wickstead (1959) has given evidence indicating that *Candacia bradyi* present in the shallow water off Singapore, feeds on chaetognaths.

The final question which arises is whether the copepods feed only on living material. The organic detritus is sometimes an important energetic source and may effectively enter the food regime of various species, but we do not know exactly what part it plays in nutrition. In addition sea water contains a certain quantity of dissolved organic matter or present in colloidal suspension. We still lack information allowing an estimate of its utilization but must mention two lines in research on this. Provasoli & Shiraishi (1959) succeeded in growing the brine shrimp *Artemia salina* in a medium containing nutrients in solution, particles of starch being added only in order to provoke ingestion of the liquid since absorption of nutrients seems to be only via the alimentary canal. Furthermore, Riley (1963) showed that dissolved organic substances in sea water can be absorbed on free interfaces (surface of the water, surface of air bubbles) and these form small particles. These particles enable *Artemia* to feed satisfactorily (Baylor & Sutcliffe, 1963). It would not, therefore, be surprising if dissolved organic matter or organic particles play a complementary role in the nutrition of plankton copepods, but this still has to be proved. It is of interest in this connection to remember Putter's (1925) theory and called after him, which considers that dissolved organic matter could be used directly as nutrients by the zooplankton; so far it has not been verified.

Other organisms of the zooplankton

In the copepods there are two principal ways of nutrition, filtration and predation which are, as has been seen, able to interact. In general, these same two methods are found in other groups of zooplankton. Sometimes the filtered particles are collected by ciliary movements and enveloped by mucous as is the case in the medusae *Aurelia* as well as in the pelagic tunicates and pteropods (Mollusca, Pteropoda). Predation also very common, is accompanied by immobilization of the prey, by injection of venom (by the nematocysts of medusae and Siphonophora) by hooks (Chaetognatha, Gymnosoma (Mollusca, Pteropoda)), by suckers (also Gymnosoma) or by adhesive cells (the colloblasts of Ctenophora).

In general, for each group of planktonic organisms we know what is the normal way of feeding and what is the food (Lebour, 1922, 1923); this has been mentioned during the systematic review, but as regards details there is still much to be learned and in spite of known nutritional methods, it is often difficult to affirm whether any given method is absolute or more or less temporary.

11-2.—QUANTITATIVE STUDY OF NUTRITION AND GRAZING

After reviewing the qualitative knowledge of nutrition we will consider quantitative studies. The first necessity is to determine the quantity of food ingested in a unit of time, hour or day, which corresponds to the daily or hourly ration of the given organism. In the case of nutrition by filtration this may also be expressed as the volume of water which the animal is able to filter and to remove its potential food during a day or an hour. Table 11-2 gives, as an example, certain determinations of the filtration rates and filtered volumes per unit of time.

TABLE 11-2

Determination of the maximum water volume filtered (M.v.f., ml/24 h) (cleared of its food) for different copepods; CV, copepodid V (after Marshall & Orr, 1962).

Species	Stage	Food	M.v.f.	Method	Author
Calanus finmarchicus & *C. helgolandicus*	♀	Flagellates < 10 μm	4	^{32}P	Marshall & Orr
Calanus finmarchicus & *C. helgolandicus*	♀	Flagellates > 10 μm	84.3	^{32}P	Marshall & Orr
Calanus finmarchicus & *C. helgolandicus*	♀	Diatoms > 10 μm	43	^{32}P	Marshall & Orr
Calanus finmarchicus & *C. helgolandicus*	♀	Natural sea water	36	analysis	Corner
Calanus finmarchicus & *C. helgolandicus*	CV	Flagellates > 10 μm	101	counting	Gauld
Calanus finmarchicus & *C. helgolandicus*	CV	Diatoms > 10 μm	4	^{32}P	Marshall & Orr
Centropages hamatus	♀	Flagellates < 10 μm	0.3	^{32}P	Marshall & Orr
Centropages hamatus	♀	Flagellates > 10 μm	15.0	^{32}P	Marshall & Orr
Centropages hamatus	♀	Diatoms > 10 μm	2.7	^{32}P	Marshall & Orr
Acartia clausi	♀	Flagellates < 10 μm	0.1	^{32}P	Marshall & Orr
Acartia clausi	♀	Flagellates > 10 μm	8.6	^{32}P	Marshall & Orr
Acartia clausi	♀	Diatoms > 10 μm	2.7	^{32}P	Marshall & Orr

Different methods are used to obtain the food ration or filtration volume. The simplest is to follow the fall in concentration of cells, a method which has been made easy by the use of particle counters. The variation in concentration in a fraction of time is proportional to the concentration (exponential law) according to the equation.

$$\frac{dC}{dt} = -C\left(\frac{nG}{V} + S\right)$$

where C is the concentration in cells, G the rate of filtration (i.e., volume filtered in a given unit of time), n the number of individuals, V the volume of the milieu and S the rate of sedimentation of the cells (Coughlan, 1969).

Integration gives,

$$C_t = C_0 \cdot \exp^{-[(nG/V)+S]t}$$

when C_0 is the concentration at time 0 and C_t is that at time t. The rate of filtration is obtained from the logarithmic relation,

$$G = \frac{V}{nt}(\log C_0 - \log C_t - S_t)$$

It is possible to facilitate the determinations by the use of graphical methods (Coughlan, 1969; Paffenhofer, 1971).

Another method consists in measuring the amount of organic matter before and after passage through the vessel which contains the experimental animals (Corner, 1961). The food may also be labelled by radioactive elements ^{32}P or ^{14}C (Rigler, 1971). In a culture of algae with skeletons which can be recognized in the faecal pellets, their rate of production also affords a method of calculating the rate of ingestion.

As a result of recent studies we are beginning to understand something of the changes in the filtration rate and food ration. For some species the filtration rate increases considerably with body weight, as was shown by Paffenhofer (1971) studying different stages of the copepod *Calanus helgolandicus* (= *pacificus*) at La Jolla, California. On logarithmic coordinates the filtration rate, expressed in ml/day is proportional to the weight expressed in μg of carbon (Fig. 11-5). Such experiments are carried out with amounts of food comparable to those in nature (50–100 μg C/l). The results are similar to those obtained for the same species by Mullin & Brooks (1970) also done in La Jolla, and for the very closely related species *Calanus finmarchicus* by Marshall & Orr (1956) and by Gauld (1951) in Scotland. The development from one stage to the next is accompanied by an increasing rate of filtration, except for adult males which have filtration rates much lower than that of females (7 times less according to Paffenhofer). Paffenhofer (1971) also observed that when all other conditions are equal, filtration rate increases with the size of the cell ingested; it was clearly higher when the cells of *Lauderia borealis* were used as food and had a dimension of 36 μm rather than 19 μm. This agrees with the observations of Richman & Rogers on *Ditylum brightwellii* noted above (Fig. 11-3). In contrast, the filtration rate decreased when the cell density was greater, as had already been shown by Mullin (1963) and Anraku (1964); this indicates some kind of regulation of filtration. If, under natural conditions, these two phenomena are able to exist together the interference will be relatively complex and the rate of filtration will be notably modified.

A number of authors, particularly Mullin (1963) have noted a decreasing filtration rate when the culture used as food is aging and some of them consider this phenomenon of biochemical origin; however, the experiments of Conover (1966) do not completely confirm any such decrease.

If we now first consider the food ration we find, on a logarithmic basis, a proportional relation, between the ration and the body weight (Fig. 11-6A,

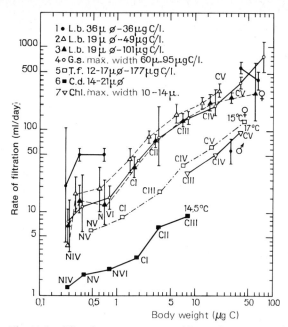

Fig. 11-5.—Filtration rate per copepod for various mean weights and various stages of *Calanus helgolandicus* (= *pacificus*) at 15°C, (1, 2, 3, 4) the concentration of food organisms is indicated in μgC/l as well as their mean size. Vertical bars give 95% confidence limits about the mean. Data of Mullin & Brooks (1970) on *Calanus helgolandicus* (= *pacificus*) (5) at 15°C, of Marshall & Orr (1956) (6) at 14.5°C and of Gauld (1951) (7) at 17°C on *Calanus finmarchicus* are added: abbreviations, L.b, *Lauderia borealis*; G.s., *Gymnodinium splendens*; T.f., *Thalassiosira fluviatilis*; C.d., *Chaetoceros decipiens*; Chl, *Chlamydomonas* sp; N, nauplius (IV, V, VI); C, copepodids (I, II, III, IV, V) (after Paffenhofer, 1971).

Paffenhofer, 1971); we have,

$\log R = k \log W + b$

where R is the ration, W the body weight while k and b are constants. This expression may also be written, in the form ($b = \log b'$):

$R = b' W^k$

The ration varies, therefore, according to a fractional power of the weight which is of the order of 0.7 in the case of *Calanus helgolandicus*. As for the ration/unit weight (R/W), or relative ration, it varies according to a negative fractional power ($R/W = b' W^{k-1}$) and so diminishes as a function of the weight as shown in Fig. 11-6B. A nauplius, at stage V, with a mean weight of 0.33 μg (in carbon) has a daily ration 340% of its weight in a suspension of *Lauderia borealis* of 19 μm diameter and containing 101 μgC/l. Under the same conditions copepodid stage III (6.94 μgC) consumes 190% of its weight and an adult female (61.3 μgC) 58%. On the other hand, the ration is a function of the available nutrients. The rate of filtration decreases with the density of food particles and as a consequence the increase of the ingested

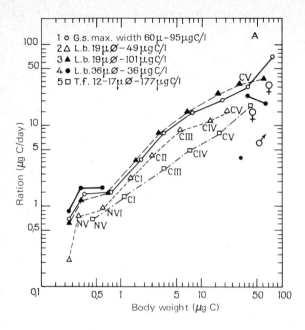

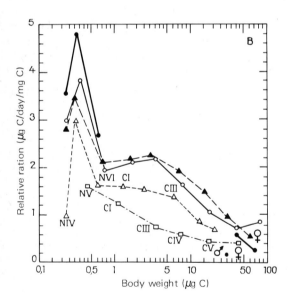

Fig. 11-6.—A, daily ration as a function of the mean body weight for different stages of *Calanus helgolandicus* (= *pacificus*) at 15°C (1, 2, 3, 4): determinations done under same condition as those of Fig. 11-5: data from Mullin & Brooks (1970) (5) are added: abbreviations as in Fig. 11-5: B, relative daily ration (ration/unit weight) as a function of the mean body weight: symbols as in A (after Paffenhofer, 1971).

food is modified when the density increases. In some cases zooplankton follow a similar relation to that established by Ivlev for fish; this has been accepted especially by Petipa (1959) by Cushing (1967). The relation was formulated by Parsons, LeBrasseur & Fulton (1967) in the following way:

$$R = R_m(1 - e^{-\delta(P-P')})$$

where R is the ingested ration for a mean concentration of phytoplankton P, R_m the maximal possible ration, δ a constant (coefficient of ingestion) and P' the concentration of phytoplankton at which the feeding began. Figs 11-7 and 11-8 show several such relations, which may also be written using the relative rations/unit weight (R' and R'_m) of which two characteristics have to be emphasized, namely the possibility of the existence of a minimum

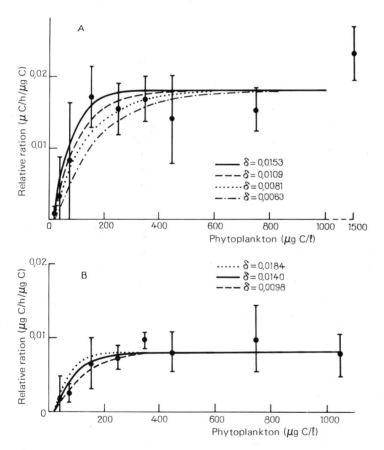

Fig. 11-7.—Relative ration R' (ration/unit weight) of *Calanus pacificus* as a function of phytoplankton concentration: P, mean values (closed circles) and standard deviation (vertical bars) A, nocturnally feeding copepods: the curve corresponds to Parsons et al. equation, with $R'm = 0.0180$, $P' = 15$ and δ varying between 0.0153 and 0.0063: B, continuously feeding copepods, the curve corresponds to the following values, $R'm = 0.0080$, $P' = 15$ δ varying from 0.0184 to 0.0098 (after McAllister, 1970).

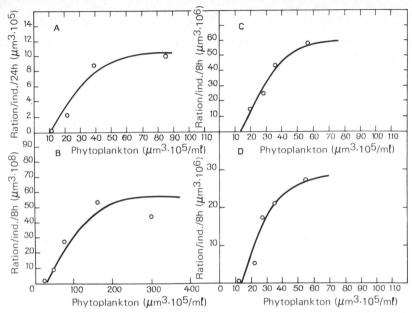

Fig. 11-8.—Variation of the alimentary ration for different concentrations of phytoplankton: A, *Pseudocalanus minutus* feeding on a mixture of *Chaetoceros socialis* and *Chaetoceros debilis* average size of 32 μm; B, *Euphausia pacifica* feeding on the same mixture; C, *Calanus pacificus* and euphausiid furcilia feeding on unidentified flagellates of average size 8 μm; D, *Calanus pacificus* and euphausiid furcilia feeding on a mixture of *Chaetoceros* (after Parsons, Le Brasseur & Fulton, 1967).

threshold and the presence of a maximum utilization in the concentration of phytoplankton. In the curves of Fig. 11-7 the ration is expressed in organic carbon which is preferable to cellular values or number of cells. Because of night respiration, the amount of organic carbon/cell decreases during the night and this is further emphasized when cell division takes place during the night as McAllister (1970) has found in his experiments. We must also remember that as regards the chemical composition of the phytoplankton, there is a decrease in the utilization of organic carbon in cells with depth.

In their experiments, neither Mullin (1963) nor Conover (1966b) found an asymptotic maximum but rather a decrease in the rate of ingestion with a high density of cells (Fig. 11-9). Rigler (1971) regarded the results of Mullin as an artefact of the method of calculating the results. As for the results of Conover, they show an important change in the values at high densities.

The interpretation of food ration and its extrapolation in calculations may sometimes need much precaution, since it is not always constant for the same organism; as emphasized by Mullin (1963) and later McAllister (1970) the ration may decrease markedly as a function of time when it is determined just after a period of fasting (Fig. 11-10). Discontinuous feeding, without doubt, exists at least for those species which migrate vertically and are found in the upper layers rich in phytoplankton only during the night.

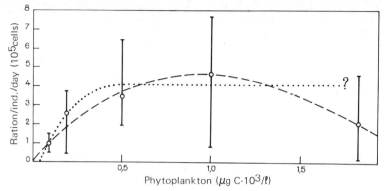

Fig. 11-9.—Changes in the ingestion rate as a function of the available food for *Calanus hyperboreus* feeding on *Thalassiosira fluviatilis*, in the dark of 6°C; broken line, curve interpreting Conover's data; dotted line, curve according to Ivlev's equation: for each density the mean and the range indicated: phytoplankton in $\mu gC \times 10^3/1$ (after Conover, 1966b).

The ingestion of phytoplankton by zooplankton makes an important contribution to the removal of the phytoplankton population; this phenomenon is called grazing. When a population of phytoplankton is subjected to grazing its rate of growth decreases and instead of the relation (Appendix, Chapter 14);

$$P_1 = P_0 \, e^{kt}$$

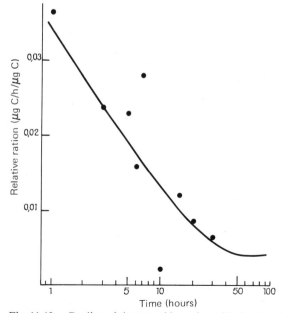

Fig. 11-10.—Decline of the rate of ingestion with the duration of feeding by *Calanus pacificus* starved prior to feeding (after McAllister, 1970).

we have,

$$P_1 = P_0 e^{(k-g)t}$$

where P_1 is the concentration of the phytoplankton at the end of the period considered, P_0 the initial concentration, k, the rate of relative growth, and g the rate of grazing (rate of relative mortality). Knowing k, calculated from a control culture, the value of g may be calculated. From the equation of the food ration McAllister (1970) made an estimate of the rate of grazing for a concentration of zooplankton equal to unity ($dP/dt = gP = ZmP$, and $mP = R'$, Z = concentration of zooplankton)

$$m = \frac{R'}{P} = \frac{R'_m}{P}(1 - e^{-\delta(P-P')})$$

This rate shows much variation with the density of phytoplankton P, with the coefficient of ingestion δ, and with the minimum threshold P' as shown in Fig. 11-11.

In conclusion, it is clear that much remains to be learned regarding zooplankton nutrition; however, it is now appreciated that a belief in the automatic filtration of the entire phytoplankton by the entire herbivores is too simplistic and the likelihood of food selection seems more certain, although it may be accompanied by a high eclecticism. As many in other biotopes, more or less clear ecological niches based on trophic relations are evident; this contributes something towards an explanation of the variation in specific composition met with in the zooplankton as a function of either time or space.

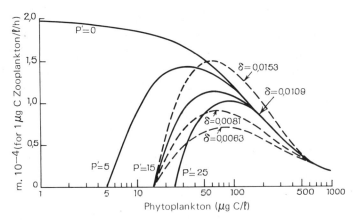

Fig. 11-11.—Relative mortality rate (m) of a concentration of zooplankton equal to 1 μgC/l (grazing rate), feeding at night, at phytoplankton concentration (P): continuous line corresponding to different values of P', 0, 5, 15, and 25, δ being constant and equal to 0.0109; the broken lines correspond to P' equal to 15 and different values of δ, R'_m is constant and equal to 0.0180 (after McAllister, 1970).

11-3.—DIGESTION AND RATE OF ASSIMILATION

Only a fraction of the ingested material is absorbed in the alimentary canal, that part not assimilated being rejected in the faeces; it is, therefore, possible to define a rate of assimilation which may vary considerably. For the determination of this rate, we must know the quantity of food consumed (C), and the quantity eliminated in the faeces (F); this rate is expressed as follows;

$$A = \frac{C - F}{C} \times 100$$

C and F may be expressed in dry weight, in organic matter (dry weight—ash weight) or in carbon and sometimes even in nitrogen or phosphate. For the determination of assimilation, radioactive tracers ^{32}P or ^{14}C have been used (Corner & Davies, 1971). The estimation of the ingested food is relatively easy, as has already been shown, but the estimation of faecal excretion is more uncertain even experimentally, since the faecal pellets may sometimes be consumed by the organisms which produce them or be more or less rapidly broken up so that their recovery without error is difficult. It is not always easy to determine the two corresponding values C and F. To avoid this difficulty Conover (1966a) proposed a method which assumes that the absolute amount of ash does not change during digestion. He used the following equation derived from the former.

$$A = \frac{(C' - F')}{(1 - F')(C')} \times 100$$

when C' is the ratio of ash-free dry weight (organic matter) to the dry weight of the ingested food and F' the same ratio in the faeces. It is then not necessary to know the absolute values of F and C but only the ratios which may be observed on partial samples. Under these assumptions an estimate of A' may even be made for the sea.

The rate of assimilation is subject to large variations and Conover (1966a) reviewing the literature gives a range of 6% to 99%. Having regard to the variety of consumed and consumers this is not surprising. The determinations of Conover (1966a, b) indicate, however, a smaller range as shown in Table 11-3 which relates to *Calanus hyperboreus* feeding on different kinds of algae; the values range from 18.8 to 89.9% and the means from 40.1 to 86.7. The flagellates such as *Exuviella* and *Dunaliella* are assimilated better than the diatoms. The rate of assimilation of *Skeletonema* varies little, and goes hand in hand with the relative constancy of its ash content. This relation between ash content and rate of assimilation of organic matter may be generalized, as shown in Fig. 11-12 which indicates an inverse relation between the two values.

Conover (1966a) also determined the rate of assimilation under natural conditions and a number of his results are given in Table 11-4. The values for *Calanus hyperboreus* are very similar to those obtained in the laboratory

TABLE 11-3

Assimilation rate (A', %, determined on organic matter) for the copepod *Calanus hyperboreus*, fed on different unicellular algae and the percentage of ash, in its composition (after Conover, 1966b).

Algae	Ash %			A' %		
	m	vr	n	m	vr	n
Diatomea						
Thalassiosira fluviatilis	27.0	11.0–43.2	47	67.3	36.7–87.9	47
Skeletonema costatum	44.3	41.8–48.3	5	40.1	36.7–44.6	5
Cyclotella nana	30.3	27.2–32.2	4	59.1	42.9–76.5	4
Coscinodiscus sp.	56.2	37.3–70.0	4	50.9	32.6–63.3	6
Ditylum brightwellii	56.7	43.7–62.2	7	50.6	39.5–66.4	4
Rhizosolenia setigera	42.0	30.3–59.4	6	56.8	18.8–73.3	9
Dinophyceae						
Exuviella sp.	10.6	9.8–15.5	6	71.6	56.3–82.9	6
Chlorophyceae						
Dunaliella sp.	7.6	4.0–14.0	3	86.7	75.1–89.9	3

m, mean; vr, variation range; n, number of observations.

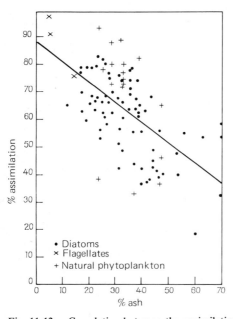

Fig. 11-12.—Correlation between the assimilation rate determined on organic matter and the ash for *Calanus hyperboreus* feeding on different sources (Table 11-3 with the exception of *Exuviella*) and for several data obtained in situ; regression equation is, $Y = 87.8 - 0.73X$, correlation coefficient $r = -0.60$; the slope is significantly different from 0; variance analysis gives $F = 55.60$ with $P < 0.001$ (after Conover, 1966,b).

TABLE 11-4

Assimilation rate (A' %) of natural particulate organic matter in the surface water, determined in situ from the organic matter and dry weight (after Conover, 1966a).

Organism	Date	A' %
Meganyctiphanes norvegica	20th February, 1964	65.0
Calanus hyperboreus	12–15th May, 1965	43.8
Calanus hyperboreus	12th May, 1965	69.5
Euchirella rostrata	10th May, 1965	17.7
Mixed zooplankton	13–18th April, 1964	32.5–65.2
Mixed zooplankton	8–9th July, 1964	75.7 and 79.5
Mixed zooplankton	12–15th May, 1965	40.0 to 80.6

(Table 11-3). As a first approximation it may be said that the rate of assimilation must lie between 40–80% (see also Table 8, page 416, in Conover & Cowey, 1968).

Beklemishev (1962) has emphasized that in periods when the phytoplankton is very abundant, notably in spring blooms, the zooplankton and particularly the copepods are capable of ingesting much more food than they can use and so the assimilation rate decreases and the faecal pellets formed are very rich in organic matter. This superfluous feeding has given rise to much argument. Conover (1966b) using far greater concentrations than the minimum of 390 µgC/l given by Beklemishev for the initiation of superfluous feeding did not find any significant decrease in the rate of assimilation. We must remember that the rate of ingestion is at a maximum when the density of food is high which results in a natural regulation.

Butler, Conover & Marshall (1970) have confirmed the absence of superfluous feeding by studying the rate of assimilation of nitrogen and phosphate. In the second fortnight of April 1969, during a spring bloom with 111 µg/l of particulate nitrogen present and using a C/N ratio of 4.4, determined for *Skeletonema* (about 488 µgC/l) they obtained a rate of assimilation of nitrogen of 62.4% for *Calanus* in the sea and 77% for phosphorus. These are high rates similar to those found experimentally.

Even in absence of superfluous feeding the faecal pellets are, however, far from being free from organic matter; on the contrary they are an appreciable source of possible food. Conover (1966b) gives the following example; *Exuviella* containing 91.1% of organic matter (% dry weight) is used by *Calanus hyperboreus* with a rate of assimilation of 68.2% and the faecal pellets still contain 76.5% organic matter (% dry weight).

There are few data on the digestive enzymes of zooplankton. Bond (1934) and Hasler (1937) found that the digestive enzymes of copepods are able to attack carbohydrates, proteins, and some lipids. In the faeces of *Calanus hyperboreus*, bacteria able to digest chitin and alginic acid were found (Vaccaro in Conover, 1966b) but we do not know whether they are able to do so in the alimentary canal and so benefit the copepod.

11-4.—UTILIZATION OF ASSIMILATED METABOLITES AND ENERGETIC BUDGET

After being assimilated, the substances originating in the food are diverted in different directions as indicated in the scheme of Warren & Davis in Fig. 11-13 (in Warren, 1971). The first fraction, corresponding to nitrogen-containing substances, has to be eliminated directly and contributes to excretion. There is, therefore, a small difference between the oxidation of proteins as determined by colorimetric methods which gives about 5.65 kcal/g, and that in vivo which does not lead to the same level of oxidation (Kersting, 1972) but to about 4.35 kcal/g; however, we do not know whether these values may be applied without some correction factor to the proteins of zooplankton or of phytoplankton.

For the determination of the calorific equivalent and the application of the notion calorie in ecology we may refer to the work of Paine (1971). Cummins & Wuychek (1970) have made an exhaustive study of the caloric equivalents of organisms of fresh water. In general, the caloric value of an organism (ash excluded) lies between a lower limit, that of glucose (3.74 kcal/g) or cellulose '(4.18 kcal/g) and an upper limit of lipids and fatty acids (9.37 kcal/g).

A certain fraction of energy must be subtracted from that of the substrates able to go into the metabolic pathways of the animals; this is

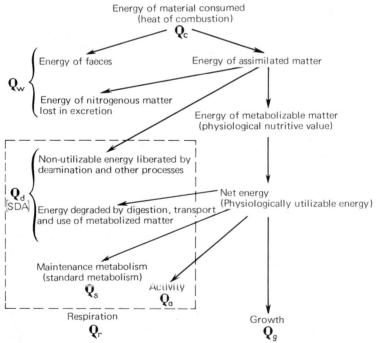

Fig. 11-13.—Fate of food consumed by an animal and the relation between the different energetic fractions (SDA = specific dynamic action).

attributed to the deamination of amino acids not incorporated into animals protein. Such a deamination will consume oxygen and this is marked by an increase in respiration after the absorption of food and is known as specific dynamic action. It is a function of the composition of the assimilated metabolites.

Only the remaining energy is used by the animal for growth, maintenance metabolism (standard metabolism), activity, and in the processes of digestion, transport, and incorporation of the metabolites. This last was added by Warren & Davis to the energy used in the specific dynamic action. The energy budget of an organism of the zooplankton may be formulated as follows:

$$Q_c - Q_w = Q_g + Q_r$$

where Q_c is the energetic value of the food consumed, Q_w the energetic value of substances eliminated in the faeces and that assimilated but not metabolized, Q_g the total change of the energetic value of the substance in the body; i.e., the energetic value of growth, Q_r the energy degraded in respiratory metabolism with, or without, benefit to the organism. Q_r may be divided to these different components:

$$Q_r = Q_s + Q_d + Q_a$$

where Q_s is the energy used when the animal is not feeding but resting (standard or maintenance metabolism), Q_d is the energy used for deamination of amino acids not used in the formation of proteins and by digestion, transport and incorporation of the assimilated metabolites and Q_a the energy used in muscular activity. Finally, Q_w may be replaced by its two components, namely, the principal one Q_f, the caloric value of the faeces, and Q_n, the caloric value of the unmetabolizable nitrogen compounds. The complete equation may be written:

$$Q_c - Q_f - Q_n = Q_g + Q_s + Q_d + Q_a$$

From nutritional studies, the food ration gives the value of Q_c, while the rate of assimilation ($A\%$) gives us $Q_c - Q_f$ as a function of Q_c, or, $Q_c - Q_f = AQ_c$. Since Q_n is neglected, we shall now examine Q_r and the respiration, then the excretion and afterwards consider the growth and Q_g.

In this energetic budget the metabolism is regarded from its caloric aspects. Some workers have tried to establish a budget based on carbon or phosphate (Corner & Davies, 1971) and budgets based on lipids, carbohydrates or proteins may also be considered.

Russian authors (e.g., Sushchenya, 1970) have used a simplified form of the energetic budget, namely,

$$R = G + T + N$$

where R is the energetic constant in the food ration, G the energy used for growth, T the energy spent in respiration, and N the energy of any unassimilated parts of the ration. From the last equation it is possible to derive the following,

$G = R - T - N = AR - T$

replacing $R - N$ (the assimilated food) by the ration multiplied by the rate of assimilation, A.

In the previous terms this may be written:

$Q_g = AQ_c - Q_r$

11-5.—RESPIRATION

The measurement of respiration can give the value of Q_r, which is the minimum energetic demand of the organism under natural conditions, when growth is excluded. This value must be subdivided, as already indicated, into several parts, Q_s, Q_d, Q_a; however, it should be stressed that, if these minimal requirements are satisfied by the food absorbed by the animal there will be a more or less important inhibition of the use of stored metabolites.

The techniques of measurement themselves give rise to a certain number of problems because of the dimensions, often small, of the zooplanktonic organisms and because of their fragility. The use of a high density of organisms in order to obtain a detectable change of oxygen may influence the results by increasing the activity of the individuals and hence their oxygen consumption; it should, therefore, be ensured that the effect of density is negligible. Capture and the various shocks accompanying it may also give rise to errors and there must be a certain acclimation time since respiration is higher immediately after collection (Fig. 11-14). In addition to the usual estimation of oxygen by chemical methods (Edmondson & Winberg, 1971) polarographic methods may be used (Nival, Nival & Leroy, 1971).

Figure 11-15, from Marshall & Orr (1964), gives the annual changes in respiration per individual. An examination of the curves show different variations in the respiration i.e., the oxygen uptake in a given time. Comparing the upper and lower curves it may be seen that temperature markedly affects the rate. Differences are also apparent according to the stage or degree of maturity, so we must examine how far this variation may be explained as a function of the weight or biochemical composition of the individual. We must also determine whether the marked seasonal variations are due to variations in weight or perhaps in nutritional state; these several points will be discussed successively.

The rate of respiration is increased by a rise of temperature up to a critical point, beyond which higher temperatures cause damage to the organism. The curve in Fig. 11-16 illustrates this general phenomenon. A value Q_{10}, is often used in order to express the effect of temperature; it is the ratio between the respiratory rates measured for two temperatures with a difference of 10°C. For female *Calanus finmarchicus* (Fig. 11-16) it is 1.43 between 0 and 10°

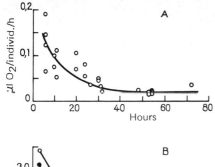
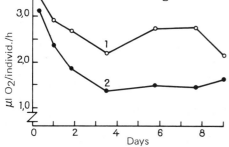

Fig. 11-14.—A, variation in the respiration rate as a function of the experimental time for the copepod *Temora longicornis*, females (at 10°C, in the dark): B, variations of the respiration rate with time after capture for (1) the euphausiid *Thysanoessa inermis* female and (2) the copepod *Calanus cristatus*, copepodid V, at 10–12°C (A, after Berner, 1962; B, after Ikeda, 1970).

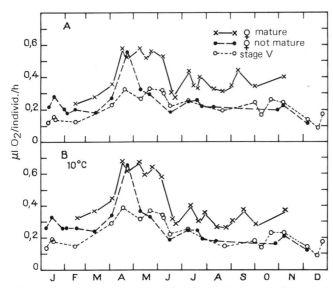

Fig. 11-15.—Seasonal variations in the respiration of *Calanus finmarchicus* mature and immature females and copepodid V: B, values obtained in the laboratory at 10°C; A, corrected values taking into account the real temperature of the sea in the Clyde Sea Area (Scotland) (after Marshall & Orr, 1964).

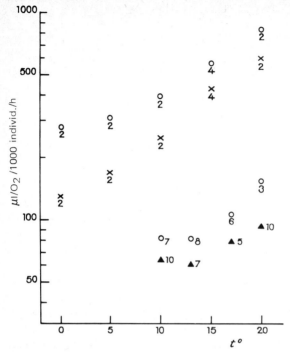

Fig. 11-16.—Variations in copepod respiration rate as a function of temperature: upper, *Calanus finmarchicus*, females (open circles) and copepodid V (crosses) in August 1931 after 18 h of acclimatization: lower, *Temora longicornis* adults (closed circles) and *Acartia clausi* adults (solid triangles) after at least 24 h of acclimatization: number of experiments indicated at each point (after data of Marshall & Orr, 1935 and of Gauld & Raymont, 1953).

(400/280) and 2.08 between 10 and 20° (830/400). Q_{10} is the temperature coefficient (related to 10°C) affecting the phenomenon and may be calculated for different temperatures when the measurements give the change with temperature. Theoretically a low Q_{10} should be an advantage in enabling an animal to compensate for the effect of temperature changes, since the effect is reduced, and should indicate a good adaptation to any thermal disturbance (eurythermia); however, good evidence for such an interpretation is still equivocal (Nicol, 1960).

In general, at a constant temperature the respiratory rate varies as a function of weight. This variation is not linear but exponential and may be expressed by the following equation,

$R = aW^b$

or in its logarithmic form,

$\log R = \log a + b \log W$

where R is the respiratory rate, W the weight and a and b are constants. Using the dry weight (DW) as a basis Conover (1959) found for neritic copepods of the Atlantic and the English Channel (at 5–10°C),

$$\log R = 1.120 + 0.856 \log DW$$

R is expressed in μlO_2/individual/h and DW in mg/individual. Nival, Nival & Palazzoli (1972) found for copepods in the Mediterranean (at 15°, with confidence intervals for $P = 0.05$)

$$\log R = 0.687 \ (\pm 0.119) + 0.837 \ (\pm 0.090) \log DW.$$

The data of Petipa et al. (in Shushkina, 1968) for the copepod *Acartia clausi* (at 16°C) allowed them to calculate

$$\log R = 0.442 + 0.81 \log DW$$

There are marked differences in the constant a which are related largely to the method of measurement; in contrast, the value of the coefficient b is reasonably constant for marine copepods, lying between 0.80 and 0.85. For the fresh water copepods this coefficient is slightly lower; Comita (1968) found a value of b ranging between 0.622 and 0.721 for the genus *Diaptomus*, with values of dry weight ranging from 3 to 300 μg.

More generally, for all aquatic organisms studied this coefficient varies between 0.66 and 1.0. At a value of 0.66 the rate of respiration is proportional to the surface of the body (in supposing $W = \alpha L^3$, $W^{2/3}$ is $\alpha' L^2$) and, therefore, follows the surface law. But the values for marine organisms do not conform exactly to this law; for small metazoans (eggs, larvae, small crustaceans) the coefficient approaches a unity (0.8–0.9) while for larger organisms the exponent is about 0.7 (Nicol, 1960).

By collecting all the data, not only for copepods but also other different planktonic groups it is possible to give a "global" relation: Nival, Nival & Palazzoli (1972) obtained for the north-western Mediterranean the general relation.

$$\log R = 0.844 + 0.706 \log DW$$

with a correlation of 0.929 for which b is very near to that obtained by Ikeda (1970) for planktonic organisms (measured at 17°C) from the north of Japan

$$\log R = 0.357 + 0.691 \log DW$$

with $r = 0.953$. The coefficient b may vary, notably as a function of the latitude (Ikeda); for the boreal species of Bering Sea the relation is:

$$\log R = 0.024 + 0.830 \log DW$$

and for tropical plankton between the Philippines and Australia:

$$\log R = 0.874 + 0.538 \log DW$$

As Nival et al. (1972) have stressed, since these measurements were done at a temperature near the natural temperature 5–12°C and 28–30°C, respectively it is not very easy to compare them.

We must draw attention to the phenomenon of temperature acclimatization (Nicol, 1960). This may well be seen from Fig. 11-17. At 0°C the respiration of crustaceans should fall along the broken line when calculated on the basis of the effect of temperature on respiration, but in reality their respiration is much higher. This adaptation to the temperature, associated here with geographical distribution, may also be observed during the various seasons and the respiration at a given temperature, when other conditions are equal, may differ as a function of the temperature to which the animals are acclimatized in their particular environment.

Since there are variations in the rate of respiration as a function of weight, it is suggested that in order to facilitate comparison, the rate of respiration should be related to a unit of weight. This relative rate of respiration (R/W), as $R = aW^b$, may be expressed as follows:

$R/W = aW^b/W = aW^{b-1}$

Following Davies (1966) and taking $R' = R/DW$ and $b' = b - 1$ the equation becomes,

$R = a\,DW^{b'}$ or in logarithms $\log R = \log a + b' \log DW$

For Mediterranean plankton (Nival et al.) we have, therefore, the relation (R' in μl O_2/mg/h, DW in mg)

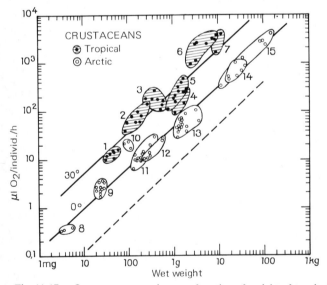

Fig. 11-17.—Oxygen consumption as a function of weight of tropical and arctic crustaceans at their normal habitat temperature; tropical forms were extrapolated down to 0°C; standard temperature curve should fall along the dotted line corresponding to a rate of 30–40 times lower than at 30°C; the metabolic rate of the arctic forms show their adaptation; the regression line corresponds to an exponent of 0.85: 1, isopod; 2, hermit crab; 3, shrimp; 4, fiddler crab (*Uca*); 5, terrestrial crab; 6, blue crab (*Calinectes*); 7, marine hermit crab; 8, fairy shrimp; 9, *Pseudalibrotus*; 10, *Lepidurus*; 11, *Gammarus locusta*; 12, *Gammarus loricatus*; 13, isopod; 14, hermit crab; 15, spider crab (after Scholander, Flagg, Walters & Irving, 1953).

$$\log R = 0.844\,(\pm 0.087) - 0.294\,(\pm 0.044)\log DW$$

Suppose that we now consider the same species at a constant temperature and with the same weight, as was almost realized by Conover & Corner (1968) using different copepods of the Atlantic coast of the United States. Figs 11-18 and 11-19 show their results for two of the most common

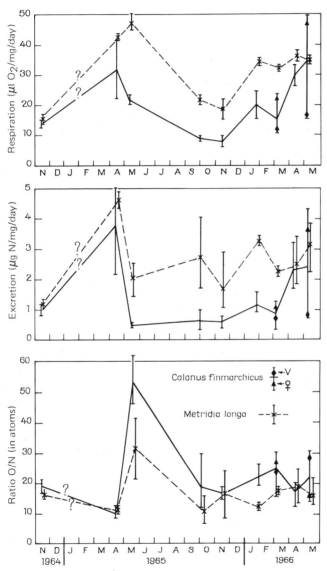

Fig. 11-18.—Seasonal changes in respiration rate and nitrogen excretion and in O:N ratio (by atoms) for the copepods *Calanus finmarchicus* and *Metridia longa* from the Gulf of Maine: points are usually the average of three or more individual observations and the vertical bars represent the range of variability (after Conover & Corner, 1968).

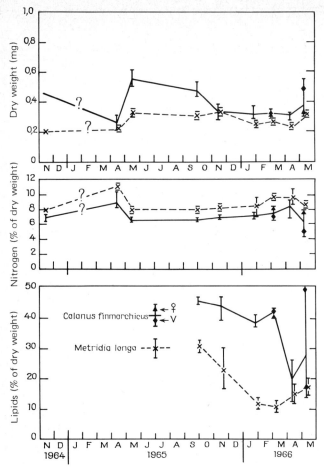

Fig. 11-19.—Seasonal changes in dry weight, nitrogen, and lipids for the copepods *Calanus finmarchicus* and *Metridia longa* from the Gulf of Maine: points are usually average of three or more individual observations and vertical bars represent the range of variability (after Conover & Corner, 1968).

copepods, *Calanus finmarchicus* is 30 $\mu l O_2$/mg/day for an animal of weight 250 μg. In May 1965 the dry weight increased to about 550 μg, with a coefficient b of 0.85, and therefore $b' = -0.15$ the respiration should be 27 $\mu l\ O_2$/mg/day but it is lower, namely, 20 μl/mg/day. According to Conover & Corner the reason for this seasonal variation in respiration intensity is the increase in nitrogen content, and hence protein which is produced in the spring. The facts were already evident in Figs 11-18 and 11-19 but the following multiple regression was found by these authors as a fit to all the data they obtained

$$\log R' = -0.03 \log W + 0.43 \log N - 0.54 \log F + 1.79.$$

when R' is in $\mu l\ O_2$/mg/day, W μg dry weight, N μg of total nitrogen, and F μg total lipids. R' is positively correlated with N. On the contrary an increase of lipids reduces the respiration. In conclusion, these seasonal variations are also related to the biochemical composition of the organism.

If these last three causes of variation, namely, temperature, weight, and season are excluded it is possible to estimate the variations which reflect specific differences; this may be shown by a comparison of *Calanus finmarchicus* and *Metridia longa* (Figs 11-18 and 11-19). The differences in weight cannot explain the higher respiratory rate of the last species. Conover & Corner relate this higher oxygen consumption of *Metridia longa* to its omnivorous rather than herbivorous diet, its lower capacity for lipid storage, and probably as a partial consequence of this the increase in density its continuous swimming activity. This last factor may be considered physiological, corresponding to differences not only between species but also between stages of the same species.

We may mention a problem which often arises in such work, namely, whether it is best or not to feed the experimental organisms before the respiration is measured. During an acclimation period prior to the experiment the animal consumes a more or less important part of its reserves. Comita (1968) compared the respiration of fed and starved individuals of different species of *Diaptomus* and found an increase in the respiratory rate with feeding in two of the species but no increase for the other species; this probably corresponds to physiological differences between them, as related to the state of their metabolic reserves. These reserves which are notably lipids may change rapidly, as shown in Table 11-5 in which data are collected from two sources, one from the calculated rates of respiration the other from direct measurements. After 24 h the decrease in lipids may reach 15–25% or even more than 50%.

We shall return now to the different components of the respiratory metabolism. By comparing *Metridia longa* and *Calanus finmarchicus* we see that it is important to be able to determine the value of Q_a, i.e., the

TABLE 11-5

Decrease in lipid reserves in starved planktonic crustaceans determined from the respiration (1–6) and directly (7–9); DW, dry weight (after Corner & Conway, 1968 and Lee, Hirota & Barnett, 1971).

Species	DW (mg)	Respiration (μlO_2/mg dry wt/day)	Lipid amount (μg/mg DW)	% of lipids used per day
Acartia clausi	0.0045	68.9	58	57.5
Metridia longa	0.25	35.0	160	10.5
Calanus finmarchicus	0.40	25.0	250	4.9
Calanus hyperboreus	2.3	12.5	350	1.7
Pareuchaeta norvegica	3.0	17.5	300	2.8
Euphausia superba	5.4	35.3	98	17.4
Gaussia princeps ♀	5–6	—	263	6.1
Megacalanus longicornis ♀	8	—	348	15.4
Gaetanus brevicornis ♀	4–5	—	333	24.8

consumption of energy due to activity and this involves considerable experimental problems. Vlyman (1970) tried to avoid this problem by calculation and estimating the energy required to overcome the resistance to propulsion; for the copepod *Labidocera trispinosa*, 1 to 2 mm in length and of 0.1–0.5 mg wet wt, he obtained an energetic output of 0.1–0.3% of the energy estimated for total respiration, and this is a very small fraction. If this estimation is correct, vertical migration should affect the energetic balance much less than had previously been thought. With regard to Q_d, the energy required for transport of the metabolites, we have no idea of its importance, although it certainly is concerned in the metabolism of animals which have just fed and should be taken into account.

Finally, we must mention the observations of Marshall & Orr (1955) in which it was shown that individuals of *Calanus* closed in a transparent vessel and placed at the surface of the water consumed much more oxygen during the day than those lowered to a depth of 5–10 m and of a control in darkness. Apart from this, down to a concentration less than 3 mlO_2/l (at 15°C) the respiratory rate was only slightly affected by the decrease in the oxygen content of the water, but below this there was a very large fall in oxygen consumption; even so, *Calanus* is found in water containing only 2 mlO_2/l.

11-6.—EXCRETION

The organic matter of zooplankton contains on an average 106 atoms of carbon for each 16 atoms of nitrogen and 1 atom of phosphorus and requires about 276 atoms of oxygen to be completely oxidized. Under these conditions one atom of nitrogen has to be excreted each time that 17 atoms of oxygen are consumed in respiration. This nitrogen is found in the form of ammonia i.e., the animals are largely ammoniotelic. A fraction of the nitrogen excreted is, however, found in the form of organic compounds; this has been the subject of some arguments since the organic nitrogen may disappear very quickly from the milieu as a result of bacterial action; it is now accepted that 13–30% of the nitrogen excreted by zooplankton is organic (Mayzaud, 1972). According to Jawed (1969) an average of 75% of this organic nitrogen consists of amino acids and urea.

The data available on nitrogen excretion are less than those dealing with measurements of respiration, but even so there is variability of the same order as found for respiratory rate. Excretion increases as a function of temperature, as the example in Fig. 11-20 shows. Its relation to weight is of the same kind as was found for respiratory rate. Conover & Corner (1968) gave the following equation for boreal crustaceans of the zooplankton,

$$\log E' = 0.51 \log W + 3.40$$

with a correlation coefficient $r = 0.67$ where E' represents the excretion in μgN/100 mg dry wt/day and W is the dry weight in μg.

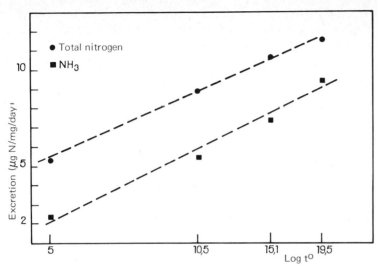

Fig. 11-20.—Variations of excretion of the amphipod *Phronima sedentaria* as a function of temperature reduced to a standard animal with a constant dry weight (36.2 mg) (after Mayzaud & Dallot, 1972).

Finally, Fig. 11-18 shows the seasonal variations and the important differences between *Calanus finmarchicus* and *Metridia longa*.

At the beginning of this discussion on excretion it was pointed out that the ratio of the oxygen consumed to nitrogen excreted should, because of the average biochemical composition of zooplankton, be about 17. The O:N ratio measured by different authors very often differs from the expected value. It may vary considerably as shown in Fig. 11-18, ranging between 10 and 50. It is extremely interesting to determine the ratio O:N in order to follow the metabolism of the organisms and their exact relation to the environment. If an organism uses only lipids or carbohydrate for its metabolism the quantity of nitrogen excreted will be zero, and the O:N ratio "infinity". A high O:N ratio indicates lipid or carbohydrate metabolism while consumption limited to protein gives an O:N ratio which may be as low as 4. From the curves in Fig. 11-18 we see that in April–May 1965 *Calanus finmarchicus* passed from a dominantly protein metabolism (O:N = 11) to a lipid-carbohydrate metabolism (O:N = 53) and this is probably due to the utilization of the phytoplankton bloom; *Metridia longa* shows the same phenomenon but less marked, with an O:N ratio usually lower than that of *Calanus finmarchicus* and which corresponds to its more omnivorous rather than herbivorous diet.

We may note that in the results of Conover & Corner (1968) since the relative rate of respiration is a function of weight by a power of -0.21 and excretion by a power of -0.51, the ratio O:N increases as a function of weight ($R'/E' = C^{te} \cdot W^{0.30}$). As the work on excretion of the zooplankton is extended, two energetic fractions must be taken into consideration, both already mentioned in the discussion of energetic budget, namely, the excretion of non-metabolizable nitrogen compounds (Q_n) and deamination of amino acids interfering in Q_d.

11-7.—GROWTH

The unexpended part of the assimilated energy increases the biomass of the animal and constitutes its growth

$$Q_g = Q_c - Q_w - Q_r$$

It is important in quantitative ecology, to estimate as a function of the absorbed food that part of the energy which is related to growth and to determine the efficiency of the growth. The gross efficiency of the growth, K_1 is, according to the terminology of Ivlev, the ratio between the growth and the food consumed,

$$K_1 = \frac{Q_g}{Q_c}$$

The net efficiency, K_2, according to Ivlev, corresponds to the ratio of the growth to the assimilated food effectively used by the organism ($Q_w \approx Q_f$):

$$K_2 = \frac{Q_g}{Q_c - Q_f} = \frac{Q_g}{AQ_c} = \frac{Q_g}{Q_g + Q_r}$$

In Table 11-6 we give a number of determinations of these efficiencies, K_1 and K_2, and the percentage of energy used in respiration (obtained by subtracting the net efficiency), as found using different bases of calculation-calories, dry weight, carbon, nitrogen, and phosphorus. According to these results the gross efficiency K_1 generally remains below 50% and the net efficiency does not exceed 60%. For comparison, the results of Hargrave & Green (1970) are also given; they studied the ingested food (Q_c) and the fraction used for respiration (Q_r) for copepods fed with natural phytoplankton. In order to calculate K_2, we must suppose a rate of assimilation (A) ranging between 40–80% and the values calculated for K_2 then range between 0 and 68%.

Our knowledge of the variations of K_1 and K_2 as a function of different ecological factors are still very fragmentary. In general the efficiency seems to decrease with age (and weight) as is shown in Table 11-6 where the data of Petipa for the copepodids I to V of *Calanus helgolandicus* show a decrease in K_1 and K_2. The results of Corner et al. and Butler et al. for *Calanus finmarchicus* also show that the efficiency is higher for the entire growth than for the growth of the latter stages. Yet the determination of K_1 by Mullin & Brooks for *Rhinocalanus nasutus* and *Calanus helgolandicus* do not agree with this low value (Table 11-8), nor do those of Hargrave & Geen for *Acartia tonsa* (Table 11-7).

A second factor, the effect of the quantity of food, is also not well

TABLE 11-6

Growth efficiencies, K_1 and K_2, and the fraction of energy degraded by respiration, Q_r (as a function of the assimilated AQ_c); CI, copepodid I, CII, copepodid II etc.; cal, calories, DW, dry weight (a, after Sushchenya, 1970; b, after Corner & Davies, 1971; c, after Lasker, 1966).

Species	Stage	Basis of calculation	$K_1(\%)$	$K_2(\%)$	$Q_r(\%)$	Author
Artemia salina (25°C)		—	9.0–18.5	23.6–27.4	72.6–76.4	Sushchenya (a)
Daphnia pulex (20°C)		—	3.9–13.2	55.4–58.7	41.3–44.6	Richman (a)
Calanus helgolandicus (10°C)		—	37.2–48.9	46.2–52.5	47.5–53.8	Corner (a)
Calanus helgolandicus (10°C)	Nauplius	Cal	34	37	63	Petipa (b)
Calanus helgolandicus (10°C)	CI	Cal	50	55	45	Petipa (b)
Calanus helgolandicus (10°C)	CII	Cal	39	43	57	Petipa (b)
Calanus helgolandicus (10°C)	CIII	Cal	28	31	79	Petipa (b)
Calanus helgolandicus (10°C)	CIV	Cal	21	23	77	Petipa (b)
Calanus helgolandicus (10°C)	CV	Cal	5	6	44	Petipa (b)
Calanus finmarchicus	Total growth	N	34	—	—	Corner et al. (b)
Calanus finmarchicus	Total growth	N	33.1	—	—	Butler et al. (b)
Calanus finmarchicus	Total growth	P	28.3	—	—	Butler et al. (b)
Calanus finmarchicus	CV and adults	N	26.8	43	57	Butler et al. (b)
Calanus finmarchicus	CV and adults	P	17.2	22.4	77.6	Butler et al. (b)
Calanus hyperboreus	CIV	Cal	5–18	23.0	77.0	Conover (b)
Calanus hyperboreus	CV	Cal	18–50	58.0	42.0	Conover (b)
Acartia clausi	Nauplius	Cal	14	17	83	Petipa (b)
Acartia clausi	CIII	Cal	23	29	71	Petipa (b)
Acartia clausi	CV	Cal	11	14	86	Petipa (b)
Penilia avirostris	Young stages	DW	40	54	46	Pavlova (b)
Euphausia pacifica	Total growth including eggs	C	28.2	18(1)	67	Lasker (c)
Sagitta elegans	Total growth	DW	11.0	—	—	Reeve (b)
Cyanea capillata Diameter 3.5 cm		DW	37.0	—	—	Fraser (b)

*Additional 15% should be added for the moults.

TABLE 11-7

Determination of Q_c, Q_r, and K_2 for different copepods feeding on natural phytoplankton at 15–20°C (Cape-Breton Island, Nova-Scotia, Canada) (after Hargrave & Geen, 1970).

	Dry weight µg	Q_c (cal/copepod/day)	Q_r (cal/copepod/day)	Q_r in % of Q_c	Q_r in % of Q_c for $A = 40\%$ and 80%	K_2 for $A = 40\%$ and 80%
Acartia tonsa						
nauplius	2.0	0.0058	0.0019	32.8	82–41	18–59
copepodid	5.0	0.0211	0.0087	41.2	100–51	0–49
adult	10.0	0.0448	0.0140	31.3	78–39	12–61
Pseudocalanus minutus and *Temora longicornis* copepodid	12–17	0.0427	0.0111	26.0	65–32	35–68
Oithona similis copepodid	3.0	0.0111	0.0034	30.6	77–38	23–62

TABLE 11-8

Gross growth efficiency, K_1 (in %), of two copepods *Rhincalanus masutus* and *Calanus helgolandicus* at two temperatures with different food sources and at different stages of development; NI, nauplius I; CI, copepodid 1, etc. (after Mullin & Brooks, 1970).

Conditions	NI to CI	CI to CIV	CIV to adult	NI to adult
Rhinocalanus				
15°C-*Ditylum*	39	32	48	45
15°C-*Thalassiosira*	21	32	40	37
10°C-*Ditylum*	39	31	34	34
15°C-*Thalassiosira*	22	55	25	30
Calanus				
15°C-*Thalassiosira*	18	26	37	34
10°C-*Thalassiosira*	21	72	30	35

established. Sushchenya (1970) applied the equation of Paloheimo & Dickie, established for fishes, namely,

$\log K_1 = -a - bR$

where R is the food ration. He concluded that K_1 decreases when the ration increases and on the contrary K_1 should be maximal with a small ration; however, as regards the last conclusion this equation is not satisfactory as was emphasized by Warren (1971).

Finally, the relation between the efficiency of growth and temperature is far from clear. It is commonly observed that the size reached by a species is an inverse function of the temperature at which it was developed, so that it is possible to consider that growth efficiency decreases as temperature increases; yet Mullin & Brooks (1970) observed that K_1 was not affected by temperature in cultures of *Rhinocalanus* and *Calanus* and this may be explained by accepting that the effect of temperature on ingestion and growth is equal, while on assimilation it is not, as observed by Conover (1966b), affected.

In adult females the metabolites not degraded may be invested not only in body growth but also in production of eggs and it is possible to calculate the gross and net efficiencies of egg production in a similar way to that of growth efficiency (Table 11-9). These values are, however, certainly underestimated relative to those reached under the natural conditions. Indeed, the calculations on *Calanus finmarchicus* agree with the results of Marshall & Orr in which an average of about 7 eggs were produced per day (250 eggs in 35 days). Paffenhöfer (1970) obtained for the same species a daily production of 44 eggs (1991 eggs in 45 days).

Besides the metabolic losses due to respiration the crustacea have a supplementary loss at their moult. The moult is sometimes not a negligible fraction of the energetic balance, and is a fraction which will contribute to the increase of particulate organic matter in the sea (tripton). Lasker (1966),

TABLE 11-9

Determination of the efficiencies of egg production as a function of the ingested (K_1) or assimilated (K_2) food: cal, calories (a, after Corner & Davies, 1971 and b, after Lasker, 1966).

Species	Basis of calculation	K_1	K_2	Author
Acartia clausi	Cal	1.8	2.0	Petipa (a)
Calanus helgolandicus	Cal	2	2.5	Petipa (a)
Calanus finmarchicus	N	14	22.5	Corner et al. (a)
Euphausia pacifica	C	7.5	9	Lasker (b)

estimated the net efficiency of moult (K_2) as 15% of the assimilated food of *Euphausia pacifica*, and the net efficiency of the growth (including the production of eggs) as 18%. Mullin & Brooks (1967, 1970) estimated the exuviae produced by one *Rhinocalanus nasutus* from egg to adult as 10 μgC (30 μg of dry tripton) while its consumption during this time is in the order of 200 μgC; this corresponds to a gross efficiency of (K_1) of about 5%.

If we return to the general equation of the energy budget

$$Q_c - Q_f = Q_r + Q_g$$

we see that the species for which it has been established are very few, and the unknown causes of any variation makes its presentation difficult. Two solutions are possible; either to make the general balance from egg to adult or to be limited to a short period and definite stages. A good example of the first is given by Lasker (1966) using his laboratory estimations and the biological data of Ponomareva on *Euphausia pacifica* from the North Pacific; expressed in carbon,

$$C_c - C_f = C_r + C_g + C_o + C_m$$

corresponding to:

$$119 - 19 = 67 + 9 + 9 + 15$$

or

$$100 - 16 = 56.1 + 7.6 + 7.6 + 12.6$$

C_o and C_m represent, respectively, the carbon in the eggs and in the exuviae.

Butler, Corner & Marshall (1970), limiting themselves to the spring growth of copepodid V, male, and female *Calanus finmarchicus* from the Clyde, studied the daily balance of nitrogen and phosphorus, where,

$$N_c - N_f = N_r + N_g$$

corresponding to:

$$100 - 37.5 = 35.7 + 26.8$$

$$P_c - P_f = P_r + P_g$$

corresponding to:

$100 - 23.0 = 59.8 + 17.2$

where N_r and P_r represent the nitrogen and phosphorus excreted in solution and N_c corresponds to 13.7% of the body nitrogen, and P_c to 17.6% of the total body phosphorus.

We must not forget that besides the main substances considered in the energy budget, certain special substances, often in very small quantities may be essential for growth. Provasoli & d'Agostino (1962) showed that in the brine shrimp *Artemia*, eight vitamins, namely, thiamine, pyridoxamine, riboflavin, nicotinic acid, panthothenic acid, biotine, putrescine, and folic acid are necessary to produce sexually mature adults. Furthermore, carnitine, inositol and choline probably take part in egg production, since in their absence the number of mature females decrease. Such needs also probably exist in the organisms of the plankton, at least in the crustaceans, but we do not yet know if the essential need of vitamins are satisfied by the algae of the phytoplankton or by the bacteria of the gut (Corner & Cowey, 1968).

CHAPTER 12

THE SECONDARY PRODUCTION OF PLANKTON

12-1.—LEVELS OF PRODUCTION

The most precise definition of secondary production is that production of organic matter by organisms which get their energy from primary production i.e., by herbivores. It is not a production of new organic matter but reorganization of it.

It is common, and natural, to speak of tertiary production, namely, the production by carnivores living on herbivores, and also possible to define a fourth level of production, that of carnivores of which their prey are predators of herbivores and so on. The herbivores are, however, far from being exclusively herbivorous and the secondary carnivores may include amongst their prey herbivores; the distinction between these various categories of production is often very uncertain. One solution is to term the total production proceeding from primary production—secondary production and to characterize within this different levels of production e.g., level 2 for the herbivores, level 3 for the carnivores, level 4 for the predators of the carnivores. In order to avoid confusion the first level is reserved for the primary production (under these conditions level 0 represents the precursor elements of the organic matter and the different minerals). If these conventions are accepted it is possible to characterize more easily the terminology without confusion: for instance, level 2-3 of secondary production will be copepods with a mixed regime of phytoplankton and of small elements of the zooplankton. The secondary production of the Euphausiacea will be dominated by level 3 (consumption of herbivorous copepods) but level 2 will be sometimes strongly represented (by consumption of phytoplankton) and also level 4 (by the consumption of carnivorous copepods and of chaetognaths). The grouping of all the higher production levels has an additional justification; the techniques for the study of higher levels of production vary slightly different amongst themselves but are very different from the methods currently used in the study of primary production which are based on the photosynthetic process.

The data on secondary planktonic production are much fewer than those on the primary production. We first shall review the principle methods so far used.

12-2.—COHORT (AGE-CLASS) METHODS: ALLEN CURVES

A cohort (age-class) is considered to be any group of individuals of the same species "hatched" during a limited period of time. Characterized by a closely similar size, these individuals may be recognized for a more or less long time as a separate group within the total population of the species. If reproduction is continuous the cohorts will be rapidly mixed since the range of their sizes overlap. If, on the contrary, reproduction is at intervals, the cohorts are clearly demarcated and may be followed over more or less long periods and so allow an estimate of production.

We consider a cohort which at the time t_1 consists of n_1 individuals with a mean weight w_1 and, at time t_2, of n_2 individuals with mean weight w_2 (Fig. 12-1A). Its biomass at time t_1 will be $n_1 w_1$ and at time $t_2, n_2 w_2$. Between t_1 and t_2 the individuals gained in weight but some were lost, either consumed by predators or died naturally and then decomposed by bacteria. We shall assume, as a first approximation, that during the relatively short interval of t_1 to t_2, the mortality and the gain in weight are uniformly distributed in time, and that the curve $O_1 O_2$ is a straight line. The production during the period t_1 to t_2 corresponds to the gain in weight of the remaining n_2, individuals together with the gain of weight of the individuals lost at before the time they disappeared: this is,

$$P = n_2(w_2 - w_1) + (n_1 - n_2)\left(\frac{w_2 - w_1}{2}\right)$$

which gives,

$$P = (n_2 + n_1)\left(\frac{w_2 - w_1}{2}\right)$$

This production is shown in Fig. 12-1A by the vertical hatching. The horizontal hatching corresponds to the biomass consumed between t_1 and t_2

$$B_c = (n_1 - n_2)w_1 + (n_1 - n_2)\left(\frac{w_2 - w_1}{2}\right)$$

which gives

$$B_c = (n_1 - n_2)\left(\frac{w_2 + w_1}{2}\right)$$

Numerical examples

1. The scheme in Fig. 12-1A is based on $n_1 = 7$, $n_2 = 3$, $w_1 = 4$ and $w_2 = 12$; under these conditions $P = 40$ and $B_c = 32$.

2. Boysen-Jensen (1919) was the first to introduce this method of calculation for a benthic population of *Solen*: in 1912 he recorded 226 individuals/m² with a mean weight of 0.15 g i.e., a biomass of 33.9 g/m². In 1913 at the same place he found 41 individuals/m² with a mean average

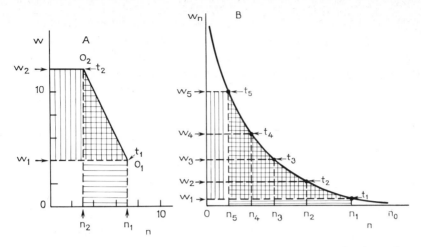

Fig. 12-1.—A, scheme illustrating the development of an age group in time interval t_1 to t_r: B, scheme representing the development of an age group counted on five occasions t_1, t_2, t_3, t_4, t_5: horizontal hatching gives the biomass consumed and vertical hatching the production (after Mann, 1969).

weight of 0.32 g, a biomass of 13.1 g/m². The calculated biomass consumed (supposing that mortality and growth are regularly progressive) is 43.5 g/m² and the production 22.7 g/m².

Now we consider an age-class in which we have more complete observations, namely, the numbers and weights at the times t_1, t_2, t_3, t_4, t_5 (Fig. 12-1B). We may generalize the preceding conclusions: from t_1 to t_5 the production corresponds to the vertical hatching, the consumed biomass to the horizontal hatching and both are given by the integral:

$$P = \int_{w_1}^{w_5} n\,dw \qquad B_c = \int_{n_1}^{n_5} w\,dn$$

If we know the entire life-history of the age-class, from n_0 to 0, the numbers, and 0 to w_n (average weights) the two hatched areas should coincide and the total production should be equal to the biomass consumed. So, knowing the variations of the average weight and of the number of individuals of the given age-class, it is possible to estimate its production. If the amount of data is sufficient it is possible to set up the curve of variations empirically and then use it in order to measure graphically the production.

According to Mann (1969) we should call these kind of curves Allen Curves, since Allen was the first to use this method to estimate the production of trout in a river in New Zealand (1951). The use of Allen Curves requires the existence of a recognizable cohort. For fish whose age can usually be identified the cohort may be followed over a long period in the form of age-classes. For benthic organisms whose movements in space are limited, the recognition of age-classes is often easy, particularly when reproduction is annual (monocyclic species). For zooplankton it is

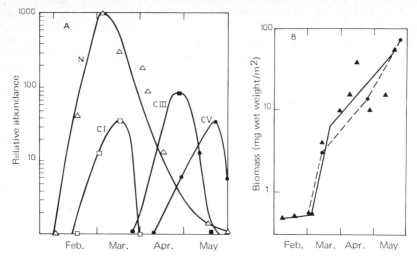

Fig. 12-2.—A, relative number of nauplii (N) and copepodid stages (CI, CIII, CV) of the copepod *Calanus plumchrus*, from February to May 1967, in the Fraser River plume (Strait of Georgia, British Columbia): B, biomass of *Calanus plumchrus* during the same period, from 0 yo 20 m depth: triangles, according to capture; circles, predicted by using the data of daily relative growth in Table 12-1 (after Parsons et al., 1969).

more difficult to obtain the conditions required to use the age-class method, mainly because of the movements of the water related to the heterogenous distribution and to the frequent existence of many generations each year (polycyclic species).

Parsons, LeBrasseur, Fulton & Kennedy (1969) have given an example for the copepod *Calanus plumchrus* in the plume of Fraser River (Strait of Georgia, British Columbia). As shown in Fig. 12-2A they succeeded in following an age-class from February to May and this enabled them to establish the stages of growth (Table 12-1) under the natural conditions. But this is a special case of the method; the number of individuals in the population remained stable and so the biomass consumed was zero and the production, therefore, reduced to $n(w_n - w_1)$. Indeed, except for the nauplii,

TABLE 12-1

In situ growth rate of *Calanus plumchrus* in the Strait of Georgia, from February to March, 1967: N, nauplius; C, copepodid (after Parsons et al., 1969).

	NI-VI	CI	CIII	CV
Date of maximal abundance	2 III	16 III	27 IV	18 V
Time interval (days)	14	44	21	
Wet weight (mg)	0.025	0.15	0.60	3.5
Relative growth rate (%)*	14.0	3.5	8.8	

*For the determination see Appendix, Chapter 14, p. 316.

the numbers of copepodids I, III, and IV remained of the same order, about 21 000 individuals/m² (at a depth of 20 m) and the production of *Calanus plumchrus* for three months, March to May was found to be about 70 g/m² wet weight (21 000 × (3.5 − 0.15)mg); which is of the order of 4 gC/m². This was possible only because of the favourable conditions, namely, abundance of phytoplankton, except at the end of April, and very low predation. In Fig.

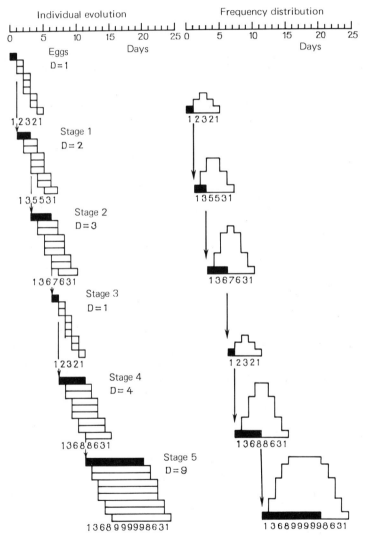

Fig. 12-3.—Diagram showing the influence of the duration of a stage on the frequency curves for a simplified age class hatched from 9 eggs: first egg and the development of the individual hatched from it is indicated in black and by an arrow: left, the development of each individual; right, frequency curve for each stage as a function of time: note the difference between the histogram of stage 3 which last 1 day and that of stage 5 which last 9 days: D, duration in days (after Edmondson in Edmondson & Winberg, 1971).

12-2B we see a good coincidence between the observed biomass and that calculated from the daily rates of relative growth, determined in Table 12-1.

It should be mentioned that in the crustaceans the use of stages and curves of their frequencies require some precautions in their interpretation. Their shape may be very different according to the duration of the stage as shown in Fig. 12-3. This may partly interfere in the relative importance of the curve for the nauplii of *Calanus plumchrus* (Fig. 12-2).

12-3.—CUMULATIVE GROWTH METHOD

The principle of this method is simple; any organism has a growth curve, relating its gain of weight to its age, and so it is possible to estimate its daily production which is its daily gain of weight, or growth, Δw. With a number, n_x of individuals of the same age the production will be: $n_x \Delta w$. The realization of such a method is illustrated in Fig. 12-4 for an imaginary population of a given species. In practice the calculation is much more complicated.

First the curve of growth, $w = f(t)$, must be obtained, and this requires experimental conditions as close as possible to the natural ones. Since temperature influences growth, a number of such curves must be established to cover the range of temperature over which a given species lives, and interpolation may then be made to the functional temperature of the population; but vertical migration may introduce variations in the ambient temperature and these are difficult to estimate and to correct the curves according to them.

A simplified method is given by several Russian authors, who use Krogh curves in order to establish, from an initial given temperature, curves for other temperatures (for instance, Fig. 12-5A, by Greze & Baldina for *Acartia clausi*). The data of Krogh (Appendix, Chapter 14) are applied to the time of development of eggs when feeding does not interfere; any extrapolation for growth should be done with care.

The theoretical calculations require a knowledge of the age-composition of the population and in practice this is not realized and it is necessary to approximate and this demands a good knowledge of the biology of the species considered. In crustaceans the determination of the different stages allow the age to be estimated when their duration is known; but this may vary as a function of the abundance of food. When other possibilities are not available a determination of size is used; since size is related to weight it is possible to use the curve $w = f(t)$ to relate to a given weight a given amount of growth; this demands some precautions and notably to be sure that the feeding is satisfactory.

Winberg et al. (in Edmondson & Winberg, 1971) gave an example of a calculation for the cladocean *Bosmina longirostris* from Lake Krivoe (North of Carelia) based on the data of Ivanova. After establishing the curve $w = f(t)$ by experiments in situ the population was divided into three

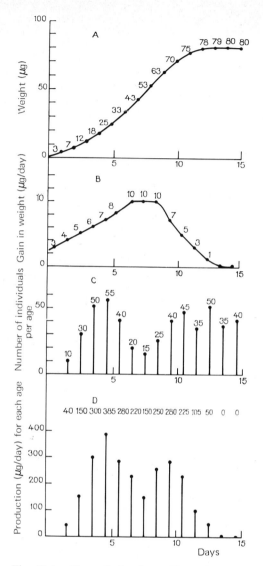

Fig. 12-4.—Theoretical scheme illustrating the cumulative growth method. The species presents a growth curve represented in A, the daily gain as a function of the age varies according to curve B. The stock of individuals of different ages is distributed in a given day according to the histogram C. The histogram D of weight gain for the given day is obtained from B and C. Summing the weight gain for the different age groups gives a production of 2.435 μg for the given day. Note that although the number of 14 and 15 day-old individuals is still high, they do not produce any more.

size-groups: I, body length less than 0.3 mm; II, body length from 0.3 to 0.4 mm; III, body length from 0.4 to 0.6 mm. The wet weight is obtained from the relation $w = 0.180 l^{0.288}$ where l is the length. If T_I, T_{II} and T_{III} are the times of growth of the three size-groups, Δw_I, Δw_{II}, Δw_{III} the corresponding

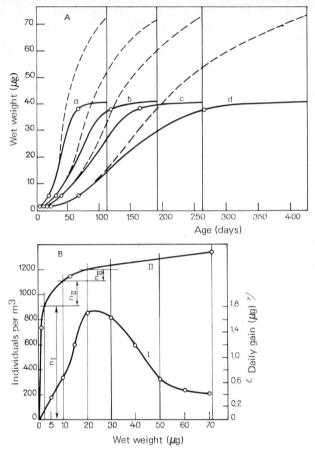

Fig. 12-5.—Curves used for the determination of the production using the cumulate growth method by the copepod *Acartia clausi* in the Black Sea. A, growth curves $w = f(t)$ at a range of mean temperature appropriate to the different seasons: a, summer, b, autumn; c, spring; d, winter: broken lines include egg production. B, curve I represents the daily weight gain (in summer) as a function of weight deduced from A; curve II from data in Table 12-3, the cumulative numbers plotted against weight: the sections of this curve are according to the weight classes with number n, n_{II}, n_{III} etc., for these classes (after Greze & Baldina, 1964 in Mann, 1969).

gain of weight and n_I, n_{II}, n_{III}, the number of individuals in the groups, then the production is given by,

$$P = n_I \frac{\Delta w_I}{T_I} + n_{II} \frac{\Delta w_{II}}{T_{II}} + n_{III} \frac{\Delta w_{III}}{T_{III}}$$

Table 12-2 gives the daily production for 18th July 1968, and is 23.58 mg/m² for a biomass of 164 mg/m² and the ratio of daily production to the biomass (daily P/B) is 0.144.

Greze & Baldina (1964) used the method of cumulative growth of a population for the copepod *Acartia clausi* from the region of Sebastopol in

TABLE 12-2

Calculation of the production (in wet weight) of the cladocerean *Bosmina longicornis* in Krivoe Lake (North Carelia) for the 18th July, 1968 from the cumulative growth method (after data of Ivanova in Winberg et al., in Edmondson & Winberg, 1971).

Size group (mm)	Number (m²)	Initial weight (mg)	Final weight (mg)	Difference (mg)	Time interval (days)	Production (mg/m²/day)
I < 0.3	5670	0.0030	0.0056	0.0026	2	7.37
II 0.3–0.4	2870	0.0056	0.0129	0.0073	4	5.18
III 0.4–0.6	2900	0.0129	0.0414	0.0285	7.5	11.03

Total biomass 164 mg/m² Daily P/B: 0.144 Total production 23.58

the Black Sea. Apart from the simple growth curve they obtained the egg laying and egg production curves at different seasons (Fig. 12-5A); from this curve the daily gain curve (Fig. 12-5B) is deduced (curve I). Since the number of nauplii, copepodids and adults are determinate (Table 12-3) curve II is obtained by accumulating these numbers and by using the mean weight of these categories. This curve allows one to obtain the number of individuals found in 6 arbitrarily chosen weight classes, namely; I, 0–2.5 μg; II, 2.5–10 μg; III, 10–20 μg etc. The production is calculated from the mean daily gain of

TABLE 12-3

Composition of the population of *Acartia clausi* in the Black Sea during different seasons (after Greze & Baldina, 1964 in Mann, 1969).

Stage	Number of individuals/m³			
	Summer	Autumn	Winter	Spring
Nauplius	730	141	571	916
Copepodids	406	36	130	316
Adults	173	20	26	128

weight of each class (Table 12-4) or;

$$P = n_I \Delta w_I + n_{II} \Delta w_{II} + n_{III} \Delta w_{III} + n_{IV} \Delta w_{IV} + n_V \Delta w_V$$

By adding the results of the four seasons Greze & Baldina estimated the annual production of *Acartia clausi* to be 66.8 mg wet weight/m³ or about 3.5 mgC/m³. Since the layer considered was at a depth of 40 m, the production will be about 140 mgC/m², and the mean daily production 0.38 mgC/m² for a mean biomass of 10.6 mgC/m²; the ratio, daily production/biomass is 0.035. Using the data of Digby (1950) for *Acartia clausi* from the Channel and assuming that the biological constants in the Black Sea are applicable, Greze & Baldina make a similar estimate which enables them to compare the two zones (Table 12-5).

TABLE 12-4

Calculation of the production (in wet weight) of *Acartia clausi* in summer, from the data in Fig. 12-5 and Table 12-3 (after Greze & Bandina, 1964 in Mann, 1969).

Weight class (μg)	Number/m^3	Daily gain/individual (μg)	Daily production
0–2.5	850	0.09	76.5
2.5–10	250	0.45	108.0
19–20	87	1.20	104.4
20–30	35	1.68	58.8
30–50	47	1.20	56.4
50–70	40	0.47	18.8
			422.9

TABLE 12-5

Comparison of the production (in wet weight) of the copepod *Acartia clausi* from the Black Sea and from the English Channel (after Greze & Baldina, 1967 in Mann, 1969).

	Black Sea	English Channel
Productive season length (days)	365	260
Mean temperature (°C)	14.9	13.0
Mean biomass (mg/m^3)	5.1	12.0
Annual production (mg/m^3)	66.8	104.8
Annual P/B ratio	13.0	8.7

TABLE 12-6

Example of the calculation of production for the copepod *Centropages typicus* from the area of Marseille: CI, CII, CIII, CIV, copepodids; I, II, III, IV, dry weight (after Gaudy, 1970).

Stage	Number	Mean weight of individual $w(10^{-4}$ mg)	Biomass nw (mg)	Intermoult period Δt (days)	Growth rate $\Delta w/\Delta t$ (10^{-4} mg/day)	Production $n \cdot \Delta w/\Delta t$ (mg/day)
CI	20930	30	62.8			1.80
				3.5	0.86	
CII	14900	33	49.2			3.98
				4.5	2.67	
CIII	2395	45	10.8			0.69
				7	2.86	
CIV	4651	65	30.2			—

Gaudy (1970) estimated the secondary production of the two copepods *Centropages typicus* and *Acartia clausi* in Marseille, using the cumulative growth method. The examination of successive samples (3 to 4 in a month) enabled him to establish five successive generations for *Centropages typicus*; the collections were made for five periods corresponding more or less to the periods of development of the five generations. The duration of development of the different stages was estimated for each generation, and in each sample the individuals of the different stages were counted. The results are given in Table 12-6. The total biomass and the total production are obtained by summing the biomasses and the production of the different stages respectively. In the estimation, the numbers corresponding to the entire samples covering one generation are then divided by the number of collections and so the mean daily production is estimated.

12-4.—TURNOVER TIME METHOD

When a population is in a steady state i.e., when its biomass and the number of individuals fluctuate very little over a period which is much longer than the duration of the life cycle, then the production may be estimated by the turnover time method.

In order to explain this method Fig. 12-6 gives a scheme for a population in a steady state, characterized by a constant number of individuals, N, equal to 4 and the life span of an adult is 4 months; each individual which disappears is immediately replaced. In one year, from January to January, the population was 3 times completely renewed: the annual rate of replacement is 3 and the time of replacement expressed in years is 1/3. The annual production is then three times the constant number of individuals i.e., 12; if w is the mean weight of an individual the constant biomass is then $12w$. The

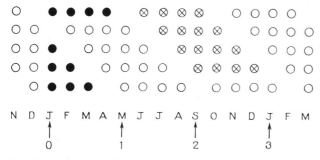

Fig. 12-6.—A scheme illustrating a population at a stable state of number equal to 4, the life of an individual is 4 months long and all the individuals that disappear are immediately replaced: individuals indicated by black in January, disappear in May; in 12 months the population is 3 times renewed (arrows 1, 2, 3): annual production P is equal to N, the constant number, multiplied by the number of the renewals, i.e., $3: P = N \cdot 3 = 12$: renewal time of 4 months expressed in years is 1/3 and we have $P = N/(1/3) = N \cdot 3 = 12$: monthly production is $P = N/4 = 1$.

following equation gives the annual production per year,

$$P = N_w/\bar{T}_N$$

where $\bar{T}_N$ is the turnover time of the population with a stable number of individuals, N. When a steady state is reached the mortality at each moment is compensated for by recruitment and the rate of mortality is equal to the rate of recruitment and the total recruitment during the turnover time $\bar{T}_N = N$. The daily recruitment is, therefore, $R = N/\bar{T}_N$ and T_N may be obtained from the estimation of the daily recruitment, R.

A method for the estimation of the recruitment is to determine the number of eggs, N_e, and to divide this by the mean time of development of an egg, T_e. When T_e is expressed in days we get the daily recruitment; $R = N_e/T_e$. The same method may be used for the estimation of recruitment of any stage; one only needs to know the number of individuals of the preceding stage and the duration of this stage; the recruitment of the stage, i, is given by:

$$R = N_{i-1}/T_{i-1}$$

The turnover time method, carried out for the recruitment was used for *Daphnia* by Stross, Neers & Hasler (1961) even in conditions where a steady state was distinctly far from being realized.

It is possible to determine directly the turnover time, T_B, for a given biomass, in this case the equation becomes,

$$P = B/\bar{T}_B$$

It should be noted that $\bar{T}_N$ and $\bar{T}_B$ need not always be equal: this will happen when the individuals disappearing have the same mean weight as that of the whole population. Furthermore, a numerical steady state may not be accompanied by steady state of the biomass.

Returning to the preceding equation we see that $P/B = 1/\bar{T}_B$. The ratio of the production to the biomass is the inverse of the turnover time. If we can obtain the ratio P/B by other methods it will be possible to calculate $\bar{T}_B$. Instead of estimating recruitment, it is also possible to determine mortality, which in a steady state is equal to recruitment. The daily mortality rate, M is equal to $(1-e^{-m})$ (see Appendix, Chapter 14); m is the coefficient of instant mortality in the equation $N_t = N_0 e^{-mt}$, which describes the decrease due to mortality of an initial population N_0 to N_t. Heinle (1966) determined this coefficient for a population of *Acartia tonsa* from the Patuxent estuary (Atlantic coast of the U.S.A.) and applied it to the study of the production by the turnover time method: for a biomass of 154 mgC/m^2 he obtained a production of 77 mgC/m^2/day for a water column 3 m during 2 summer months. The value of P/B obtained for the daily production was about 0.50 which is very high.

12-5.—PHYSIOLOGICAL METHOD

The method was introduced by Winberg (1965, 1966) and applied to planktonic organisms by Shushkina (1968). Referring to the energetic budget, studied earlier, the energy invested by an organism in its growth is given by $Q_g = AQ_c - Q_r$. The growth represents the production of the organism. The net efficiency of growth, K_2, may be written as follows:

$$K_2 = \frac{Q_g}{Q_g + Q_r} \quad \text{or:} \quad Q_g = Q_r \frac{K_2}{1 - K_2}$$

So it should be possible to calculate the production of one individual, knowing K_2 and Q_r: Q_r the energy degraded by respiration is directly related to the consumption of oxygen, by the oxycaloric coefficient K_c, a relation that may be determined. Since the respiratory rate is itself a function of the weight of an organism according to the relation:

$$R = aW^b$$

we may write:

$$Q_r = K_c aW^b$$

Finally, if P_e is the elementary production, Q_g, of an organism of weight W, it is equal to:

$$P_e = K_c aW^b \cdot \frac{K_2}{1 - K_2} = pW^b$$

where p is a coefficient and a function of K_c, a, and K_2; it is possible to calculate the total production by knowing the number of individuals of each weight W. In practice, the growth rate, C_w, is used instead of P_e (C_w being the growth per unit time and unit weight) or:

$$C_w = \frac{dW}{dt} \cdot \frac{1}{W} = pW^{b-1}$$

and the general equation:

$$P = \Sigma (C_w)_i W_i n_i$$

where n_i is the number of individuals in the ith group, w_i their mean weight and $(C_w)_i$ their growth rate estimated by physiological measurements.

Shushkina (1968) compared the results obtained by this method with those obtained by the accumulated growth method for the fresh water copepod *Cyclops*, and obtained very similar values (Table 12-7). He also applied this method to the copepod *Haloptilus longicornis* from the Fiji Islands but because of the lack of knowledge of the biology of this species the results obtained are speculative. This method may be used for estimation of little known populations and also as a method of cross-checking in order to determine the position of a population within the ecosystem.

TABLE 12-7

Comparison of P/B ratio of the monthly production to the mean biomass for a copepod of the genus *Cyclops*, in Batorine Lake, a eutrophic lake in Bielorussia, calculated by the cumulative growth method (CGM) and by the physiological method (PM), 1962: relation used for the oxygen consumption is $R = 0.165 W^{0.8}$ at 20°C; the coefficient a (0.165) was corrected for temperature, two values of K_2 were tried (after Shushkina, 1968).

Month	T_c^0	Corrected a	Monthly P/B		
			CGM	PM($K_2 = 0.25$)	PM($K_2 = 0.30$)
May–June	16	0.115	1.8	2.0	2.7
June–July	18	0.137	2.9	2.5	3.2
July–August	19	0.151	2.6	2.8	3.6
August–Sept	15	0.105	1.2	1.1	2.2

12-6.—USE OF MODELS

A different method for the estimation of secondary production is to construct models which represent the different parameters of the pelagic ecosystem; when these models reflect satisfactorily the development of the ecosystem, it is possible to deduce from them values of the secondary production. Riley (1946, 1947) was the pioneer of this kind of research in his work on George's Bank at Cape Cod (Atlantic coast of the USA). He expressed the biomass of the herbivores, which compose the main part of the zooplankton, by the equation:

$$H_t = H_0 e^{(a-r-c-d)t}$$

H_t is the biomass of the herbivore at time t, H_0 the biomass at the time $t = 0$ and a, r, c, d are the instant coefficients of assimilation of the phytoplankton by the herbivores, of degradation by respiration, of predation by carnivores, and of death of the herbivores, from other causes respectively. This equation only expresses the fact that changes in the biomass of the herbivores are equal to the difference between the increase of the biomass by addition of assimilated organic matter and that due to respiration and death (by predation or other ways). Fig. 12-7 shows the development of these different activities during the year. Fig. 12-8A gives the results and Fig. 12-8B gives a comparison of the observed and calculated zooplankton biomass. As a first approximation there is a satisfactory agreement between the two. Some of Riley's premises are no longer completely accepted, notably the continuous "automatic" filtration of herbivorous copepods and the superfluous nutrition which cuts the curve of assimilation from the end of March to the end of May. In general, however, this curve seems closely to represent the facts, since the growth of the zooplankton so deduced fits the data quite well. Knowing the biomass and having a satisfactory estimation of growth, predation, and

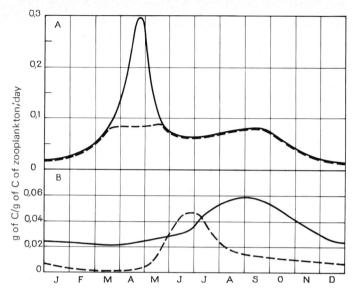

Fig. 12-7.—Zooplankton population on Georges Bank; changes during the year: A, of the grazing rate (continuous line) and assimilation rate (broken line): B, rate of respiratory degradation (continuous line) and loss by predation (broken line) (after Riley, 1946).

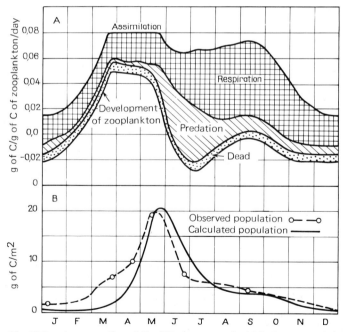

Fig. 12-8.—A, summation of the different rates of which the seasonal changes are given in Fig. 12-7, and the variation of growth rate of a population of zooplankton determined by plotting these rates: B, comparison between the variation of the standing biomass observed in Georges Bank (broken line) and the calculated biomass (continuous line) obtained by integration of the variations in the growth of a zooplankton population indicated in A (after Riley, 1976).

natural death it is possible to calculate production; this was realized by Mullin (1969). For a mean biomass of 6400 mgC/m² the daily mean production on George's Bank was 200 mgC/m²/day, i.e., the daily ratio P/B is 0.03 or 3%.

Cushing (1959) developed a model for studying the production of *Calanus* in the North Sea. By the application of computers and the improvement in our knowledge of the energy transfer within the food web, the use of models should lead to further progress. Fig. 12-9 gives an example of an energy budget elaborated in a form which enables secondary production to be easily calculated. This method may, however, only be properly exploited when our knowledge of the different energetic transfers (feeding, assimilation,

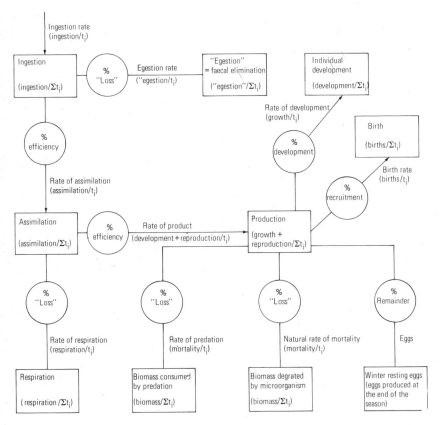

Fig. 12-9.—Energy budget represented by compartments established on the fresh water cladocerem *Leptodora* studied by Cummins et al. (1969): rates measured for each time interval t_i are summed for a long period (Σt_i), year or season of the development of a species and the totals obtained are presented in the rectangular compartments. The circles represent the percentage which can pass from a compartment to a closer one (loss is given in inverted commas since it represents only a relative loss for the species considered). The nitrogen excretion is supposed to be only ammonia and the corresponding energy spent is included in the respiration (after Cummins in Edmondson & Winberg, 1971).

respiration etc.) is certain. The McAllister model given in Fig. 12-10 is very instructive. From measurements of the primary production and of the algal biomass at the point P, in the North-east Pacific, the instant coefficients of phytoplankton growth were determined throughout the year. Knowing its theoretical development as a function of its growth and its real development, the mortality coefficient due to grazing is estimated and this enables the ration available for the zooplankton to be calculated. The ration enables a calculation of secondary production to be made after correcting for assimilation and respiration. Depending upon the way in which the grazing is realized and on the respiratory rate adopted, the result obtained at the end of the year for the cumulative secondary production varies extremely.

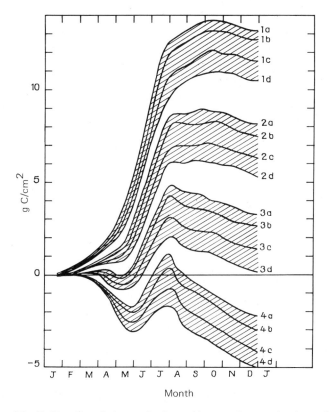

Fig. 12-10.—Cumulative totals of monthly secondary production estimated from field data from Ocean Station P (North-East Pacific) on the assumption of different grazing schemes, and zooplankton respiration rates: curves 1, 2, 3 and 4 show the production estimated on the assumption of zooplankton respiration rates of 4, 6, 8, and 10% of the body weight per day, respectively: curves marked a and b estimated assuming exclusively nocturnal grazing each with two different rates of decline of the feeding rate (see Fig. 11-10): curve c assumes nocturnal grazing at a constant rate (equal to the mean of the variable rate): curve d assumes continuous grazing (after McAllister, 1970).

12-7.—RESULTS OBTAINED BY THESE METHODS

The data on planktonic secondary production are much fewer than on primary production since the methods described involve a considerable amount of work. Table 12-8 brings together the principal results which are now available. Apart from a few values for total zooplankton and those for *Euphausia pacifica*, all the data concern copepods.

It is difficult to come to any general conclusions from these data, but we may assume that secondary production of zooplankton from temperate seas is of the order of 50 to 100 mgC/m^2/day i.e., about 10–30% of the net primary production is transferred to secondary production. By an examination of the daily P/B ratio we see that except for the very high values given by Heinle for *Acartia tonsa* from the Chesapeake estuary, the values lie between 0.002 and 0.23. The existence of a relatively narrow range of variation of P/B ratios has been confirmed by Zaika (1968) for three copepods from the Sea of Azov as may be seen from Table 12-9. In *Acartia clausi*, for example, the mean biomass and the annual production vary considerably, but the annual P/B ratio fluctuates very little about an average of 15.8, i.e., a daily P/B ratio of 0.043. According to Zaika (1968), the small variation of the P/B ratio should bear an inverse relation to the mean individual weight of the population ($\log P/B = a - b \log \bar{W}$) and a direct relation to the temperature ($\log P/B = p + q \log t^0$).

12-8.—TOTAL COMPARISON OF THE PRODUCTION

We may consider the metabolites produced by photosynthesis in primary production as divided into three fractions; the non-assimilated fraction either not ingested or passing through the alimentary canal of the herbivores without being absorbed, the fraction oxidized in the respiration of the herbivores, and the fraction involved in secondary production, in growth, or in egg production. Using the same units for primary production and the respiratory consumption of herbivorous planktonic organisms—even when it is difficult to estimate the production because of a lack of knowledge of the rate of assimilation—we can at least indicate its upper limits. Such a procedure will only have meaning if it integrates the values obtained over a long enough time, and over a large enough area. Several cases may be considered

1. If the metabolites resulting from photosynthesis are completely consumed in the respiration of the zooplankton the final secondary production is zero, whatever may be the number of production levels involved.

2. If, on the contrary, the oxidation is incomplete we may consider there is net production by the zooplankton.

3. If the oxidation exceeds the local photosynthetic supply, it must be concluded that there exists a lateral exchange with the surrounding water masses.

TABLE 12-8

Estimation of the planktonic secondary production, daily P/B ratio, and ratio between the secondary production and the net primary production (P_{II}/P_I) (after Mullin, 1969 completed after Parsons et al., 1969, Gaudy, 1970 and Razouls, 1972).

Organism or group	Geographical area	Period	P mgC/m²/day	Daily P/B	P_{II}/P_I	Reference
Zooplankton	Georges Bank	year	200	0.03	0.25	Riley, 1947
Zooplankton	English Channel	year	75	0.10	0.30	Harvey, 1950
Zooplankton	Long Island Sound	year	166	0.17	0.30	Conover, 1956[c]
Zooplankton	North Sea (North)	Apr–Sept	180	0.048	0.58	Steele, 1958[c]
Euphausia pacifica	Pacific (NE)	year	0.9	0.008	0.0048	Lasker, 1966
Herbivorous copepods	North Sea	Jan–June	4.9	0.08	0.14	Cushing, 1969
Copepods (mainly *Calanus*)	North Sea	Mar–June	46	0.10	0.20	Cushing & Vucetic, 1963[c]
Copepods (mainly *Calanus*)	North Sea	Mar–June	15	0.033	0.043 0.065	Mullin, 1969[a]
Calanus finmarchicus	Barents Sea (East)	year	7.8	0.002	0.03	Kamshilov, 1958[c]
Calanus helgolandicus	Black Sea	June	28	0.15	0.07	Petipa, 1967
Calanus plumchrus	Strait of Georgia	Mar–May	51	—	—	Parsons et al., 1969
Acartia clausi	Black Sea (Bay)	June	15	0.17	—	Petipa, 1967
Acartia clausi	Black Sea	June	6.6	0.23	—	Petipa, 1967
Acartia clausi	Black Sea	year	0.38	0.035	0.08	Greze & Baldina, 1964
Acartia clausi	Mediterranean (NW)	—	0.33[b]	0.05	0.001	Gaudy, 1970
Acartia tonsa	Chesapeake Bay	summer	77	0.50	—	Heinle, 1966
Centropages kroeyeri	Black Sea	summer	0.19	0.077	0.05	Greze & Baldina, 1954
Centropages typicus	Mediterranean (NW)	—	0.9[b]	0.05	0.0002	Gaudy, 1970
Centropages typicus	Mediterranean (NW)	year	7.75	0.06	—	Razouls, 1972
Temora stylifera	Mediterranean (NW)	June–Nov	2.04	0.05	—	Razouls, 1972

[a] Re-calculated by Mullin from data of Cushing & Vucetic; primary production determined according to two different methods (growth rate and phosphate).
[b] Aware that the plankton tows were made half at 0 m and half at 25 m and corresponding to 370 m³, is representative for the column 0.50 m.
[c] Not given in the bibliography, for detailed references consult Mullin, 1969.

TABLE 12-9

Variation of the biomass (B, mg/m^3), production (P, mg/m^3), and annual P/B ratio (all in wet weight) for populations of three copepods in the Sea of Azov, from 1963 to 1965 (after Zaika, 1968).

Acartia clausi			Calanipeda aquae-dulcis			Centropages kroeyeri		
B	P	P/B	B	P	P/B	B	P	P/B
4	54	14.1	22	877	40.2	1.3	30.4	23.4
18	299	16.3	42	1389	32.6	5.8	71.7	12.3
39	631	16.2	55	2055	37.0	7.6	192.4	25.3
55	936	17.0	374	9898	26.5	10.9	182.6	16.7
214	3031	14.2	575	17953	31.2	40.9	671.8	16.4
219	3676	16.8	686	14696	21.4	88.1	985.1	11.2
55*	63	1.2	31	20	1.9	70	33	2.0

*Ratio of the maximum and minimum values for each parameter.

In this way a knowledge of primary production and the oxygen consumption of the whole zooplankton enables a "synthetic" investigation of the secondary production. Menzel & Ryther (1961) gave an example of this method, in the Sargasso Sea, 15 miles off the south east of Bermuda. The primary production was estimated over 30 months using the ^{14}C method and the zooplankton was collected by a net (mesh 0.366 mm) towed at the surface, 500 m, or 2000 m; the dry weight was determined. The respiration was measured on a sample of unsorted zooplankton and expressed in dry weight, which enables the respiration per m^2 to be calculated. The development of the primary production and the respiratory combustion of the zooplankton is given in Fig. 12-11. It appears at a first glance, that the primary production is almost always consumed by the zooplankton and only a small amount of organic carbon remains available for the higher levels of secondary production.

This kind of research is still very approximate, but it will improve as more knowledge on the respiration accumulates; it represents a very interesting point of view.

12-9.—CHEMICAL COMPOSITION OF THE ZOOPLANKTON

We shall end this chapter on production by gathering together certain data on the chemical composition of the zooplankton. Table 12-10 gives data based on Curl (1962a, b) and Omori (1969) for the dry weight, ash, carbon, and nitrogen of some planktonic species analyzed separately. Table 12-11 (after Beers, 1966), gives the total analysis of different groups collected in Sargasso Sea over a whole year.

Considering dry weight in terms of wet weight we see that it varies considerably, with a minimum less than 5% in the Cnidaria, Ctenophora, and

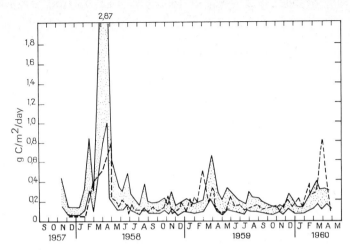

Fig. 12-11.—Relation between net primary production (broken line) and the metabolic requirements of zooplankton represent as the stippled envelope: lower limit based on measured zooplankton population between 0 and 500 m depth, the upper limit of estimated zooplankton population between 0 and 2000 m: Sargasso Sea 15 miles S.E. of Bermuda (after Menzel and Ryther, 1961).

Tunicata and a maximum of 30% in the Pteropoda. Within the Copepoda it varies between 12 and 22% and this is confirmed by the data of Mazza (1964) who gives 12.0% to 21.2%. There may be seasonal variations and the value obtained in May for *Calanus cristatus* 33.9% is particularly remarkable relative to the values for July and December (16.7 and 15.5).

The amount of ash and, therefore, of inorganic matter, expressed relative to dry weight, varies considerably according to the group. Even within the same group there are notable differences, for instance in the Copepoda and Euphausiacea. The discrepancies do not seem to be due to the methods used; Curl's results which are higher than those of Omori, as are those of Raymont, Srinivagasam & Raymont (1971) obtained a variation of ash weight in different samples of 6.9 to 25% in the euphausiid *Meganyctiphanes norvegica*, over one year. The monthly mean oscillates between 11.7 in September and 21.0% in January with a general mean of 16.1%.

In Cnidaria, Ctenophora, and Tunicata, not only the dry weight as a percentage of wet weight, but also carbon as a percentage of dry weight is relatively very low, of the order of 10% and so 100 g of wet material contain 4 g dry weight and 0.4 g of carbon, while for copepods the values are, 100, 17 and 8.5 g, respectively.

Finally, the amount of nitrogen, is also very low in Cnidaria, Ctenophora, and Tunicata being less than 3%. In Chaetognatha and Crustacea it ranges between 5 and 13%. For copepods, Conover & Corner (1968) showed a marked seasonal variation from 6 to 9% for *Calanus finmarchicus* and 8 to 11% for *Metridia longa* (Fig. 11-19); the maximum corresponds to the utilization of the phytoplankton bloom.

TABLE 12-10

Chemical composition of zooplankton: DW, dry weight; WW, wet weight (after Curl, 1962 and Omori, 1969).

	DW (% of WW)	% of DW				
		Ash	Org. Mat.	C	N**	Ref.
Cnidaria						
Aequorea vitrina	0.9	49.1	50.9	17.8–22.5	0.5	Curl
Pelagia noctiluca	4.1	69.0	31.0	8.2–9.9	1.4	Curl
Cyanea capillata	4.6	63.0	37.0	11.6	0.4–1.4	Curl
Ctenophora						
Beroe cucumis	4.7	70.3	29.7	—	1.1	Curl
Mnemiopsis sp.	5.0	75.0	25.0	6.4	0.2	Curl
Pteropoda						
Limacina retroversa & sp.	18.7	64.2	35.8	28.3*	4.1	Curl
Limacina inflata	36.4	46.6	53.4	17.0	1.5	Omori
Limacina helicina juv.	25.0	28.5	71.5	29.0	6.0	Omori
Euclio pyramidata	31.1	42.8	57.2	20.3	2.9	Omori
Clione limacina	9.0	33.3	66.7	26.3	2.2–5.0	Curl
Copepoda						
Calanus finmarchicus	10.2	17.6	82.4	35.7–41.7	4.7–5.9	Curl
Calanus pacificus	20.5	3.6	96.4	52.2	9.5	Omori
Calanus cristatus (July)	16.7	2.1	97.9	60.9	6.3	Omori
Calanus cristatus (December)	15.5	3.4	96.6	39.0	7.6	Omori
Calanus cristatus (May)	33.9	2.9	97.1	59.0	5.9	Omori
Calanus cristatus (average)	22.0	2.8	97.2	53.0	6.6	Omori
Rhincalanus nasutus	13.5	3.4	96.6	52.2	9.9	Omori
Pareuchaeta birostrata ♀	18.5	2.1	97.9	58.4	7.1	Omori
Metridia okhotensis ♀	18.8	2.7	97.3	63.5	5.8	Omori
Pleuromamma xiphias	12.2	3.3	96.7	47.4	13.1	Omori
Candacia aetiopica	14.3	3.7	96.3	46.6	12.6	Omori
Labidocera acutifrons	11.3	5.7	94.3	45.8	12.9	Omori
Pontellina plumata	16.5	6.4	93.6	44.3	12.2	Omori
Centropages typicus & *hamatus*	15.8	22.8	77.2	32.5–38.5	5.2–7.1	Curl
Amphipoda						
Parathemisto japonica	18.4	13.4	86.6	48.4	8.2	Omori
Mysidacea						
Siriella aequiremis	18.7	10.2	89.8	42.4	11.0	Omori
Euphausiacea						
Euphausia pacifica	20.2	8.0	92.0	38.7	10.7	Omori
Euphausia pacifica juv.	20.7	8.5	91.5	39.6	10.1	Omori
Euphausia krohnia	20.0	18.6	81.4	35.8	6.8	Curl
Meganyctiphanes norvegica	19.0	22.4	77.6	33.4–37.0	5.2–7.1	Curl
Chaetognatha						
Sagitta elegans	10.6	21.6	78.4		7.8	Curl
Sagitta elegans	14.1	4.8	95.2	47.7	10.7	Omori
Sagitta nagae	11.6	4.2	95.8	43.5	11.1	Omori
Tunicata						
Salpa fusiformis	4.0	77.1	22.9	7.2–10.6	0.5–1.5	Curl
Pyrosoma sp.	4.1	71.2	31.0	9.4	0.3–0.4	Curl

*1.74% in the form of carbonate. **Curl's values are under-estimated.

TABLE 12-11

Estimation of the amount of C, N, P, and of the percentage of dry weight for different groups of zooplankton from samples taken in the Sargasso Sea (0–500 m) off south to southeast Bermuda; mean values obtained from monthly values; extreme values in parentheses: DW, dry weight; WW, wet weight (after Beers, 1966).

Group	Carbon (% DW)	Nitrogen (% DW)	Phosphorus (% DW)	DW (% WW)
Copepoda	41.6 (35.2–47.6)	9.62 (8.16–11.17)	0.79 (0.69–0.87)	13.5
Euphausiacea and Mysidacea	40.7 (35.4–43.4)	9.96 (9.43–10.46)	1.48 (1.39–1.60)	15.9
Other Crustacea	36.9 (32.9–41.7)	7.83 (6.95–8.85)	1.26 (1.10–1.48)	18.2
Chaetognatha	28.3 (21.9–34.3)	7.84 (6.28–9.36)	0.63 (0.46–0.71)	6.8
Fish and fish larvae	37.9 (32.6–41.7)	9.68 (8.27–10.65)	1.40 (0.94–1.84)	14.0
Polychaeta	29.9 (15.9–43.9)	8.92 (4.37–11.18)	0.99*(0.44–1.80)	9.7
Siphonophora	10.9 (3.0–16.0)	2.97 (0.98–4.36)	0.14 (0.05–0.18)	4.0
Hydromedusae	7.2 (5.4–10.4)	2.89 (1.34–6.92)	0.17 (0.12–0.44)	4.4
Pteropoda	22.7*(20.8–25.4)	3.25*(2.69–4.20)	0.30*(0.23–0.38)	25.6

*Mean values obtained from less than 8 monthly values.

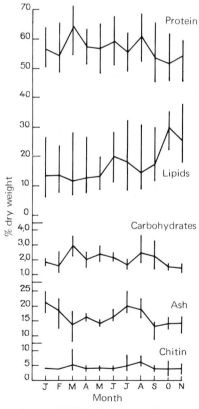

Fig. 12-12.—Changes during the year 1968 in the amount of proteins, lipids, carbohydrates, ash and chitin, expressed as a function of dry weight (monthly means and limits of variation) in the euphausiid *Meganyctiphanes norvegica* in the Firth of Clyde (Scotland) (after Raymont, Srinivagasam & Raymont, 1971).

As for the biochemical composition of the zooplankton organisms we again refer to the detailed study on *Meganyctiphanes norvegica* carried out over one year by Raymont et al. (1971) and illustrated in Fig. 12-12. The protein or more precisely the amino acids, free or bound, represent the largest fraction of the dry weight, from 50 to 60% as was also found for *Calanus* by Cowey & Corner (1963a) who give 50% of the dry weight as amino acids, 80% as proteins and 20% as free amino acids. The amino acids contain the largest fraction of the nitrogen, up to 90%; the remainder is in the form of trimethylamine oxide, betaine, or polymerized glucosamine in chitin.

Cowey & Corner (1963b) studied the abundance of different amino acids in the adult females of *Calanus*. Table 12-12 gives their results for particulate organic matter (phytoplankton, nanoplankton, and detritus) and for the diatom *Skeletonema costatum*. Except for taurine, the amino-acid composition of *Calanus* is similar to that of its food.

After proteins, lipids constitute the most important fraction; their seasonal variations are considerable, ranging from 10 to 30% in *Meganyctiphanes norvegica* (Fig. 12-12) and from 20 to 50% in *Calanus finmarchicus* (Fig. 11-19). A detailed study of the different fractions of lipids has been made by Lee, Hirota & Barnett (1971); Table 12-13 gives their results for four species of copepods. According to their results, triglyceride tends to be replaced by fatty esters in copepods of deep water and in those of cold shallow water. During starvation experiments the triglycerides disappear first and the fatty esters may serve rather as reserve. A comparison of fatty acids in phytoplankton, a copepod, and a euphausiacean (Corner & Cowey, 1968; Ackman et al., 1970; Jeffries, 1970) has shown, on the other hand certain

TABLE 12-12

Amino-acid composition (g amino-acid nitrogen/100 g of total amino-acid nitrogen) of adult females of the copepod *Calanus helgolandicus* and of seston (0.5–1.0 μm) collected off Plymouth in 1961 and a comparison with the diatom *Skeletonema costatum* grown in culture: seston (zooplankton previously removed by filtration through a fine gauze) was obtained by filtration of water through membrane with pore size of 0.5–1.0 μm and includes phytoplankton nanoplankton and detritus (tripton): mean values for *Calanus* and the seston from 9 determinations during the year (after Cowey & Corner 1963b).

Amino acid	*Calanus*	Seston	*Skeletonema*	Amino acid	*Canalus*	Seston	*Skeletonema*
Arginine	17.9	15.9	12.2	Serine	3.9	8.2	6.4
Glycine	11.9	14.3	10.2	Threonine	3.9	4.6	4.8
Lysine	11.2	9.3	9.1	Proline	3.9	3.6	4.6
Alanine	9.1	7.3	8.1	Iso-leucine	3.7	4.3	4.7
Glutamic acid	8.1	7.7	8.2	Histadine	3.6	2.2	3.5
Aspartic acid	7.1	7.7	9.1	Tyrosine	2.5	1.9	1.7
Leucine	6.0	6.3	6.8	Phenylalanine	2.4	2.7	3.7
Valine	5.7	5.7	6.0	Taurine	1.7	—	—
				Methionine	1.2	0.8	0.7
				Cystine/2*	0.6	0.1	0.4

*Both cysteine (C–S–H) and cystine (C–S–S–C) are included here; they are not differentiated by the method used.

TABLE 12-13

Lipid composition of four copepods (% total lipid): DW, dry weight (after Lee, Hirota & Barnett, 1971).

Lipid fraction	Gaetanus brevicornis	Amallothrix sp.	Gaussia princeps	Rhincalanus nasutus
Hydrocarbons	3	2	1	6
Wax esters	58	74	49	69
Triglycerides	13	13	18	9
Polar lipids*	9	3	15	5
Phospholipids	17	8	17	11
Total lipids (% of DW)	30	59	26	76

*Including mainly sterols with some free fatty acids and pigments.

differences in composition, contrary to what was found for the amino acids. The percentage of long chain fatty acids (notably C22-6) is higher in the zooplankton. The lipids (Table 12-13) contain a small amount of hydrocarbons; the most important being pristane (2, 6, 10, 14-tetramethylpentadecane) (Corner & Cowey, 1968).

Finally, carbohydrates constitute only a small part of the organic matter of zooplankton: in *Meganyctiphanes norvegica* they represent only 1 to 3% of the dry weight (Fig. 12-12).

It should be mentioned that certain organisms of the zooplankton contain important quantities of Vitamin A, notably the euphausiid, *Meganyctiphanes norvegica* with 1 to 10 μg of vitamin A/g wet weight and this is concentrated mainly in the eyes (Mauchline & Fisher, 1969).

CHAPTER 13

THE PLANKTON IN THE MARINE ECOSYSTEM

13-1.—THE PLANKTON WITHIN THE SESTON

The seston consists of the entire particulate matter in suspension in the water and includes, therefore, the plankton, i.e., the living "particles" and the tripton, the non-living particles. The latter may be divided into the inorganic and organic tripton, in which we may include the particles formed from dissolved organic matter, by condensation at the surface film or on gas bubbles; this is the neo-formed organic tripton. We shall examine in detail the relations between the plankton and the tripton in the total seston.

The plankton contributes to the formation of tripton in various ways; in addition to dead animals, the excretion of the faeces of planktonic animals notably in the form of faecal pellets makes an important contribution to the organic particles. The exuviae of moulting planktonic crustaceans, is also important and has been particularly stressed in Euphausiacea. To these the residual "houses" of appendicularia may be added. These elements of the tripton containing the remains of more or less recognizable organisms are added to detritus of other origin, including the debris of benthic algae, fibres from terrestrial plants, starch particles. Numerous other particles of the tripton clearly do not correspond to the detritus of living organisms and they appear as particles whose most frequent mean dimension lies between 25 and 50 μm but reaching 1 mm and more; some have the structure of film-like loose flakes resembling films of mucus or bacterial films; others are present as smaller, semi-transparent flakes, less than 50 μm in size (Riley, 1970). The origin of these particulate flakes is still uncertain; fragmentation of larger particles, coalescence of small particles or condensation of dissolved organic matter may all take place.

Beers & Stewart (1969, 1971) measured the dry weight of the seston and its different constituents in shallow water and in the open sea off California between San Diego and Guadalupe Island. Table 13-1 shows their results. It appears that the phytoplankton or zooplankton represent only a small fraction of the entire seston, namely, 10 to 20%; this is similar to the results obtained elsewhere (Riley, 1970; Finenko & Zaika, 1970). It is only at the time of the phytoplankton bloom that the plankton exceeds 50% or even 90% of the total seston. Beer & Stewart (1971) found low values (6–17%), during the Eastropac Cruise in the North Pacific (10°38′ N to 12°21′ S on 105°W, March 1968), the proportion of sestonic carbon varying from 10 to 28% (mean, 21%) of the dry weight. Taking the mean amount of carbon in organic matter as 40%, the percentage of organic matter in the seston is 25–70%. Because

TABLE 13-1

The amount of seston (/0.45 μm) in the euphotic zone (3 times the disappearance depth of Secchi disc) at 4 stations off San-Diego (California) in February 1967, and the estimation of the phytoplankton (by chlorophyll) of the zooplankton (by volume) and the tripton (by the difference): DW, dry weight (after Beers & Stewart, 1969).

Station	Total depth (m)	Depth of photic zone (m)	Seston DW (μg/l)	Tripton DW (μg/l)	Tripton % of Seston	Phytoplankton DW (μg/l)	Microzooplankton DW (μg/l)
I	1170	66	160	123	77	32.3	4.5
II	1800	84	140	125	89	12.9	2.5
III	2200	108	120	102	85	14.5	3.4
IV	3600	108	110	96	87	11.3	2.6

such a small fraction of the plankton is represented in the seston, a notable quantity of organic matter must belong to the tripton.

Seston values are often expressed directly in carbon measured by dry combustion or by wet oxidation of the material collected on a filter with pores of 0.2 μm to several μm. The amount in the euphotic zone varies from tenths to several hundreds of μgC/l. A clear maximum is found in the surface proper; Wangersky & Gordon (1965) found 300 to 800 μgC/l at the surface and 55–200 μgC/l at a depth of 13–15 m in Guinea Current while at depths greater than 1000 m, the amount fell to not more than several tenths of μgC/l.

Table 13-2 shows the contribution to the seston of different size fractions; it is particularly important in the fine fraction, less than 35 μm, and of the order of a 100 μg/l dry weight, in the waters studies. Above 35 μm and 103 μm it represents only several μg/l, and here zooplankton is an important fraction (30 to 80%). These observations in the open sea agree with those of Kenchington (1970) for very near coastal waters in the Menai Strait, North Wales, where the seston collected on a net of 190 μm mesh, fluctuated from 5 to 15 μg/l (dry weight) during the year. The zooplankton included in the seston varied from 25 to 90% (Fig. 13-1).

The tripton of small dimensions (nanotripton), because it is so abundant, represents, when it contains organic matter, a possible source of food for zooplankton. In two samples from the surface Parsons & Strickland (1962a) found 202 μgC/l as tripton, corresponding to about 400 μg/l of organic matter of which 115 μg/l was carbohydrates, calculated as glucose, and 219 μg/l was protein as estimated by multiplying the quantity of nitrogen by 6.25; but this organic material, mainly in the deep water tripton, may be relatively resistant and difficult to be used by any organisms (Hobson & Menzel, 1964).

Besides the particulate organic matter there is a variable quantity of organic matter in solution or at which, at least, passes through filter; it

TABLE 13-2

Distribution of seston (0.45 μm) in the euphotic zone, at the same station as in Table 13-1, in the different size-classes and relative proportions of the microzooplankton, phytoplankton, and tripton (after Beers & Stewart, 1969).

Station	Seston DW (μg/l)	Seston % of the total (1+2+3)	1, <35 μm			2, <35 μm				3, 103 μm and 202 μm			
			Micro-zooplankton (%)	Phyto-plankton (%)	Tripton (%)	Seston DW (μg/l)	Micro-zooplankton (%)	Phyto-plankton (%)	Tripton (%)	Seston DW (μg/l)	Micro-zooplankton (%)	Phyto-plankton (%)	Tripton (%)
I	140	90	0.4	24	76	11.6	7.6	6.5	86	4.4	71	1.4	28
II	135	97	0.3	11	89	2.6	46	2.8	51	1.7	54	1.6	44
III	111	96	0.7	12	87	2.0	29	4.1	67	2.4	82	1.4	17
IV	104	97	0.3	11	89	2.1	76	3.0	21	1.7	51	1.1	48

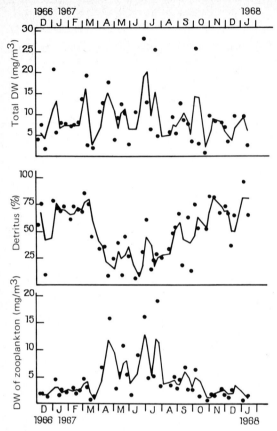

Fig. 13-1.—Variations in the amount of seston (/190 μm) on the eastern edge of Menai Straits; sampled with a high speed net, a modification of "Gulf III net": weekly measurements (points): line drawn between the means of two successive points (after Kenchington, 1970).

originates from marine organisms, notably phytoplankton, and benthic algae. Williams (1969) found 400 to 1000 μg/l of organic carbon, between 0–300 m in the open sea off San Diego. Besides the possibility that this organic matter may condense into particulate organic matter Khailov & Finenko (1970) considered another way in which it could be utilized by filter feeding animals of the zooplankton; they consider that the dissolved macromolecules such as proteins and polysaccharides may be attached to particles, so enriching them with organic matter and increasing their nutritive value to the zooplankton. Another way in which it might be returned to the plankton was studied by Parsons & Strickland (1962b); they found that at a concentration of 10^{-7} mole, or even less, glucose or acetate allowed the growth of heterotrophic bacteria. If we allow that 1–10% of the dissolved organic matter is biologically utilisable at the same level as glucose, the amounts observed enable us to consider the existence of 10^{-7} mole (1.2 μgC/l) are quite capable of supporting bacterial growth. With absorption on the particulate matter the minimal concentrations of utilisable organic matter are perhaps able to be more easily realised and the growth of bacteria facilitated (Riley, 1970). It is also possible that by heterotrophic nutrition the enigmatic pigmented cells found in the depth by Fournier (1970) can also multiply.

13-2.—THE PLANKTON IN THE FOOD WEB

The plankton, as we have already indicated, is situated at the lowest level of production, phytoplankton corresponding to the most important part of the primary production of the oceans. The zooplankton is found at the second, third, and fourth levels and constitutes a considerable part of the secondary production. We shall try to assemble data which will enable us to understand the place of the plankton in the whole of the production of the marine ecosystem.

Riley in 1970 in his report on particulate organic matter gave the following picture for the Sargasso Sea—a region not very productive but particularly well studied. Taking into account criticisms of the various methods used he estimated the primary production of phytoplankton in the upper hundred meters, as 320–380 $mgC/m^2/day$; this value includes the net production and the organic matter released to the exterior of the cell. Of this primary production 75 to 80% is used in the shallow and upper medium layers by the zooplankton and by the heterotrophic fraction of the ultraplankton, which according to Pomeroy & Johannes (1968) contains organisms less than 10 μm such as bacteria, heterotrophic algae, and protozoans. Then, 15% is transported to lower medium layers, as deep as the oxygen minimum (about 800 m) by the zooplankton and heterotrophic organisms. Less than 5% is used in the bathypelagic water and on the bottom.

It is difficult to measure the part played by higher, non-planktonic animals in the consumption and degradation of the organic matter but it may be deduced from energetic considerations; it is not greater than 5%. Without doubt we must investigate relations of the same type in other regions where they will vary—more or less—with locality; in shallow water the consumption by the benthos must increase considerably.

Margalef (1967) has presented a quantitative scheme of the circulation of energy and of material for a marine ecosystem corresponding to a coastal zone in the western Mediterranean; it is shown in Fig. 13-2. By grouping them and by numerical data it brings to light the different relations studied in this book. The importance of the dissolved or particulate organic matter in regard to the plankton is particularly well stressed. On the other hand, in these coastal waters the bottom and the benthos receive directly or indirectly an important fraction, one quarter to one fifth of the phytoplankton production. Regarding pelagic fishes whose place in the ecosystem is difficult to assess because of migration, their production is of the order of 1 to 2% of the primary production. By estimating from the data

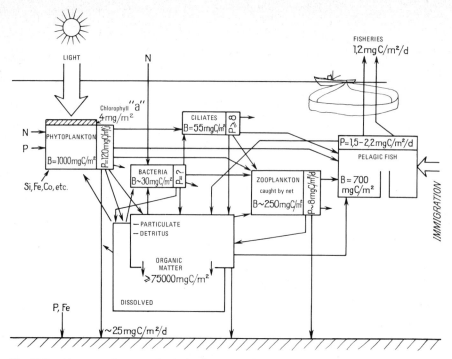

Fig. 13-2.—Energy and matter circulation in an exploited marine ecosystem corresponding to a coastal area in the Western mediterranean: biomass is expressed in mg organic carbon under a surface of 1 m^2: production in mg organic carbon/m^2/day: arrows indicate the direction of flow: in the exchange with the environment only few elements, apart from carbon, are represented (after Margalef, 1967).

of Lasker (1970) for the Pacific sardine that the part used for growth and therefore for production is 5 to 10% of the energy consumed, the consumption of the pelagic fishes would represent some 10 to 40% of the primary production. If this value seems high, relative to the values for the Sargasum Sea, it should be noted that the trophic level of pelagic fish does not correspond exactly to that of a carnivorous regime but also contains an herbivorous fraction of the second level, as has already been seen with regard to anchovy of Peru.

Apart from the global study of the food web, we shall now give an example of a detail of a web taken from the work of Parsons & LeBrasseur (1970) and made in Saanich Inlet, a kind of fjord situated at the south east of Vancouver Island in British Columbia (Fig. 13-3). Initially in one cubic meter of water in the photic zone, there may be approximately 1000 *Pseudocalanus minutus*, (1 mm) 400 *Calanus plumchrus* (3–4 mm) and 20 *Euphausia pacifica* (20 mm). The Euphausiacea produced a large number of eggs in June on the appearance of *Chaetoceros* bloom (32 μm) which was only little used by *Pseudocalanus* and *Calanus* but consumed abundantly and almost exclusively by the Euphausiacea. The euphausiid larvae hatching from the

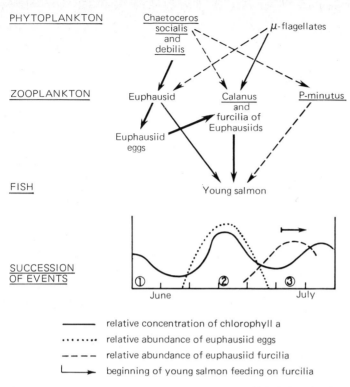

Fig. 13-3.—Tentative food web in Saanich Inlet (British Columbia, Canada) June–July, 1966: arrows indicate energy transfer and their thickness represents the importance of the transfer (heavy—important, fine—mean, broken—low transfer): major phytoplanktonic species causing bloom are indicated: 1, *Distephanus* sp. (Silicoflagellata); 2, *Chaetoceros socialis* and *Ch. debilis* (Diatomea); 3, microflagellates (after Parsons & LeBrasseur, 1970).

eggs then found a very favourable food situation with the appearance of a flagellate bloom (<10 μm). Finally the young pink salmon (*Oncorhynchus gorbuscha*), about 90 mm, were able to feed on the *Calanus* and in particular on the furcilia larvae (3.5 mm), of the euphausiids, two kinds of food perfectly suitable for them and quite different from *Pseudocalanus*. This example illustrates clearly that in the relation between plankton and necton not only quantitative relations but also variables associated with space and time interact.

13-3.—THE PLANKTON AND MAN

Knowing that the biomass of the plankton is much higher than that of its consumers by a factor of approximately 100 for the phytoplankton, and 10 for the zooplankton, it is tempting to consider its possible use in order to get direct supplementary resources from the sea. Theoretically nothing prevents this but the practical difficulties are great since it is necessary, ultimately, to be sure that the operation will be economically valid. In 1957 Jackson in a report on this problem, reached the conclusion that the collection of plankton could be an economic proposition only if zones with high population densities could be localized, or produced artificially; otherwise, the only alternative would be for a new "fishery" method to be developed. Very recently, Parsons (1972) examined the question again and presented a much more optimistic assessment of the possibilities.

So far the exploitation which has been effectively realised concerns particularly the Euphausiacea, which traditionally were caught as bait at Monaco during the winter and are an important fishery at present on certain coasts of Japan. This is possible due to the concentration of Euphausiacea in shallow shoals during a part of the year. Fig. 13-4 show the periods of abundance during which the harvest is made in the Island of Kyushu and Fig. 13-5 illustrates the method of fishery. Komaki (1967) indicates a production of 432 tons in 1953, 1420 tons in 1959 (a very good year) for the area of Kinkazan onlythe harvesting season is between the end of March and May. The Russians have tried to harvest Euphausiacea on an industrial scale; Fig. 13-6 represents a horizontal net used for this purpose. Using a pelagic trawl, the research vessel "Akademik Knipovitch" captured 4 to 6 tons per hour (Parsons, 1972).

Among other organisms exploited we may mention the Mysidacea in the near coastal waters of Thailand and Japan and the *Pleurocondes*

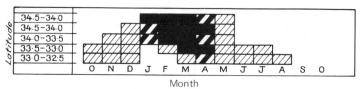

Fig. 13-4.—Seasonal shoaling at the surface of the euphausiid *Euphausia pacifica* in the waters off Kyusha Island, south west of Japan: black rectangles, months in which *Euphausia* most often shoaling; heavy shading, periods of moderate intensity; fine shading, months in which swarming is low (after Komaki, 1967).

Fig. 13-5.—Diagram showing a euphausiid net being operated from a small motor boat off the coast of Kinkazan, in the north east of Hondo Island, Japan (after Parsons, 1972).

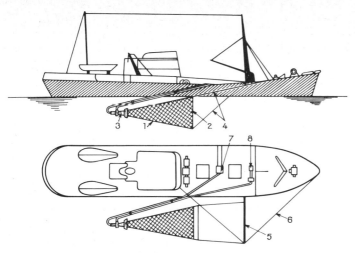

Fig. 13-6.— Side trawl used for euphausiid harvesting on Soviet vessels: 1, net; 2, brace (5m × 5m) with sinker; 3, hydraulic lift; 4, cables; 5, swing boom; 6, guy wire; 7, water separator, separating the euphausiids caught from the water; 8, pump (after Parsons, 1972).

(Galatheidea) on the Chilean coast; these are, however, not typical planktonic forms but partly benthic.

Any extension to a direct exploitation of other planktonic groups needs to detect the existence of a concentration. This is probably possible for certain copepods for which it has been shown that they can be recorded by an echosounder and correspond to scattering layers (Fig. 10-24). Essentially important catches (150 kg/tow) of the copepod *Calanus plumchrus* were obtained in the Strait of Georgia (Canada, B.C.) (Parsons, 1972); however repercussions of this type of exploitation must be expected—after a more or less short delay—on the fisheries themselves.

The exploitation of the plankton by man, directly or indirectly, cannot be done without a particularly thorough knowledge of its ecology and the food web in which it takes part. We have already seen that much progress has been made in this direction but numerous gaps still remain and new ones are found with increasing knowledge. The interference by man can only be satisfactory if our knowledge of ecology is thorough. The first result of the importance of nutrient minerals for primary production leads to the possibility of enrichment of a restricted area, but the results are less favourable than expected, often giving rise to unwanted eutrophication. Even so, relying on numerical data and our knowledge of the mechanisms involved, Parsons and his collaborators have recently made a very satisfactory fertilization operation in the Great Central Lake in Vancouver Island. Adding carefully nitrogen and mineral phosphate they succeeded, in doubling the primary production compared with the preceding year, multiplying the biomass of the zooplankton by eight, and increasing the mean weight of young salmon by about 40% (Parsons, Stephens & Takahashi, 1972; McAllister, LeBrasseur & Parsons, 1972).

CHAPTER 14

APPENDIX

14-1.—SYSTEMATIC PLACE OF THE PRINCIPAL GENERA OF THE MARINE PHYTOPLANKTON

(According to Parke & Dixon, 1968 and Hendey, 1964)

All the classes of marine algae are reported. Only the orders, or sub-orders, which contain planktonic forms are mentioned. Some are not truly planktonic genera but are very commonly cultured in laboratories and are, therefore, mentioned. (O = order, SO = sub order)

Cyanophyceae
    O Nostocales   *Oscillatoria (Trichodesmium) Nostoc, Richelia*

Rhodophyceae

Cryptophyceae
    O Cryptomonadales   *Hemiselmis, Cryptomonas*

Dinophyceae
    O Prorocentrales   *Exuviella, Prorocentrum*
    O Dinophysiales   *Amphisolenia, Dinophysis, Phalacroma, Ornithocercus*
    O Peridinales   *Amphidinium, Cochlodinium, Gymnodinium, Gyrodinium, Warnowia, Polykrikos, Oxyrrhis, Noctiluca, Glenodinium, Peridinium, Gonyaulax, Pyrodinium, Ceratium, Goniodoma, Oxytoxum, Cladopyxis, Podolampas, Oobdinium, Blastodinium*
    O Phytodiniales   *Pyrocystis*

Haptophyceae
    O Isochrysidales   *Dicrateria, Isochrysis*
    O Prymnesiales   *Chrysochromulina, Prymnesium, Phaeocystis, Coccolithophores: Acanthoica, Calciosolenia, Calyptrosphaera, Cricosphaera, Crystallolithus, Pontosphaera, Syracosphaera, Coccolithus, Cyclococcolithus, Rhabdosphaera*

Chrysophyceae
    O Ochromonadales   *Dinobryon*
    O Chromulinales   *Chromulina, Monochrysis, Pseudopedinella*
    O Dictyochales   *Dictyocha* (silicoflagellates)

Craspedophyceae

Xanthophyceae
    O Heterochloridales   *Olisthodiscus, Phacomonas*

Bacillariophyceae
    O Bacillariales
        SO Coscinodiscineae   *Coscinodiscus, Cyclotella, Porosira, Skeletonema, Stephanopyxis, Thalassiosira*
        SO Biddulphineae   *Biddulphia, Cerataulina, Ditylum, Eucampia, Hemiaulus, Chaetoceros, Bacteriastrum*
        SO Rhizosolenineae   *Corethron, Dactyliosolen, Lauderia, Leptocylindrus, Schroederella, Guinardia, Rhizosolenia*
        SO Fragilariineae   *Asterionella, Thalassionema, Thalassiothrix*
        SO Naviculineae   *Nitzschia, Phaeodactylum*

Phaeophyceae

Prasinophyceae
    O Pyramimonadales   *Pyramimonas, Platymonas, Tetraselmis*
    O Halosphaerales   *Halosphaera*

Chlorophyceae
    O Volvocales   *Dunaliella, Carteria, Chlamydomonas*
    O Chlorococcales   *Chlorella, Nannochloris*

Euglenophyceae   *Euglena*

14-2.—SYSTEMATIC PLACE OF THE PRINCIPAL ELEMENTS OF THE ZOOPLANKTON

Based on the classification of Grasse, Poisson & Tuzet (1961) and of Rothschild (1961).

Abbreviations: P, Phylum; C, Class; SC, Sub-class; O, order; bw, brackish water organism; fw, fresh water organism; ?, uncertain classification (*incertae sedis*); between quotation marks planktonic larval forms.

All the classes are cited (except in the Vertebrata) as well as the sub-classes of the Crustacea; only orders containing planktonic forms are indicated. All the genera referred to in the text, are mentioned. Among the parasitic groups a large number is secondarily planktonic, parasites of planktonic organisms.

P Protozoa
   C Flagellata
      SC Phytoflagellata (= Phytomastigina)
         O Dinoflagellata (= Peridineae)(*Noctiluca, Oxyrrhis*)
   C Rhizopoda
      O Foraminifera (*Globigerina, Globigerinella*)
   C Actinopoda
      O Acantharia (*Acanthochiasma, Acanthometra, Amphilonche, Lithoptera*)
      O Radiolaria (*Aulacantha, Hexacontium, Thalassicolla, Collozoum*)
      O Heliozoa (fw. bw.: *Actinophrys, Acanthocystis*)
   C Sporozoa
   C Cnidosporidia
   C Ciliata
      SC Holotricha
         O Peritrichida (*Zoothamnium*)
      SC Spirotricha
         O Tintinnida (*Tintinnopsis, Proplectella, Favella, Dictyocysta, Eutintinnus*)

P Porifera "larvae", "gemmules".
   C Calcarea
   C Demospongiae
   C Hexactinellida

P Cnidaria
   C Hydrozoa
      O Athecata (= Gymnoblastea + Anthomedusae)(*Sarsia, Zanclea, Vellella, Cladonema, Rathkea, Pandea*): "Actinula"
      O Thecata (= Calyptoblastea + Leptomedusae)(*Laodicea, Phialidium, Tiaropsis, Obelia, Aequorea*)
      O Limnomedusae (*Olindias, Gonionemus*)
      O Trachymedusae (*Geryonia, Liriope, Aglaura, Persa, Rhopalonema*)
      O Narcomedusae (*Solmaris, Solmissus, Aegina, Cunina*)
      O Siphonophora* (*Forskalia, Nanomia, Physalia, Sulculeolaria, Chelophyes, Muggiaea, Abylopsis*)
   C Scyphozoa
      O Cubomedusae (*Carybdea*)
      O Coronatae (*Atolla, Nausithoe, Periphylla*)
      O Semaeostomae (*Pelagia, Chrysaora, Cyanea, Aurelia*)
      O Rhizostoma (*Rhizostoma, Cotylorhiza*)
   C Anthozoa
      O Octocorallia*: "*planula*"
      O Hexacorallia*: "*arachnatis*", "*cerianthula*", "*edwardsia*", "*halcampula*" "ectoparasite larvae of *Peachia*"

---

*Considered as a class in Grassé, Poisson & Tuzet.

P Ctenophora
  O Tentaculata* (*Pleurobranchia, Bolina, Mnemiopsis, Cestus*)
  O Nuda (*Beroe*)

P Platyhelminthes
  C Turbellaria (*Alaurina*): "Larva of Mueller"
  C Temnocephaloidea
  C Cestodaria
  C Monogenea
  C Cestoda
  C Trematoda
  ? C Acanthocephala

P Mesozoa
  C Dictyemida
  C Orthonectida

P Nemertina: "*pilidium*"
  C Anopla
  C Enopla (*Pelagonemertes*)

P Aschelminthes**
  C Rotifera (bw: *Synchaeta, Asplanchna*; fw: *Keratella*)
  C Gastrotricha
  C Echinoderida
  C Priapulida
  C Nematomorpha
  C Nematoda

P Mollusca
  C Aplacophora: "barrel larvae"
  C Polyplacophora: "trochophora"
  C Monoplacophora
  C Gastropoda
    Prosobranchia: "veliger", "echinospira".
      O Archaegastropoda
      O Mesogastropoda (*Janthina*; Heteropoda: *Atlanta, Carinaria, Pterotrachea*)
      O Neogastropoda
        Opisthobranchia: "veliger"
      O Tectibranchia
      O Pteropoda, Thecosoma: (*Creseis, Limacina, Spiratella, Euclio, Cymbulia*)
        Gymnosoma: (*Pneumodermopsis, Clione*)

---

*A union of the orders Filictenida and Platyctenea of Grassé, Poisson & Tuzet.
**Grassé, Poisson & Tuzet do not justify this phylum and prefer to divide it into three smaller phyla: Rotifera, Nematorhyncha (Nematoda, Nematomorpha). Priapulida is *incertae sedis*.

O Nudibranchia (*Phylliroe*)
Pulmonata
C Scaphopoda: "veliger"
C Bivalvia: "veliger"
C Cephalopoda

P Annclida
C Polychaeta (*Alciope, Tomopteris*): "trochophora", "nectocheta", "mitraria"
C Oligochaeta
C Hirudinea
C Sipunculoidea (*Pelagosphaera*)
C Echiuroidea

P Lophophora
C Entroprocta "*trochophora*"
C Phoronida "*actinothrocha*"
C Ectoprocta "*cyphonautus*"
C Brachiopoda "bivalvia larvae"

P Arthropoda
C Merostomacea
C Arachnida
C Pycnogonida
C Crustacea
Entemostraca "*nauplius*", "*metanauplius*"
SC Branchiopoda
O Anostraca (*Artemia* (not planktonic))
O Cladocera (*Podon, Evadne, Peniia,* fw: *Daphnia, Bosmina, Leptodora*)
SC Ostracoda
O Myodocopa (*Cypridina, Conchoecia, Gigantocypris*)
O Podocopa (*Cythere*)
O Platycopa (*Cytherella*)
SC Copepoda
O Calanoida (*Calanus, Pseudocalanus, Paracalanus, Microcalanus, Clausocalanus, Megacalanus, Rhinocalanus, Euchaeta, Pareuchaeta, Pleuromamma, Isias, Labidocera, Anomalocera, Candacia, Gaussia, Tortanus, Pontellina, Metridia, Haloptilus, Gaetanus, Euchirella, Temora, Centropages, Acartia*; fw: *Diaptomus, Calanipeda*)
O Cyclopoida (*Oithona*; fw: *Cyclops*)
O Harpacticoida (*Euterpina, Oncaea,* ? *Corycaeus,* ? *Sapphirina*)
SC Branchiura
SC Mystacocarida
SC Cirripedia: "*nauplius*", *cypris*"
Malacostraca: "*nauplius*", *metanauplius*"
SC Leptostraca

SC Hoplocarida
O Stomatopoda: "*pseudozoea*", "*antizoea*", "*erichthus*", "*alima*"
SC Syncarida
SC Peracarida
O Mysidadacea (*Mysis, Siriella, Gnathophausia, Eucopia*)
O Cumacea (*Diastylis*)
O Isopoda (*Munnopsis*) "praniz", "epicard larvae"
O Amphipoda (*Hyperidea*: *Phronima, Hyperia, Parathemisto, Euthemisto, Vibilia*)
SC Eucarida
O Euphausiacea (*Meganyctiphanes, Nyctiphanes, Euphausia, Nematoscelis, Stylocheiron, Thysanopoda, Thysanoessa*): "*calyptopis*", "*cyrtopia*", "*furcilia*".
O Decapoda (*Pleurocondes*): "*mysis*", "*zoea*", "*metazoea*", "*megalopa*", "*elaphocaris*", "*acanthosoma*", "*mastigopus*", "*tracheliper*", "*eretmocaris*", "*glaucothoe*", "*phyllosoma*"
C Myriapoda
C Insecta: fw: "larvae of *Chaoborus*"
C Onychophora
C Tardigrada
C Pentastomida

? P Chaetognatha (*Sagitta, Eukrohnia*)

P Echinodermata
C Crinoidea "barrel like larva"
C Holothuroidea (*Pelagothuria*): "*auricularia*", "*doliolaria*", "*pentactula*"
C Echinoidea: "pluteus"
C Asteroidea: "bipinnaria", "brachiolaria"
C Ophiuridea: "ophiopluteus"

P Stomatochordata
C Enteropneusta: "tornaria"
C Pterobranchia
C Planctospheridea (*Planctosphaera*)

P Pogonophora

P Tunicata
C Thaliacea (*Pyrosoma, Doliolum, Salpa*)
C Appendicularia (*Oikopleura, Fritillaria*)
C Ascidiacea "tadpole of ascidian"

P Cephalochordata "Larvae of amphioxus"

P Vertebrata
C Pisces
SC Neopterygii "eggs and larvae"

14-3.—SOME IDEAS CONCERNING GROWTH

(a) *Growth of unicellular algal cultures**

Assuming a population of the size "n" its growth i.e., its positive change as a function of the time, will be,

$$\frac{dn}{dt} = K_n n \tag{1}$$

It may be seen that growth is a function of n; this may also be written,

$$K_n = \frac{dn}{dt} \cdot \frac{1}{n} \tag{2}$$

where K_n, the instantaneous growth coefficient, is the rate of growth related to the size of the population: it is a relative growth rate (also called the specific growth rate) with the dimension t^{-1}.

In practice this growth rate is measured over a finite interval of time Δt. Care must be taken that the value of Δn does not increase very much during the interval Δt, so that it would affect the calculation. If the increase is not $> 20\%$, it may be satisfactorily corrected by adding to n $\Delta n/2$, using the equation

$$K_n = \frac{\Delta n}{\Delta t} \cdot \frac{1}{n + \Delta n/2} \tag{3}$$

Instead of estimating the growth of a cell population or of individuals, the growth of a particular cell substance may be estimated and the relative growth of such a substance (lipid, chlorophyll, for example) may be defined in the same way as the preceding one:

$$\frac{dX_a}{dt} \cdot \frac{1}{X_a} = K_a \tag{4}$$

when X_a is the amount of substance a. It is then possible to compare different relative rates of growth, K_n, K_a, K_b etc. ...

Because of practical considerations it is also convenient to express the rate of relative growth of the material, a, as a function of another material, say, b. This is the reason why the rate of carbon fixation per unit of chlorophyll a is often used;

$$\frac{dX_c}{dt} \cdot \frac{1}{X_{chl_a}} = r \tag{5}$$

the coefficient r is related to K_c:

$$K_c = r \frac{X_{chl_a}}{X_c} \tag{6}$$

---

*We follow here the work of Eppley & Strickland (1968).

If the coefficient K_n remains constant during the time $t - t_0$, equation (1) may be integrated to give:

$$n_t = n_{t_0} e^{K_n(t-t_0)} \tag{7}$$

where n_t is the size of the population at time t and n_{t_0} the size at time t_0; from this,

$$K_n = \frac{1}{t - t_0}(\ln n_t - \ln n_{t_0}) \tag{8}$$

This expression of K_n is preferable to that in (3) when growth is $>20\%$ of the initial size of the culture or of the natural population. K_n will be different according to the units, days or hours of t; K_n (day^{-1}) = 24 K_n (hr^{-1}).

Certain authors, rather than using K_n, prefer to use the doubling time t_g. In (8) $t - t_0 = t_g$ and $\ln n_t = \ln 2nt_0$ which enables one to calculate t_g:

$$t_g = \frac{\ln 2}{K_n} \tag{9}$$

which is equal to $0.69/K_n$ (day^{-1}) if t_g is expressed in days or $0.69/K_n$ (hr^{-1}) if t_g is expressed in hours.

It is also possible to calculate the number of divisions/day which is equal to K_n (day^{-1})/0.69 (doublings/day).

As an example see Table 7-9 where some values of K_n (day^{-1}) are given for a few species of phytoplankton.

(b) *Calculation of the daily rate of relative growth*

Considering the growth of an organism during the exponential phase, i.e., before there is any limitation and which corresponds to the characteristic section of a logistic curve, its growth dw/dt is proportional to the weight, or $dw/dt = Kw$ and at the end of an interval of a given time-unit its weight will be $w + Kw$ or $w(1 + K)$; at the end of two intervals of time it becomes $w(1 + K)^2$ and at the end of t intervals, $w(1 + K)^t$. K is the instantaneous growth coefficient (rate of relative growth, rate of specific growth). Supposing that we know the initial weight, w_1 and the weight at the end w_2 and that the coefficient of growth is constant over the time interval t, we have,

$$w_2 = w_1(1 + K)^t$$

from which it may be deducted that $(1 + K)^t = w_2/w_1$ and in its logarithmic form,

$$\log(1 + K) = \frac{\log w_2 - \log w_1}{t}$$

According to the definition of the logarithm of a number (the power by which the base has to be raised in order to obtain the number) we have in decimal logarithms:

$$1 + K = 10 \frac{\log w_2 - \log w_1}{t}$$

and:

$$K = 10 \frac{\log w_2 - \log w_1}{t} - 1$$

or in natural logarithms:

$$K = e^{\ln w_2 - \ln w_1/t} - 1$$

By expressing t in days we obtain the daily rate of relative growth, which is a finite rate, calculated for a time of which the limits are given supposing that the instantaneous coefficient remains constant over the interval of time considered.

(c) *Calculation of the finite rate of growth populations*

Assuming the relation $N_t = N_0 e^{gt}$ expressing the growth of a population of which the instantaneous growth coefficient (relative growth rate, specific growth rate) is g the finite growth rate corresponds to the relative increase of the population in a limited interval of time. For example for 1 day we have:

$$\frac{N_t - N_0}{N_0} = \frac{N_t}{N_0} - 1 = e^g - 1$$

for 10 days we have

$$\frac{N_t - N_0}{N_0} = \frac{N_t}{N_0} - 1 = e^{10g} - 1$$

etc. The finite rate should therefore be expressed only when accompanied by the interval of time considered (see also Edmondson & Winberg, 1971, p. 302).

14-4.—DERIVATION OF THE SVERDRUP EQUATION $D_{cr} = \frac{\bar{I}_e}{KI_c}$

Sverdrup (1953) started from the following assumptions:

1. There exists a thoroughly mixed upper layer of thickness D_m;
2. Within the top layer the turbulence is sufficient to distribute the plankton organisms evenly through the layer.
3. Within the upper layer the production is not limited by a lack of nutrient salts.
4. Within the upper layer the extinction coefficient, K, of radiation is considered as constant.
5. The energy considered is at wave-lengths not shorter than 420 nm and not longer than 560 nm. For other wave lengths, the extinction coefficient is so

large that the radiation is rapidly absorbed by the water. The energy considered represents about 20% of the total energy of the incident radiation for the fraction effectively active in photosynthesis (efficient fraction).

6. The production of organic matter by photosynthesis is considered as proportional to the energy of the radiation at the depth under consideration. This is not absolutely exact but may be considered to be fulfilled below a depth of a few meters.

7. The energy, I_c, at the compensation depth is known; for *Coscinodiscus excentricus* it is equal to 0.13 cal/cm²/h (Jenkin); for a mixed population a value of 0.15 was found in Gullmar Fjord, Sweden.

If I_0 is the total radiation in cal/cm²/h, of which a fraction is lost by reflection from the surface, the irradiance after penetration through the water is $I_w = (1-a)I_0$ and the effective energy I_e, is, therefore, $I_e = 0.20 I_w$.

At a depth z (the z axis vertical) we have,

$$I_z = I_e e^{Kz} = 0.2(1-a)I_0 e^{Kz} \tag{2}$$

according to the definition of the coefficient K.

Supposing that production at the depth z is proportional to the available irradiance energy, I_z, the production, dp, during a time interval dt is then $dp = mI_z \cdot dt$, where m is a factor depending upon the character of the population and the temperature of the water; the loss of matter by respiration over the same time interval is $dr = ndt$, where n is another factor depending on the same conditions as m. At the compensation depth, D_c, by definition $dp = dr$ which means that $m \cdot I_c dt = ndt$ and, therefore, $I_c = n/m$. For the interval of time T the total production by photosynthesis in the mixed superficial layer of thickness D_m is (since $z = -D_m$):

$$P = \int_0^T \int_{-D_m}^0 m I_e e^{Kz} dz dt \tag{3}$$

The total loss in the same layer is:

$$R = \int_0^T \int_{-D_m}^0 n \, dz \, dt = nTD_m \tag{4}$$

On the other hand, calling $\bar{I}_e$ the algebraic mean of the effective irradiance in the time T we have:

$$\bar{I}_e = \frac{1}{T} \int_0^T I_e dt \tag{5}$$

The calculation of D_{cr}, the critical depth, is obtained, and gives, when $P = R$,

$$\frac{D_{cr}}{1 - e^{-KD_{cr}}} = \frac{\bar{I}_e}{KI_c} \tag{6}$$

Finally accepting that $1 - e^{-KD_{cr}}$ is very close to 1:

$$D_{cr} \approx \frac{\bar{I}_e}{KI_c} \tag{7}$$

14-5.—CONSTRUCTION OF PLANKTON NETS

We may rely on the excellent monograph on zooplankton sampling published by UNESCO (Anonymous, 1968) and particularly on the work of Tranter & Smith.

The efficiency of filtration or filtration coefficient, F, of a net is defined as the ratio of the volume of water filtered by the net to the volume of water swept by the net mouth. That is,

$$F = \frac{W}{A \cdot D}$$

where A is the mouth area of the net and D is the distance towed. Multiplied by 100 it is expressed in percentage, or it may be expressed as,

$$F = \frac{AV't}{AVt} = \frac{V'}{V}$$

where t is the time of towing, V is the velocity of towing and V' is the mean velocity of the water through the net mouth.

Whatever is the filtering gauze, most often silk or nylon, or sometimes metallic gauze, it consists of pores enclosed by strands and it is characterized by a certain porosity, β, which may be calculated by the equation:

$$\beta = \frac{m^2}{(d+m)^2}$$

where m is the mesh width and d is the diameter of the strand in the meshwork. The porosity may readily be estimated from the curve of Fig. 14-1.

From the porosity, β, it is possible to calculate the opening area ratio, R, given by the equation

$$R = \frac{a\beta}{A}$$

where a is the total porous area and A is the mouth area of the net.

Experience shows that when the opening area ratio is >3, the filtration

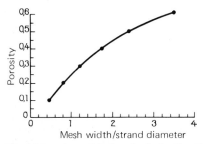

Fig. 14-1.—A graphical method of estimating porosity from the ratio of the mean mesh width to the mean strand diameter (after Tranter & Smith, 1968).

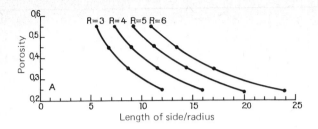

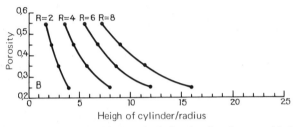

Fig. 14-2.—A, a graphical method of estimating the cone side length for nets with opening area ratio of 3 to 6 where the gauze porosity is between 0.25 and 0.55: units on the abscissa are to be multiplied by the radius to determine side length: B, graphical method for estimating the cylinder height of the reserve mesh to be added to a net for opening area ratios of 2 to 8: units on the abscissa are to be multiplied by the radius to determine cylinder height (after Tranter & Smith, 1968).

coefficient increases slowly. For a conical net it is higher than 85% above an opening area ratio of 3; it does not reach 95% for a ratio higher than 5. Fig. 14-2 for an opening area ratio between 3 and 6, enables one to obtain the ratio between the side length of the net and its mouth radius as a function of the porosity. The radius is determined by different considerations such as the volume to be sampled and the avoiding reaction of the organisms sought.

The filtration coefficient may, however, be markedly changed by clogging when the porosity is progressively reduced by particles and planktonic organisms which adhere to the strands of the gauze during filtration. It is, necessary, therefore, to include a filtration "reserve" and this may be done by adding a cylinder of gauze to the terminal cone. The graphs in Fig. 14-2 allow one to determine the height of the supplementary cylinder. For a gauze with mesh width > 333 μm, an overall open area ratio of 5.3 in the cone and 2 in the cylinder are considered satisfactory. Data on the importance and speed of clogging are still few; it is interesting to give the data of Smith, Counts & Clutter (Table 14-1). The extrapolation of these data (for nets with mesh width of 0.333 mm) shows that a net of opening area ratio of 7.8 could withstand a tow of 750 m^3 in a neritic zone ("green" water) before the filtration coefficient falls below 85%. It is also possible to add, in front of the filtration cone, a reduction cone which reduces the opening area of the net; this is particularly used in high-speed nets. The principal effect of a reduction cone is to increase the filtration coefficient which may be over 100% as shown in Fig. 14-3. Finally, if the net is encased in a protective cylinder, the filtration may be markedly modified.

TABLE 14-1

The effect of opening area ratio (R) on sustained filtration efficiency (F) for a net with nylon gauze of mesh width 333 μm (after Smith, Counts & Clutter in Anonymous, 1968).

R	Volume filtered (m³ at $F>85\%$)	
	"Green" water	"Blue" water
3.2	49	390
4.8	123	1172
6.4	300	2564

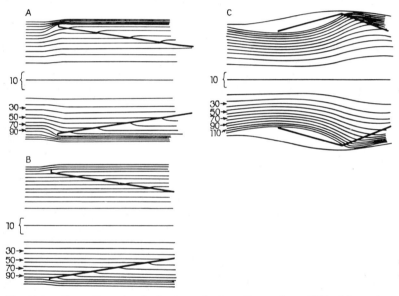

Fig. 14-3.—Streamlines ahead of nets; each streamline encloses 10% of the water entering a circular net: a, conical net which is accepting 75% of the water presented to it; b, conical net which accepting 95% of the water presented to it; c, conical net, with a mouth reduction cone, which is accepting 125% of the water presented to it (after Tranter & Smith, 1968).

It is interesting to know the resistance of the net during towing; its variation may be estimated from the curves in Fig. 14-4. The total resistance to the free passage of the net, defined as drag, D, is related to the square of the towing velocity, V, and is given by the equation:

$$D = C_D \cdot \tfrac{1}{2}\rho V^2 A$$

ρ is the density of the fluid, in our case water, A is the opening area and C_D is the drag coefficient of the net. We have only poor estimations of the drag; for

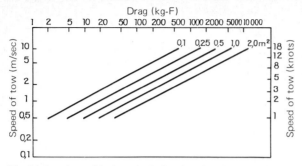

Fig. 14-4.—A graphical method of estimating the drag of nets with mouth area of 0.1 to 2.0 m², towed at velocities of 0.5 m/sec (1 kt) to 9.50 m/sec (18 kt) assuming a drag coefficient of 1.33 (after Motoda in Tranter & Smith, 1968).

the Indian Ocean Standard net and the WP2 net it should be of the order of 1.33.

Finally, it is useful to know two velocities, namely v the approach velocity which is the effective velocity at which the water approaches the immediate environ of the filtering surface, before its acceleration through the meshes, and it is given by the equation,

$$v = V \cdot F(A/a)$$

and the mesh velocity v' which is the mean velocity at which the water discharges through the meshes and is calculated from the equation,

$$v' = \frac{V \cdot F \cdot A}{a\beta} = V \cdot F/R$$

The decrease in pressure due to the flow of the water through the plankton net, Δp, is related to the approach velocity as follows,

$$\Delta p = K \tfrac{1}{2} \rho v^2$$

where K is the resistance coefficient of the gauze. The damage caused to organisms, when a net is towed with increasing velocity, is attributed to this fall in pressure.

Except for small quantitative nets, such as the Clarke–Bumpus net, where the entire opening of the net corresponds to a flowmeter, it is difficult to estimate precisely the quantity of water filtered by the net. Fig. 14-5 shows that flowmeters, the size of which is small in relation to the net opening, may give different results according to their position. The relative stability of the filtration coefficient >85%, enables one to overcome the difficulty. It should be ensured that the mesh does not clog during towing. This is possible by placing 2 flow meters, one of them outside the net, not influenced by it and which will record the towing velocity, the second, inside the net placed in the mean flow. An increase in the difference between the two meters enables one to estimate the beginning of clogging (Fig. 14-6). As long as the net is not clogged, the product of the opening area A and the distance of towing D,

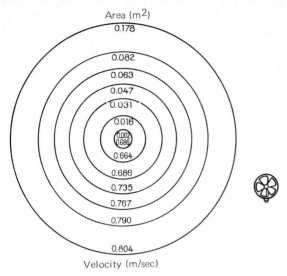

Fig. 14-5.—Velocity profile of the flow at different points of the mouth of a Hensen egg net, the diameter of the net being 73 cm and the diameter of the flow meter 7.6 cm: water velocity in the test channel was 0.718 m/sec (1.4 kt): under these conditions the flow meter situated 20 cm from the centre gives the most correct results: area presented in the drawing corresponds to different rings: assuming a radial symmetry of flow the volume filtrated by the net can be calculated by multiplying the velocity by the area it is intended to represent (after Tranter & Smith 1968).

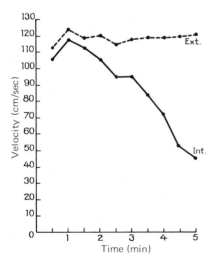

Fig. 14-6.—The clogging rate of an early model of the WP-2 net towed at 120 cm/sec in neritic water off San Diego, California; clogging indicated by the progressive decrease in the velocity through the mouth of the net, measured by an internal flow meter (Int.): external flowmeter (Ext.) does not show a decrease in the velocity (after Tranter & Smith, 1968).

TABLE 14-2
Characteristics of some plankton nets (after Tranter & Smith, 1968).

		Mouth area (m²)	Form*	Mesh width (mm)	Porosity (β)	Open area ratio (R)	Filtration efficiency** (F)	Approach velocity** (v)(cm/sec)	Reference
Low-speed nets (<3 knots)	**I Coarse gauze (>0.4 mm)**								
	Bongo net	0.38	Cone	0.51	0.51	6.8	0.96	7	McGowan & Brown
	FAO-Larval tuna	0.79	Cyl+Cone	0.51	0.51	4.8	0.97	10	Matsumoto
	CALCOFI standard	0.79	Cone	0.55	0.36	3.2	0.91	10	Smith et al.
	II Medium gauze (0.2–0.4 mm)								
	CalCOFI Anchovy Egg	0.20	Cyl+Cone	0.33	0.46	7.8	0.95	6	Smith et al.
	Australian Clarke-Bumpus	0.012	Cone	0.27	0.44	5.3	0.88	7	Tranter
	Indian Ocean Standard	1.00	Cyl+Cone	0.33	0.46	4.3	0.96	10	Currie
	III Fine gauze (≤0.2 mm)								
	WP-2	0.25	Cyl+Cone	0.20	0.45	6.0	0.94	7	Fraser
	Tropical Juday-Reg	0.50	Red+Cone	0.17	0.32	4.2	—	—	Bogorov
	Bé net MPS	0.25	Pyramid	0.20	0.45	2.7	0.88	15	Bé
	IV Mixed gauzes								
	International Standard	0.20	Cone	0.23	0.36	2.6	—	—	Ostenfeld & Jespersen
				0.08	0.20				
	N 70	0.37	Cone	0.37	0.34	2.4	—	—	Foxton
				0.17	0.32				
High-speed nets (>3 knots)	Miller high speed I	0.0081	Red+Cone	0.26	0.44	22.0	—	—	Miller
	Gulf III modified	0.0320	Red+Cone+Enc	0.38	0.44	13.1	—	—	Bridger
	Hardy recorder	0.00016	Red+Disc+Enc	0.22	0.37	11.8	—	—	Glover

*Cyl: Cylinder; Red: Reduction cone. Enc: Encased gauze section; Disc: filter disc.
**Towing velocity: 100 cm/sec.

multiplied by the coefficient of filtration, allows one to estimate the volume of water effectively filtered. Examples and data are given in Table 14-2 for the characteristics of operation of some nets in current use and which have international acceptance.

14-6.—EQUIVALENTS AND CORRESPONDENCIES

Some data used in the calculations are grouped in the following paragraph.

(a) *Light energy*

1 g-cal/cm^2/min or: 1 cal/cm^2/min = 0.07 watts/cm^2, since 1 cal = 4.185 joules and 1 watt = 1 joule/sec. 1 kilolux of solar radiation in the visible part of the spectrum (380–720 nm) is equal to:

0.006 cal/cm^2/min
1×10^{-4} cal/cm^2/sec
4.185×10^{-4} joule/cm^2/sec
0.4185 mW/cm^2

1 quantum of light energy (1 photon/sec) = $(19.86 \times 10^{-17}W)/\lambda$; λ is the wave length considered expressed in nanometers (10^{-9} m). 1 foot candle = 10.764 meter-candles or lux.

(b) *Chemical composition of the plankton*

Dry weight (DW) as a function of wet weight (WW). According to Beers & Stewart, 1969:

Phytoplankton: DW = 0.35 WW.

According to Mullin, 1969:

Zooplankton: DW = 0.13 WW.

According to Margalef & Vives 1967:

	Phytoplankton	Zooplankton	Total plankton
Dry Weight	100	100	100
Ash	23	7	15
Organic carbon	35	60	45
Nitrogen	6	10	8
Phosphorus	0.7	1.1	0.9
Chlorophyll	0.2–0.3		

According to Mullin, 1969:

Zooplankton: Organic carbon = 0.40 of dry weight
Organic carbon = 0.50 of dry weight without ash

According to Strickland, 1960:

Phytoplankton: Organic carbon = 60 chl *a*.

All these values are approximate and should be used with precaution. See also paragraphs 7-14 and 12-9.

For the estimation of carbon in the phytoplankton as a function of the cell volume or area, see Mullin, Sloan & Eppley, 1966.

(c) *Energetic equivalents*

According to Brody, 1945:

1 g of protein = 4.6 kcal or 0.95 l of O_2
1 g of glucose = 3.8 kcal or 0.745 l of O_2
1 g of lipids = 9.45 kcal or 2.14 l of O_2

According to Platt, 1971:

1 g of phytoplankton carbon = 15.8 kcal

According to Richman, 1958:

1 g of dry weight of fresh water phytoplankton = 5.269 kcal

Caloric equivalent of oxygen:
According to Sushchenya, 1970:

1 ml O_2 = 5 cal

According to Hargrave & Geen, 1970:

1 mg O_2 = 3.38 cal
or
1 ml O_2 = 4.83 cal

(d) *Equivalents for the time of development*

According to Winberg, 1956 in Edmondson & Winberg, 1971: Accepting that the variation of time of development as a function of the temperature follows the normal curve of Krogh, it is possible to obtain the time of development x_2, knowing the time of development at the temperature x_1 from the equation:

$t_{x_2} q_{x_1} = t_{x_1} q_{x_2}$

q_{x_1} and q_{x_2} are given in the following table:

Temperature (°C)	5	6	7	8	9	10	11
q	5.19	4.55	3.98	3.48	3.05	2.67	2.40
Temperature (°C)	12	13	14	15	16	17	
q	2.16	1.94	1.74	1.57	1.43	1.31	
Temperature (°C)	18	19	20	21	22	23	24
q	1.20	1.09	1.00	.920	.847	.779	.717
Temperature (°C)	25	26	27	28	29	30	
q	.659	.609	.563	.520	.481	.444	

Example for calculation: the time of development at 15°C(t_{15}) is equal to 3 days; the time of development at 25°C(t_{25}) is given by:

$$t_{25}q_{15} = t_{15}q_{25}$$

or

$$t_{25} \times 1.57 = 3.0 \times 0.659$$

$$t_{25} \equiv \frac{3.0 \times 0.659}{1.57} = 1.25 \text{ days.}$$

BIBLIOGRAPHY

Ackman, R. G., Eaton, C. A., Sipos, J. C., Hooper, S. N. & Castell, J. D., 1970.—Lipids and fatty acids of two species of north Atlantic Krill (*Meganyctiphanes norvegica* and *Thysanoessa inermis*) and their role in the aquatic food web. *J. Fish. Res. Bd. Can.*, **27**, 513–533.

Allen, K. R., 1951.—The Horokiwi stream. A study of a trout population. *N. Z. mar. Dep. Fish. Bull.*, **10**, 1–233.

Alvariño, A., 1964.—Bathymetric distribution of Chaetognaths. *Pacif. Sci.*, **18**, 64–82.

Alvariño, A., 1965.—Chaetognaths. *Oceanogr. mar. Biol. A. Rev.*, **3**, 115–194.

Anderson, G. C., 1969.—Subsurface chlorophyll maximum in the northeast Pacific ocean. *Limnol. Oceanogr.*, **14**, 386–391.

Anderson, G. C. & Zeutschel, R. P., 1970.—Release of dissolved organic matter by marine phytoplankton in coastal and offshore areas of the northeast Pacific ocean. *Limnol. Oceanogr.*, **15**, 402–407.

Anonymous, 1966.—*Determination of photosynthetic pigment in sea-water*. Monogr. oceanogr. Methodol. Unesco, **1**, 69 p.

Anonymous, 1968.—*Zooplankton sampling*. Monogr. oceanogr. Methodol., Unesco, **2**, 174 p.

Anraku, M., 1964.—Some technical problems encountered in quantitative studies of grazing and predation by marine plankton copepods. *J. oceanogr. Soc. Japan*, **20**, 221–231.

Armstrong, F. A. J., 1957.—The iron content of sea water. *J. mar. biol. Ass. U.K.* **36**, 509–517.

Armstrong, F. A. J., 1965.—Silicon. In *Chemical Oceanography*, ed. Riley & Skirrow, Academic Press, Lond. & New York, **1**, 409–432.

Armstrong, F. A. J., Butler, E. I. & Boalch, G. T., 1970.—Hydrographic and nutrient chemistry surveys in the western English Channel during 1961 and 1962. *J. mar. biol. Ass. U.K.*, **50**, 883–906.

Armstrong, F. A. J. & Tibbits, S., 1968.—Photochemical combustion of organic matter in sea water, for nitrogen, phosphorus and carbon determination. *J. mar. biol. Ass. U.K.*, **48**, 143–152.

Arnon, D. I., 1958.—Some functional aspects of inorganic micro-nutrients in the metabolism of green plants. In *Perspectives in marine biology*, ed. Buzzati-Traverso, Univ. Calif. Press, Berkeley & Los Angeles, 351–383.

Arrhenius, G., 1952.—Sediment cores from the East Pacific. *Rep. swed. Deep-Sea Exped.*, **5**, 122–227.

Arthur, C. R. & Rigler, F. H., 1967.—A possible source of error in the 14C method of measuring primary productivity. *Limnol. Oceanogr.*, **12**, 121–124.

Atkins, W. R. G., 1922.—The hydrogen ion concentration of open sea water and its variation with depth and season. *J. mar. biol. Ass. U.K.*, **13**, 754–771.

Bachrach, E. & Lucciardi, N., 1932.—Influence de la concentration en ion hydrogène (pH) sur la multiplication de quelques diatomées marines. *Rev. Algol.*, **6**, 251–261.

Bainbridge, R., 1957.—The size, shape and density of marine phytoplankton concentrations. *Biol. Rev.*, **32**, 91–115.

Bainbridge, R., 1961.—Migrations. In *The physiology of Crustacea*, II, ed. Waterman, Academic Press, Lond. & New York, 431–463.

Baker, A. de C., 1970.—The vertical distribution of euphausiids near Fuertaventura, Canary Islands ("Discovery" SOND Cruise, 1965). *J. mar. biol. Ass. U.K.*, **50**, 301–342.

Ballantine, D., 1953.—Comparison of the different methods of estimating nannoplankton. *J. mar. biol. Ass. U.K.*, **32**, 129–147.

Ballantine, D., 1956.—Two new marine species of *Gymnodinium* isolated from the Plymouth area. *J. mar. biol. Ass. U.K.*, **35**, 467–474.

Barber, R. T., Dugdale, R. L., MacIsaac, J. J. & Smith, R. L., 1971.—Variations in phytoplankton growth associated with the source and conditioning of upwelling water. *Inv. pesq.*, **35**, 171–193.

Barber, R. T. & Ryther, J. H., 1969.—Organic chelators: factors affecting primary production in the Cromwell Current upwelling. *J. exp. mar. Biol. Ecol.*, **3**, 191–199.

Barker, H. A., 1935.—The culture and physiology of marine dinoflagellates. *Arch. Mikrobiol.*, **6**, 157–181.
Barnes, H., 1957.—Nutrient elements. In *Treatise on marine Ecology and Paleoecology*, ed. Hedgdeth, Geol. Soc. America, Mem., **67**, 297–344.
Barnes, H. & Tranter, D. J., 1965.—A statistical examination of the catches, numbers and biomass taken by three commonly used plankton nets. *Aust. J. mar. Freshwat. Res.*, **16**, 293–306.
Bary, M. B., Stefano, J. G. de, Forsyth, M. & Kerkhof, J. Van den, 1958.—A closing, high-speed plankton catcher for use in vertical and horizontal towing. *Pacif. Sci.*, **12**, 46–59.
Baylor, E. R. & Sutcliffe, W. H., 1963.—Dissolved organic matter in sea water as a source of particulate food. *Limnol. Oceanogr.*, **8**, 369–371.
Bé, A. W. H., 1967.—Foraminifera. Families *Globigerinidae* and *Globorotaliidae*. *Fiches identif. zooplancton, Cons. Explor. Mer*, n° 108.
Bé, A., 1968.—Measuring total plankton biomass. In *Zooplankton sampling. Monogr. oceanogr. Methodol.*, Unesco, **2**, 173–174.
Beauchamp, P. de, 1960.—Classe des Chétognathes. In *Traité de Zoologie*, éd. Grassé, Masson, Paris, **5** (2), 1500–1520.
Beers, J. R., 1966.—Studies on the chemical composition of the major zooplankton groups in the Sargasso Sea off Bermuda. *Limnol. Oceanogr.*, **11**, 520–528.
Beers, J. R. & Stewart, G. L., 1969.—Microzooplankton and its abundance relative to the zooplankton and other seston components. *Mar. Biol.*, **4**, 182–189.
Beers, J. R. & Stewart, G. L., 1971.—Microzooplankters in the plankton communities of the upper waters of the eastern tropical Pacific. *Deep-Sea Res.*, **18**, 861–883.
Beeton, A. M., 1959.—Photoreception in the opossum shrimp, *Mysis relicta* Loven. *Biol. Bull.*, **116**, 204–216.
Beeton, A. M., 1960.—The vertical migration of *Mysis relicta* in Lakes Huron and Michigan. *J. Fish. Res. Bd Can.*, **17**, 517–539.
Beklemishev, C. W., 1962.—Superfluous feeding of marine herbivorous zooplankton. *Rapp. Cons. Explor. Mer*, **153**, 108–113.
Belser, W. L., 1959.—Bioassay of organic materials in sea water. *Int. Ocean. Congress Reprints*, A.A.A.S. Washington, 908–909.
Bernard, F. & Lecal, J., 1960.—Plancton unicellulaire récolté dans l'Océan Indien par le Charcot (1950) et le Norsel (1955–1956). *Bull. Inst. oceanogr. Monaco*, 1166, 1-59.
Berner, A., 1962.—Feeding and respiration in the copepod *Temora longicornis* (Mueller). *J. mar. biol. Ass. U.K.*, **42**, 625–640.
Berrill, N. J., 1948.—The gonads, larvae and budding of the polystyelid ascidians *Stolonica* and *Distomus*. *J. mar. biol. Ass. U.K.*, **27**, 633–650.
Bhaud, M., 1969.—Étude de la migration verticale quotidienne des larves de *Mesochaetopterus sagittarius* à Nosy-Bé (Madagascar). *Mar. biol.*, **4**, 28–35.
Bhaud, M., 1971.—Aspects systématiques et biogéographiques de l'étude des larves planctoniques d'Annélides polychètes. *Thèse d'État, Univ. Paris VI*, 464 p. manusc.
Black, C. C., 1971.—Ecological implications of dividing plants into groups with distinct photosynthetic production capacities. *Adv. ecol. Res.*, **7**, 87–114.
Blanc, F., Leveau, M. & Szekielda, K. H., 1969.—Effets eutrophiques au débouché d'un grand fleuve (Grand Rhône). *Mar. Biol.*, **3**, 233–242.
Blanke, W., 1956.—*The sea fisheries of North-west Europe. Structure and problems.* Kröger, Hamburg, 611 p.
Blaxter, J. H. S. & Currie, R. I., 1967.—The effects of artificial lights on acoustic scattering layers in the oceans. *Symp. zool. Soc. Lond.*, **19**, 1–14.
Blumer, M., Guillard, R. R. L. & Chase, T., 1971.—Hydrocarbons of marine phytoplankton. *Mar. Biol.*, **8**, 183–189.
Boden, B. P. & Kampa, E. M., 1967.—The influence of natural light on the vertical migrations of an animal community in the sea. *Symp. zool. Soc. Lond.*, **19**, 15–26.
Bogorov, B. G., 1946.—Peculiarities of diurnal vertical migration of zooplankton in Polar Seas. *J. mar. Res.*, **6**, 25–32.

Bogorad, L., 1962.—Chlorophylls. In *Physiology and Biochemistry of Algae*, ed. Lewin, Academic Press, Lond. & New York, 385–408.

Bond, R. M., 1934.—Digestive enzymes of the pelagic copepod *Calanus finmarchicus*. *Biol. Bull.*, **67**, 461–465.

Bottazzi Massera, E. & Nencini, G., 1969.—*Acantharia*. Order: *Holacantha*. Family: *Acanthochiasmidae*. *Fiches identif. zooplancton, Cons. Explor. Mer.*, n° 114.

Bougis, P., 1962.—Le cuivre en écologie marine. *Pubbl. Sta. zool. napoli*, 32 (Suppl.), 497–514.

Bougis, P., 1964.—Sur le développement des plutéus *in vitro* et l'interprétation du test de Wilson, *C.R. Acad. Sci. Paris*, **259**, 1250–1253.

Bougis, P., 1967.—*Le plancton. Coll.* "Que sais-je?", n° 1241. Presses univ. Fr., Paris, 128 p.

Bourdillon, A., 1971.—L'échantillonnage du zooplancton marin. In *Problèmes d'écologie: l'échantillonnage des peuplements animaux des milieux aquatiques*. Masson, Paris 109–184.

Bowen, R. A., Onge, J. M., Colton, J. B. & Price, C. A., 1972.—Density gradient centrifugation as an aid to sorting planktonic organisms. I. Gradient materials. *Mar. Biol.*, **14**, 242–247.

Boyd, C. M., 1967.—The benthic and pelagic habitats of the red crab, *Pleurocondes planipes*. *Pacif. Sci.*, **21**, 394–403.

Boysen-Jensen, D., 1919.—Valuation of the Limfjord. I. *Rep. dan. biol. Sta.*, **26**, 1–24.

Braarud, T., 1966.—Initial stock versus invasion, grazing and sinking. In *Marine Biology*, II, ed. Oppenheimer, Acad. Sci., New York, 178–195.

Braconnot, J. C., 1967.—Sur la possibilité d'un cycle court de développement chez le Tunicier pélagique *Doliolum nationalis* Borg. *C.R. Acad. Sci., Paris*, **264**, 1434–1437.

Braconnot, J. C., 1968.—Distribution du tunicier pélagique *Salpa fusiformis* Cuvier à Villefranche-sur-Mer. *Rapp. Comm. int. Mer Médit.*, **19**, 481–482.

Brand, T., Rakestraw, N. W. & Renn, C. E., 1939.—Further experiments on the decomposition and regeneration of nitrogenous organic material in sea water. *Biol. Bull.*, **77**, 285–296.

Brinckmann-Voss, A., 1970.—Anthomedusae-Athecatae (Hydrozoa, Cnidaria) of the Mediterranean. I. Capitata. *Fauna e Flora del Golfo di Napoli*, **39**, 1–96.

Brinton, E., 1967.—Vertical migration and avoidance capability of Euphausiids in the California Current. *Limnol. Oceanogr.*, **12**, 451–483.

Brody, S., 1945.—*Bioenergetics and growth*. Reinhold, New York, 1 023 p.

Brongersma-Sanders, 1948.—The importance of upwelling water to the Vertebratae Paleontology and Oil-Geology. *Verh. Khl. Nederl. Med. Amsterdam*, sect. 2, 45, n° 4, 1–112.

Brouardel, J., 1971.—Variations, dans le temps, de la production primaire de la mer au voisinage de Monaco. *Mem. Inst. océanogr. Monaco*, **3**, 1–20.

Brouardel, J., 1973.—Influence de la température sur la fixation du carbone à l'obscurité par le plancton. Étude expérimentale. *Mem. Inst. oceanogr. Monaco*, **6**, 1–10.

Brouardel, J. & Rinck, E., 1963.—Mesure de la production organique en Méditerranée dans les parages de Monaco, à l'aide du 14C. *Ann. Inst. océanogr.*, **40**, 109–164.

Brown, A. H., 1953.—The effect of light on respiration using isotopically enriched oxygen. *Am. J. Botany*, **40**, 719–731.

Brown, A. H. & Webster, G. C., 1953.—The influence of the light on the rate of respiration of the blue-green algae *Anabaena*. *Am. J. Botany*, **40**, 753–756.

Buch, K., 1951.—Das Kohlensäure gleichgewichtsystem in Meerwasser. *Havsforskn. Inst. Skr.*, n° 151, 1–18.

Budd, J. A., 1969.—Catabolism of trimethylamine by a marine bacterium, *Pseudomonas* NCMB 1154. *Mar. Biol.*, **4**, 257–266.

Bunt, J. S. & Lee, C. C., 1970.—Seasonal primary production in Antarctic Sea Ice at McMurdo Sound in 1967. *J. mar. Res.*, **28**, 304–320.

Butler, E. I., Corner, E. D. & Marshall, S. M., 1970.—On the nutrition and metabolism of zooplankton. VII. Seasonal survey of nitrogen and phosphorus excretion by *Calanus* of the Clyde Sea area. *J. mar. biol. Ass. U.K.*, **50**, 525–560.

Caperon, J., 1968.—Population growth response of *Isochrysis galbana* to nitrate variation at limiting concentrations. *Ecology*, **49**, 866–872.

Carlucci, A. F. & Schubert, H. R., 1969.—Nitrate reduction in sea water of the deep nitrite maximum off Peru. *Limnol. Oceanogr.*, **14**, 187–193.

Carlucci, A. F. & Silbernagel, S. B., 1966.—Bioassay of sea water. III. Distribution of vitamin B12 in the northeast Pacific Ocean. *Limnol. Oceanogr.*, **11**, 642–646.

Cassie, R. M., 1962.—Frequency distribution models in the ecology of plankton and other organisms. *J. anim. Ecol.*, **31**, 65–92.

Cassie, R. M., 1963.—Microdistribution of plankton. *Oceanogr. mar. Biol. A. Rev.*, **1**, 223–252.

Cassie, R. M., 1968.—Sample design. In *Zooplankton sampling*. Monogr. oceanogr. Methodol., Unesco, **2**, 105–121.

Chamberlain, W. & Shapiro, J., 1969.—On the biological significance of phosphate analysis; comparison of standard and new methods with a bioassay. *Limnol. Oceanogr.*, **14**, 921–927.

Champalbert, G., 1969.—Microdistribution d'un Pontellidae dans le Golfe de Marseille: *Anomalocera patersoni*. *Mar. Biol.*, **2**, 346–349.

Clarke, G. L., 1934.—Further observations on the diurnal migration of copepods in the Gulf of Maine. *Biol. Bull.*, **67**, 432–448.

Clutter, R. I. & Anraku, M., 1968.—Avoidance of samplers. In *Zooplankton sampling*. Monogr. oceanogr. Methodol., Unesco, **2**, 57–76.

Colebrook, J. M., 1964.—Continuous plankton records: a principal component analysis of the geographical distribution of zooplankton. *Bull. mar. Ecol.*, **6**, 78–100.

Colebrook, J. M., 1965.—The analysis of variation in the plankton, the environment and the fisheries. *Spec. Publ. int. Comm. Nthw. Atlant. Fish.*, **6**, 291–302.

Colebrook, J. M., 1969.—Variability in the plankton. *Progr. oceanogr.*, **5**, 115–125.

Comita, G. W., 1968.—Oxygen consumption in *Diaptomus*. *Limnol. Oceanogr.*, **13**, 51–57.

Conover, R. J., 1959.—Regional and seasonal variation in the respiratory rate of marine copepods. *Limnol. Oceanogr.*, **4**, 259–268.

Conover, R. J., 1960.—The feeding behavior and respiration of some marine planktonic crustacea. *Biol. Bull.*, **119**, 399–415.

Conover, R. J., 1961.—The turnover of phosphorous by *Calanus finmarchicus*. *J. mar. biol. Ass. U.K.*, **41**, 484–488.

Conover, R. J., 1966a.—Assimilation of organic matter by zooplankton. *Limnol. Oceanogr.*, **11**, 338–345.

Conover, R. J., 1966b.—Factors affecting the assimilation of organic matter by zooplankton and the question of superfluous feeding. *Limnol. Oceanogr.*, **11**, 346–354.

Conover, R. J., 1966c.—Feeding on large particles by *Calanus hyperboreus* (Kröyer). In *Some contemporary studies in marine science*, ed. H. Barnes, Allen & Unwin, Lond., 187–194.

Conover, R. J. & Corner, E. D. S., 1968.—Respiration and nitrogen excretion by some marine zooplankton in relation to their life cycle. *J. mar. biol. Ass. U.K.*, **48**, 49–75.

Cooper, L. H. N., 1935.—Liberation of phosphate in sea water by the breakdown of plankton organisms. *J. mar. biol. Ass. U.K.*, **20**, 197–200.

Cooper, L. H. N., 1948.—The distribution of iron in the waters of the western English Channel. *J. mar. biol. Ass. U.K.*, **27**, 297–325.

Cooper, L. H. N., 1955.—Deep water movements in the north Atlantic as a link between climatic changes around Iceland and biological productivity of the English Channel and Celtic Sea. *J. mar. Res.*, **14**, 347–362.

Corner, E. D. S., 1961.—On the nutrition and metabolism of zooplankton. I. Preliminary observations on the feeding of the marine copepod, *Calanus helgolandicus* (Claus). *J. mar. biol. Ass. U.K.*, **41**, 5–16.

Corner, E. D. S. & Cowey, C. B., 1968.—Biochemical studies on the production of marine zooplankton. *Biol. Rev.*, **43**, 393–426.

Corner, E. D. S. & Davies, A. G., 1971. Plankton as a factor in the nitrogen and phosphorus cycles in the sea. *Adv. mar. Biol.*, **9**, 101–204.

Coughlan, J., 1969.—The estimation of filtering rate from the clearance of suspensions. *Mar. Biol.*, **2**, 356–358.

Cowey, C. B., 1956.—A preliminary investigation of the variation of vitamin B12 in oceanic and coastal waters. *J. mar. biol. Ass. U.K.*, **35**, 609–620.

Cowey, C. B. & Corner, E. D. S., 1963a.—Amino acid composition and some other nitrogenous constituents in *Calanus finmarchicus. J. mar. biol. Ass. U.K.*, **43**, 485–493.

Cowey, C. B. & Corner, E. D. S., 1963b.—On the nutrition and metabolism of zooplankton. II. The relationship between the marine copepod *Calanus helgolandicus* and particulate material in Plymouth sea water in terms of amino acid composition. *J. mar. biol. Ass. U.K.*, **43**, 495–511.

Craigie, J. S. & McLachlan, J., 1964.—Excretion of colored ultraviolet absorbing substances by marine algae. *Can. J. Botany*, **42**, 23–33.

Crouzet, P., 1972.—Contribution à la connaissance de la physicochimie et de la production primaire du lac nord de Tunis. *Thèse de spécialité, Univ. Paris VI*, 72 p. manusc.

Culver, D. A. & Brunskill, G. J., 1969.—Fayetteville Green Lake. V. Studies of primary production and zooplankton in a meromictic marl lake. *Limnol. Oceanogr.*, **14**, 862–873.

Cummins, K. C. & Wuychek, J. C., 1970.—Caloric equivalents for investigations in ecological energetics. *Verh. int. Ver. Limnol.*, **18**, 1–158.

Curl, H., 1962a.—Standing crops of carbon, nitrogen and phosphorus and transfer between trophic levels, in continental shelf waters south of New York. *Rapp. Cons. Explor. Mer*, **153**, 183–189.

Curl, H., 1962b.—Analyses of carbon in marine plankton organisms. *J. mar. Res.*, **20**, 181–188.

Curl, H. & McLeod, G. C., 1961.—The physiological ecology of a marine diatom, *Skeletonema costatum* (Grev.) Cleve. *J. mar. Res.*, **19**, 70–88.

Curl, H. & Small, L., 1965.—Variations in photosynthetic assimilation ratios in natural, marine phytoplankton communities. *Limnol. Oceanogr.*, 10 (Suppl.), R62-R67.

Cushing, D. H., 1951.—The vertical migration of planktonic Crustacea. *Biol. Rev.*, **26**, 158–192.

Cushing, D. H., 1959.—On the nature of production in the sea. *Fish. Invest. Lond.*, Ser. 2, **22**, 1–40.

Cushing, D. H., 1964.—The work of grazing in the sea. In *Grazing in terrestrial and marine environments*, ed. D. J. Crisp, Blackwell, Oxford & Edinburgh, 207–225.

Cushing, D. H., 1973.—*The detection of fish.* Pergamon Press, Oxford, 200 p.

Cushing, D. H. & Nicholson, H. F., 1966.—Method for estimating algal production rates at sea. *Nature*, **212**, 310–311.

Dallot, S. & Ducret, F., 1969.—Un Chaetognathe mésoplanctonique nouveau: *Sagitta megalophthalma* sp. n., *Beaufortia*, **17**, 13–20.

Daumas, R. & Fiala, M., 1969.—Évaluation de la matière organique vivante dans les eaux marines par la mesure de l'adénosine triphosphate. *Mar. Biol.*, **3**, 243–246.

David, P. M., 1967.—Illustrations of oceanic neuston. *Symp. zool. Soc. Lond.* **19**, 211–213.

Davies, P. S., 1966.—Physiological ecology of *Patella*. I. The effect of body size and temperature on metabolic rate. *J. mar. biol. Ass. U.K.*, **46**, 647–658.

Davis, C. C., 1955.—*The marine and freshwater plankton.* Univ. Press Michigan, 541 p.

Davis, C. C., 1967.—Circulation of matter versus flow of energy in production studies. *Arch. hydrobiol.*, **63**, 250–255.

Deacon, G. E. R., 1933.—A general account of the hydrology of the south Atlantic ocean. *Discovery Rep.*, **7**, 171–238.

Deevey, G. B., 1971.—The annual cycle in quantity and composition of the zooplankton of the Sargasso sea off Bermuda. I. The upper 500 m. *Limnol. Oceanogr.*, **16**, 219–240.

Digby, P. S. B., 1950.—The biology of the small planktonic copepods of Plymouth. *J. mar. biol. Ass. U.K.*, **29**, 393–438.

Droop, M. R., 1955.—A pelagic marine diatom requiring cobalamin. *J. mar. biol. Ass. U.K.*, **34**, 229–231.

Droop, M. R., 1958a.—Requirement for thiamine among some marine and supralittoral protista. *J. mar. biol. Ass. U.K.*, **37**, 323–329.

Droop, M. R., 1958b.—Optimum relative and actual ionic concentrations for growth of some euryhaline algae. *Verh. int. Ver. Limnol.*, **13**, 722–730.

Droop, M. R., 1961.—Vitamin B12 and marine ecology: the response of *Monochrysis lutheri. J. mar. biol. Ass. U.K.*, **41**, 69–76.

Droop, M. R., 1970.—Nutritional investigation of phagotrophic Protozoa under axenic conditions. *Helgoländer wiss. Meeresunters.*, **20**, 272–277.

Dugdale, R. C., Goering, J. J. & Ryther, J. H., 1964.—High nitrogen fixation rates in the Sargasso sea and the Arabian sea. *Limnol. Oceanogr.*, **9**, 507-510.
Dussart, B. H., 1965.—Les différentes catégories de plancton. *Hydrobiologia*, **26**, 72-74.
Dussart, B. H., 1965.—*Limnologie. L'étude des eaux continentales.* Gauthier-Villars, Paris, 618 p.
Duursma, E. K., 1961.—Dissolved organic carbon, nitrogen and phosphorus in the sea. *Neth. J. Sea Res.*, **1**, 1-147.
Edmondson, W. T. & Winberg, G. C., 1971.—*A manual on methods for the assessment of secondary productivity in freshwaters.* I.B.P. Handbook n° 17, Blackwell, Oxford & Edinburgh, 358 p.
Einarsson, H., 1945.—*Euphausiacea.* I. Northern Atlantic species. *Dana Rep.*, **27**, 1-185.
Eppley, R. W., Coastworth, J. L. & Solorzano, L., 1969.—Studies on nitrate reductase in marine phytoplankton. *Limnol. Oceanogr.*, **14**, 194-205.
Eppley, R. W., Holm-Hansen, O. & Strickland, J. D. H., 1968.—Some observations on the vertical migrations of dinoflagellates. *J. Phycol.*, **4**, 333-340.
Eppley, R. W., Rogers, J. N. & McCarthy, J. J., 1969.—Half-saturation constants for uptake of nitrate and ammonium by marine phytoplankton. *Limnol. Oceanogr.*, **14**, 912-920.
Eppley, R. W. & Strickland, J. D. H., 1968.—Kinetics of marine phytoplankton growth. *Adv. Microbiol. Sea*, **1**, 23-62.
Eppley, R. W. & Thomas, W. H., 1969.—Comparison of half-saturation "constants" for growth and nitrate uptake of marine phytoplankton. *J. Phycol.*, **5**, 365-369.
Fee, E. J., 1969.—A numerical model for the estimation of photosynthetic production integrated over time and depth, in natural waters. *Limnol. Oceanogr.*, **14**, 906-911.
Fenaux, L., 1962.—Maturation des gonades et cycle saisonnier des larves chez *A. lixula*, *P. lividus* et *P. microtuberculatus* (Échinides) à Villefranche-sur-Mer. *Vie Milieu*, **19**, 1-52.
Fenaux, L., 1968.—Aspects écologiques de la reproduction chez les Échinides et les Ophiurides de Villefranche-sur-Mer. *Thèse d'État, Fac. Sci. Univ. Paris*, 193 p. manusc.
Fenaux, R., 1963.—Écologie et biologie des Appendiculaires méditerranéens (Villefranche-sur-Mer). *Vie et Milieu*, Suppl., **16**, 142 p.
Fenaux, R., 1967.—*Les appendiculaires des mers d'Europe et du Bassin méditerranéen.* Faune Europe Bassin méditerranéen 2, Masson, Paris, 116 p.
Finenko, Z. Z. & Zaika, V. E., 1970.—Particulate organic matter and its role in the productivity of the sea. In *Marine Food Chains*, ed. J. H. Steele, Oliver & Boyd, Edinburgh, 32-44.
Fitzgerald, G. P., 1968.—Detection of limiting or surplus nitrogen in algae and aquatic weeds. *J. Phycol.*, **4**, 121-126.
Fleming, R. H., 1940.—The composition of plankton and units for reporting population and production. *Proc. sixth pacif. Sci. Congr. Calif.*, **3**, 535-540.
Fournier, R. O., 1970.—Studies on pigmented microorganisms from aphotic marine environments. *Limnol. Oceanogr.*, **15**, 675-682.
Fraser, J. H., 1939.—Distribution of Chaetognatha in Scottish waters in 1937. *J. Cons.*, **14**, 25-34.
Fraser, J. H., 1961.—The survival of larval fish in the northern north sea according to the quality of sea water. *J. mar. biol. Ass. U.K.*, **41**, 305-312.
Fraser, J., 1962.—*Nature adrift. The story of marine plankton.* Foulis, London, 178 p.
Fraser, J. H., 1969.—Variability in the oceanic content of plankton in the scottish area. *Progr. Oceanogr.*, **5**, 149-159.
Fraser, J. H., 1970.—The ecology of the ctenophore *Pleurobranchia pileus* in scottish waters. *J. Cons.*, **33**, 149-168.
Fretter, V. & Pilkington, M. C., 1970.—*Prosobranchia.* Veliger larvae of *Taenioglossa* and *Stenoglossa.* Fiches identif. zooplancton, Cons. Explor. Mer, n° 129-132.
Frontier, S., 1969.—Sur une méthode d'analyse faunistique rapide du zooplancton. *J. exp. mar. Biol. Ecol.*, **3**, 18-26.
Frontier, S., 1971.—Étude statistique de la dispersion du plancton. In Zooplancton d'une baie eutrophique tropicale. *Doc. sci. Nosy-Bé, Orstom*, **24**, 55-95.
Frontier, S., 1972.—Calcul de l'erreur sur un comptage de zooplancton. *J. exp. mar. Biol. Ecol.*, **8**, 121-132.

Gaudy, R., 1970.—Contribution à la connaissance du cycle biologique et de la physiologie des Copépodes du golfe de Marseille. *Thèse d'état, Univ. Marseille* II, 294 p. manusc.

Gauld, D. T., 1951.—The grazing rate of planktonic copepods. *J. mar. biol. Ass. U.K.*, **29**, 695–706.

Gauld, D. T. & Raymont, J. E. G., 1953.—The respiration of some planktonic copepods. II. The effect of temperature. *J. mar. biol. Ass. U.K.*, **31**, 447–460.

Gerloff, G. C. & Skoog, F., 1954.—Cell contents of nitrogen and phosphorus as a measure of their availability for growth of *Microcystis aeruginosa*. *Ecology*, **35**, 348–353.

Gilmartin, M., 1964.—The primary production of a British Columbia fjord. *J. Fish. Res. Bd. Can.*, **21**, 505–538.

Giraud, G., 1959.—Sur les rapports entre le spectre d'absorption et l'intensité photosynthétique de *Rhodosorus marinus* Geitler (Rhodophycée). *C.R. Acad. Sci., Paris*, **248**, 277–280.

Glover, R. S., 1967.—The continuous plankton recorder survey of the north Atlantic. *Symp. zool. Soc. Lond.*, **19**, 189–210.

Glover, R. S., Robinson, G. A. & Colebrook, J. M., 1970.—Plankton in the north Atlantic. An example of the problems of analysing variability in the environment. FAO technical conference on Marine Pollution. *FAO MP/70/E* **55**, 1–14.

Goering, J. J., Dugdale, R. C. & Menzel, D. W., 1966.—Estimates of *in situ* rates of nitrogen uptake by *Trichodesmium* sp. in the tropical Atlantic ocean. *Limnol. Oceanogr.*, **11**, 614–620.

Gold, K., 1970.—Cultivation of marine ciliates (Tintinnida) and heterotrophic flagellates. *Helgöländer wiss. Meeresunters.*, **20**, 264–271.

Goldberg, E. D., Walker, T. J. & Whisenand A., 1951.—Phosphate utilization by diatoms. *Biol. Bull.*, **101**, 274–284.

Gostan, J., 1968.—Contribution à l'étude hydrologique du bassin liguro-provençal entre la Riviera et la Corse: distribution et variations saisonnières de la température, de la salinité, de l'oxygène et des phosphates minéraux dissous dans les masses d'eau superficielles, intermédiaires et profondes. *Thèse d'État, Fac. Sci. Univ. Paris*, 206 p. manusc.

Goy, J., 1964.—Note sur l'écologie de l'hydroméduse *Persa incolorata* MacCrady devant Villefranche-sur-Mer. *Vie et Milieu*, **15**, 565–568.

Grall, J. R., 1966.—Détermination de la production de matière organique en Manche occidentale à l'aide du carbone 14. *C.R. Acad. Sci., Paris*, **262**, 2514–2517.

Grall, J. R., 1972.—Recherches quantitatives sur la production primaire du phytoplancton dans les parages de Roscoff. *Thèse d'État, Univ. Paris VI*, 247 p. manusc.

Grall, J. R. & Jacques, G., 1964.—Étude dynamique et variations saisonnières du phytoplancton de la région de Roscoff. I.B. Phytoplancton. *Cah. Biol. mar.*, **5**, 432–455.

Grassé, P. P., 1952.—*Traité de Zoologie.* T. 1-1, Phylogénèse, Protozoaires: Généralités, Flagellés. Masson, Paris, 1071 p.

Grassé, P. P., Poisson, R. A. & Tuzet, O., 1961.—*Zoologie. I. Invertébrés.* Masson, Paris, 919 p.

Greve, W., 1970.—Cultivation experiments on North Sea Ctenophores. *Helgöländer wiss., Meeresunters.*, **20**, 304–317.

Greze, V. N. & Baldina, E. P., 1964.—Population dynamics and the annual production of *Acartia clausi* and *Centropages kroeyeri* in the neritic zone of the Black Sea. *Trud. Sevastopol biol. Sta.*, 18. Fish. Res. Bd. Translation, 893.

Gueredrat, J. A. & Friess, R., 1971.—Importance des migrations nycthémérales de Copépodes bathypélagiques. *Cah. Orstom, Ser. Océanogr.*, **9**, 187–196.

Hamilton, D. H., 1969.—Nutrient limitation of summer phytoplankton growth in Cayuga lake. *Limnol. Oceanogr.*, **14**, 579–590.

Handa, N., 1969.—Carbohydrate metabolism in the marine diatom *Skeletonema costatum*. *Mar. Biol.*, **4**, 208–214.

Handa, N. & Yanagi, K., 1969.—Studies on water-extractable carbohydrates of the particulate matter from the northwest Pacific ocean. *Mar. Biol.*, **4**, 197–207.

Hansen, W. J. & Dunbar, M. J., 1970.—Biological causes of scattering layers in the Arctic Ocean. *Proc. int. Symp. biol. sound scattering in the ocean*, ed. Brooke, Report 005, Maury Center, Washington, 508–526.

Hardy, A. C., 1935.—The plankton community, the whale fisheries and the hypothesis of animal exclusion. *Discovery Rep.*, **11**, 273-370.

Hardy, A., 1958.—*The open sea. Its natural history. Part I, The world of plankton*, 2ᵉ ed., Collins, London, 335 p.

Hargrave, B. T. & Geen, G. H., 1970.—Effects of copepod grazing on two natural phytoplankton populations. *J. Fish. res. Bd. Can.*, **27**, 1395-1403.

Harris, E., 1959.—The nitrogen cycle in Long Island Sound. *Bull. Bingham oceanogr. Coll.*, **17**, 31-65.

Harris, E. & Riley, G. A., 1956.—Oceanography of Long Island Sound, 1952-1954. VIII: Chemical composition of the plankton. *Bull. Bingham oceanogr. Coll.*, **15**, 315-323.

Harvey, H. W., 1934.—Measurement of phytoplankton population. *J. mar. biol. Ass. U.K.*, **19**, 761-773.

Harvey, H. W., 1937.—Notes on selective feeding by *Calanus. J. mar. biol. Ass. U.K.*, **22**, 97-100.

Harvey, H. W., 1947.—Manganese and the growth of phytoplankton. *J. mar. biol. Ass. U.K.*, **26**, 562-579.

Harvey, H. W., 1950.—On the production of living matter in the sea off Plymouth. *J. mar. biol. Ass. U.K.*, **29**, 97-137.

Harvey, H. W., 1955.—*The chemistry and fertility of sea waters.* Cambridge Univ. Press. 224 p.

Harvey, H. W., Cooper, L. H. N., Lebour, M. V. & Russell, F. S., 1935.—Plankton production and its control., *J. mar. biol. Ass. U.K.*, **20**, 407-441.

Hasler, A. D., 1937.—The physiology of digestion in planktonic crustacea. II. Further studies on the digestive enzymes of *Daphnia* and *Polyphemus, Diaptomus* and *Calanus. Biol. Bull.*, **72**, 290-298.

Haxo, F. T. & Blinks, L. R., 1950.—Photosynthetic action spectra of marine algae. *J. gen. Physiol.*, **33**, 389-422.

Head, P. C., 1971.—Observations on the concentration of iron in sea water, with particular reference to Southampton waters. *J. mar. biol. Ass. U.K.*, **51**, 891-903.

Heinle, D. R., 1966.—Production of a Calanoid Copepod, *Acartia tonsa*, in the Patuxent River Estuary. *Chesapeake Sci.*, **7**, 59-74.

Hendey, N. I., 1954.—Note on the Plymouth *Nitzschia* culture. *J. mar. biol. Ass. U.K.*, **33**, 335-339.

Hendey, N. I., 1964.—An introductory account of the smaller algae of British coastal waters. Part V: *Bacillariophyceae* (Diatoms). *Fish. Invest., Lond.*, Ser. 4, 1-317.

Heron, A. C., 1968.—Plankton gauze. In *Zooplankton sampling.* Monogr. oceanogr. Methodol., Unesco, **2**, 19-25.

Herrera, J. & Margalef, R., 1961.—Hidrografia y fitoplancton de las costas de Castellon, de julio de 1958 a junio de 1959. *Inv. pesq.*, **20**, 17-63.

Hobson, L. A. & Menzel, D. W., 1969.—The distribution and chemical composition of organic particulate matter in the sea and sediments off the east of South America. *Limnol. Oceanogr.*, **14**, 159-162.

Hoffmann, C., 1956.—Untersuchungen über die Remineralisation des Phosphors im Plankton. *Kieler Meeresforsch.*, **12**, 25-36.

Holmes, R. W., Williams, P. M. & Eppley, R. W., 1967.—Red water in La Jolla bay, 1964-1966. *Limnol. Oceanogr.*, **12**, 503-512.

Holm-Hansen, O. & Booth, C. R., 1966.—The measurement of adenosine triphosphate in the ocean and its ecological significance. *Limnol. Oceanogr.*, **11**, 510-519.

Holm-Hansen, O., Lorenzen, C. J., Holmes, R. W. & Strickland, J. D. H., 1965.—Fluorometric determination of chlorophyll. *J. Cons.*, **30**, 3-15.

Ikeda, T., 1970.—Relationship between respiration rate and body size in marine plankton animals as a function of the temperature of habitat. *Bull. Fac. Fish. Hokkaido*, **21**, 91-112.

Ivanoff, A., 1972.—*Introduction à l'Océanographie. Propriétés physiques et chimiques des eaux de mer.* Vuibert, Paris, I, 206 p.; II, à paraître.

Ivanoff, A. & Morel, A., 1970.—Terminologie concernant l'optique océanographique. *Cah. océanogr.*, **22**, 457-468.

Jackson, P., 1954.—Engineering and economic aspects of marine plankton harvesting. *J. Cons.*, **20**, 167–174.
Jacques, G., 1970.—Aspects quantitatifs du phytoplancton de la région de Banyuls-sur-Mer (Golfe du Lion) 1965–1969. *Thèse d'État, Univ. Paris VI*, 210 p. manusc.
Jawed, M., 1969.—Body nitrogen and nitrogenous excretion in *Neomysis rayii* Murdoch and *Euphausia pacifica* Hansen. *Limnol. Oceanogr.*, **14**, 748–754.
Jeffries, H. P., 1970.—Seasonal composition of temperature plankton communities: fatty acids. *Limnol. Oceanogr.*, **15**, 419–426.
Jenkin, P. M., 1937.—Oxygen production by the diatom *Coscinodiscus excentricus* Ehr. in relation to submarine illumination in the English Channel. *J. mar. biol. Ass. U.K.*, **22**, 301–343.
Jerlov, N. G., 1951.—Optical studies of ocean waters. *Rep. swed. Deep-Sea Exped.*, **3**, 1–59.
Jitts, H. R., Morel, A. & Saijo, Y., 1975.—The relation of marine primary production to available photosynthetic irradiance. *Austr. J. mar. Freshwater Res.*, in press.
Johannes, R. E., 1964.—Phosphorus excretion and body size in marine animals: microzooplankton and nutrient regeneration. *Science*, **146**, 923–924.
Johannes, R. E., 1968.—Nutrient regeneration in lakes and oceans. *Adv. Microbiol. Sea.* **1**, 203–213.
Johnstone, J., Scott, A. & Chadwick, H. C., 1924.—*The marine plankton.* Univ. Press, Liverpool, 194 p.
Kain, J. M. & Fogg, G. E., 1958a—Studies on the growth of marine phytoplankton. I. *Asterionella japonica* Gran. *J. mar. biol. Ass. U.K.*, **37**, 397–413.
Kain, J. M. & Fogg, G. E., 1958b.—Studies on the growth of marine phytoplankton. II. *Isochrysis galbana*. *J. mar. biol. Ass. U.K.*, **37**, 781–788.
Kain, J. M. & Fogg, G. E., 1960.—Studies on the growth of marine phytoplankton. III. *Prorocentrum micans* Ehrenberg. *J. mar. biol. Ass. U.K.*, **39**, 33–50.
Kamatani, A., 1971.—Physical and chemical characteristics of biogenous silica. *Mar. Biol.*, **8**, 89–95.
Kenchington, R. A., 1970.—An investigation of the detritus in Menai Straits plankton samples. *J. mar. biol. Ass. U.K.*, **50**, 489–498.
Kersting, K., 1972.—A nitrogen correction for caloric values. *Limnol. Oceanogr.*, **17**, 643–644.
Ketchum, B. H., 1939.—The absorption of phosphate and nitrate by illuminated cultures of *Nitzschia closterium*. *Amer. J. Bot.*, **26**, 399–407.
Ketchum, B. H. & Redfield, A. C., 1949.—Some physical and chemical characteristics of algae grown in mass cultures. *J. cell. comp. Physiol.*, **33**, 281–300.
Khailov, K. M. & Finenko, Z. Z., 1970.—Organic macromolecular compounds dissolved in sea-water and their inclusion into food chains. In *Marine food Chains*, ed. J. H. Steele, Oliver & Boyd, Edinburgh, 6–18.
Kiefer, D. & Strickland, J. D. H., 1970.—A comparative study of photosynthesis in sea water samples incubated under two types of light attenuator. *Limnol. Oceanogr.*, **15**, 408–412.
Kimball, H. H., 1928.—Amount of solar radiation that reaches the surface of the earth on the land and on the sea, and the methods by which it is measured. *Mon. Weath. Rev.*, **56**, 393–398.
Komaki, Y., 1967.—On the surface swarming of Euphausiid Crustaceans. *Pacif. Sci.*, **21**, 433–448.
Krause, H. R., 1964.—Zur Chemie und Biochemie der Zersetzung von Süsswasser-organismen, under besonderer Berücksichtigung des Abbaues der organischen Phosphorkomponenten. *Verh. int. Ver. Limnol.*, **15**, 549–561.
Kreps, E. & Verbinskaia, N., 1930.—Seasonal changes in the phosphate and nitrate content and in hydrogen ion concentration in the Barents sea. *J. Cons.*, **5**, 329–346.
Kuenzler, E. J., 1970.—Dissolved organic phosphorus excretion by marine phytoplankton. *J. Phycol.*, **6**, 7–13.
Kuenzler, E. J. & Ketchum, B. H., 1962.—Rate of phosphorus uptake by *Phaeodactylum tricornutum*. *Biol. Bull.*, **123**, 134–145.
Kuenzler, E. J. & Perras, J. P., 1965.—Phosphatases of marine algae. *Biol. Bull.*, **128**, 271–284.
Laevastu, T., 1962.—The adequacy of plankton sampling. *Rapp. Cons. Explor. Mer*, **153**, 66–73.

Lasker, R., 1966.—Feeding, growth, respiration and carbon utilization of a Euphausiid Crustacean. *J. Fish. Res. Bd. Can.*, **23**, 1291–1317.

Lasker, R., 1970.—Utilization of zooplankton energy by a Pacific sardine population in the California current. In *Marine Food Chains*, ed. J. H. Steele, Oliver & Boyd, Edinburgh, 265–284.

Laval, Ph., 1968.—Observations sur la biologie de *Phronima curvipes* Voss. (Amphipode Hypéride) et description du mâle adulte. *Cah. biol. mar.*, **9**, 347–362.

Laval, Ph., 1972.—Comportement, parasitisme et écologie d'*Hyperia schizogeneios* Stebb (Amphipode hypéride) dans le plancton de Villefranche-sur-Mer. *Ann. Inst. oceanogr., Paris*, **48**, 49–74.

Laval, Ph., 1974.—Un modèle mathématique de l'évitement d'un filet à plancton, son application pratique, et sa vérification indirecte en recourant au parasitisme de l'amphipode hypéride *Vibilia armata* Bovallius. *J. exp. mar. Biol. Ecol.*, **14**, 57–87.

Lebour, M. V., 1922.—The food of plankton organisms. *J. mar. biol. Ass. U.K.*, **12**, 644–677.

Lebour, M. V., 1923.—The food of plankton organisms II. *J. mar. biol. Ass. U.K.*, **13**, 70–92.

Lebour, M. V., 1925.—*The dinoflagellates of Northern seas.* Mar. biol. Ass. U.K., 250 p.

Lebour, M. V., 1944.—The larval stages of *Portumnus (Crustacea Brachyura)* with notes on some other genera. *J. mar. biol. Ass. U.K.*, **26**, 7–15.

Lee, R. F., Hirota, J. & Barnett, A. M., 1971.—Distribution and importance of wax esters in marine copepods and other zooplankton. *Deep-Sea Res.*, **18**, 1147–1165.

Lee, R. F., Neuvenzel, J. C. & Paffenhöfer, G. A., 1971.—Importance of wax esters and other lipids in the marine food chain: phytoplankton and copepods. *Mar. Biol.*, **9**, 99–108.

Legendre, L. & Watt, W. D., 1970.—The distribution of primary production relative to a cyclonic gyre in Baie des chaleurs. *Mar. Biol.*, **7**, 167–170.

Lewin, J. C., Lewin, R. A. & Philpot, D. E., 1958.—Observations on *Phaeodactylum tricornutum*. *J. gen. Microbiol.*, **18**, 418–426.

Lincoln, R. J., 1970.—A laboratory investigation into the effects of hydrostatic pressure on the vertical migration of planktonic Crustacea. *Mar. Biol.*, **6**, 5–11.

Lohmann, H., 1920.—Die Bevölkerung des Ozeans mit Plankton nach den Ergebnissen der Zentrifugenfänge während der Ausreise der "Deutschland" 1911. *Arch. Biont.*, **4**, 1–617.

Lorenzen, C. J., 1966.—A method for the continuous measurement of *in vivo* chlorophyll concentrations. *Deep-Sea Res.*, **13**, 223–228.

McAllister, C. D., 1970.—Zooplankton rations, phytoplankton mortality and estimation of marine production. In *Marine food chains*, ed. J. H. Steele, Oliver & Boyd, Edinburgh, 419–457.

McAllister, C. D., LeBrasseur, R. J. & Parsons, T. R., 1972.—Stability of enriched aquatic ecosystems. *Science*, **175**, 562–564.

McAllister, C. D., Parsons, T. R. & Strickland, J. D. H., 1960.—Primary productivity and fertility at Station "P" in the northeast Pacific ocean. *J. Cons.*, **25**, 240–259.

McAllister, C. D., Shah, N. & Strickland, J. D. H., 1964.—Marine phytoplankton photosynthesis as a function of light intensity, a comparison of methods. *J. Fish. Res. Bd. Can.*, **21**, 159–181.

McCarthy, J. J., 1972.—The uptake of urea by natural populations of marine phytoplankton. *Limnol. Oceanogr.*, **17**, 738–748.

McGinn, M. P. & Gold, K., 1969.—Axenic cultivation of *Noctiluca scintillans*. *J. Protozool.*, **16** (suppl.).

McGowan, J. A. & Fraudorf, V. J., 1964.—A modified heavy fraction zooplankton sorter. *Limnol. Oceanogr.*, **9**, 152–155.

MacIsaac, J. J. & Dugdale, R. C., 1969.—The kinetics of nitrate and ammonia uptake by natural populations of marine phytoplankton. *Deep-Sea Res.*, **16**, 45–57.

McIsaac, J. J. & Olund, R. K., 1971.—An automated extraction procedure for the determination of ammonia in sea water. *Inv. pesq.*, **35**, 221–232.

McLaren, I. A., 1963.—Effects of temperature on growth of zooplankton and the adaptive value of vertical migration. *J. Fish. Res. Bd. Can.*, **20**, 685–727.

McLeod, G. C., 1958.—Delayed light action spectra of several algae in visible and ultraviolet light. *J. gen. Physiol.*, **42**, 243–250.
Maddux, W. S. & Kanwisher, J. W., 1965.—An *in situ* particle counter. *Limnol. Oceanogr.*, **10** (suppl.), 162–168.
Mann, K. H., 1969.—The dynamics of aquatic ecosystems. *Adv. ecol. Res.*, **6**, 1–81.
Margalef, R., 1957.—Nuevos aspectos del problema de la suspension en los organismos planctonicos. *Inv. pesq.*, **7**, 105–116.
Margalef, R., 1958.—Temporal succession and spatial heterogeneity in phytoplankton. In *Perspectives in marine biology*, ed. Buzzati-Traverso, Univ. Calif. Press, Berkeley & Los Angeles, 323–349.
Margalef, R., 1961.—Corrélations entre certains caractères synthétiques des populations de phytoplancton. *Hydrobiologia*, **18**, 155–164.
Margalef, R., 1963.—Algunas regularidades en la distribucion a escala pequeña y media de las problaciones marinas de fitoplancton y en sus caracteristicas funcionales. *Inv. pesq.*, **23**, 169–230.
Margalef, R., 1967a—El ecosistema. In *Ecologia marina*. Monografia n° 14, Fund. La Salle de Ciencias Naturales, Caracas, 377–453.
Margalef, R., 1967b—Las algas inferiores. In *Ecologia marina*. Monografia n° 14, Fund. La Salle de Ciencias Naturales, Caracas, 230–272.
Margalef, R. & Estrada, M., 1971.—Simple approaches to a pattern analysis of phytoplankton. *Inv. pesq.*, **35**, 269–297.
Margalef, R. & Vives, F., 1967.—La vida suspendita en las aguas. In *Ecologia marina*. Monografia n° 14, Fund. La Salle de Ciencias Naturales, Caracas, 493–562.
Marshall, P. T., 1958.—Primary production in the Arctic. *J. Cons.*, **23**, 173–177.
Marshall, S. M., 1933.—The production of microplankton in the Great Barrier Reef region. *Sc. Rep. brit. Mus. Great Barrier Reef Exped., 1923–1929*, **2**, 111–157.
Marshall, S. M., Nicholls, A. & Orr, A. P., 1935.—On the biology of *Calanus finmarchicus*. VI. Oxygen consumption in relation to environmental conditions. *J. mar. biol. Ass. U.K.*, **20**, 1–28.
Marshall, S. M. & Orr, A. P., 1930.—A study of the spring diatom increase in Loch Striven. *J. mar. biol. Ass. U.K.*, **16**, 853–878.
Marshall, S. M. & Orr, A. P., 1955.—*The biology of a marine copepod, Calanus finmarchicus (Gunnerus)*. Oliver & Boyd, Edinburgh, 188 p.
Marshall, S. M. & Orr, A. P., 1956.—On the biology of *Calanus finmarchicus*. IX. Feeding and digestion in the young stages. *J. mar. biol. Ass. U.K.*, **35**, 587–603.
Marshall, S. M. & Orr, A. P., 1961.—On the biology of *Calanus finmarchicus*. XIII. The phosphorus cycle; excretion, egg production, autolysis. *J. mar. biol. Ass. U.K.*, **41**, 463–488.
Marshall, S. M. & Orr, A. P., 1962.—Food and feeding in copepods. *Rapp. Cons. Explor. Mer*, **153**, 92–98.
Marshall, S. M. & Orr, A. P., 1964.—Grazing by copepods in the sea. In *Grazing in terrestrial and marine environments*. ed. D. J. Crisp, Blackwell, Oxford & Edinburgh, 227–238.
Martin, J. H., 1970.—Phytoplankton-zooplankton relationships in Narragansett Bay. IV. The seasonal importance of grazing. *Limnol. Oceanogr.*, **15**, 413–418.
Massuti, M. & Margalef, R., 1950.—*Introduccion al estudio del plancton marino*. Inst. Biol. Apl., Barcelona, 182 p.
Mauchline, J. & Fisher, L. R., 1969.—The biology of Euphausiids. *Adv. mar. Biol.*, **7**, 1–454.
Mayzaud, P., 1972.—Respiration and nitrogen excretion of zooplankton. II. Studies of the metabolic characteristics of starved animals. *Mar. Biol.*, **21**, 19–28.
Mayzaud, P. & Dallot, S., 1973.—Métabolisme du zooplancton: respiration et excrétion azotée. I. Étude des niveaux métaboliques de quelques espèces. *Mar. Biol.*, **19**, 307–314.
Mazza, J., 1964.—Premières observations sur les valeurs de poids sec chez quelques copépodes de Méditerranée. *Rev. Trav. Inst. Pêch. marit.*, **28**, 293–301.
Menzel, D. W. & Ryther, J. K., 1961.—Zooplankton in the Sargasso Sea off Bermuda and its relation to organic production. *J. Cons.*, **26**, 250–258.

Menzel, D. W. & Spaeth, J. P., 1962.—Occurrence of vitamin B12 in the Sargasso sea. *Limnol. Oceanogr.*, **7**, 151–154.
Michel, A. & Grandperrin, N. R., 1971.—Traitement des récoltes micronectoniques. *Mar. Biol.*, **8**, 238–242.
Miller, C. B., 1970.—Some environmental consequences of vertical migration in marine zooplankton. *Limnol. Oceanogr.*, **15**, 727–741.
Minas, H. J., 1970.—La distribution de l'oxygène en relation avec la production primaire en Méditerranée nord-occidentale. *Mar. Biol.*, **7**, 181–204.
Moore, H. B. & Corwin, E. G., 1956.—The effect of temperature, illumination and pressure in the vertical distribution of zooplankton. *Bull. mar. Sci. Gulf Caribb.*, **6**, 273–287.
Morel, A. & Caloumenos, L., 1972.—Analyse spectrale de l'énergie disponible pour la photosynthèse dans diverses régions marines typiques. *33ᵉ Assemblée Comm. int. Explor. Mer Médit.*
Morris, I., Yentsch, C. M. & Yentsch, C. S., 1971.—Relationship between light carbon dioxide fixation and dark carbon dioxide fixation by marine algae. *Limnol. Oceanogr.*, **16**, 854–858.
Morton, J. E., 1954.—The biology of *Limacina retroversa*. *J. mar. biol. Ass. U.K.*, **33**, 297–312.
Morton, J. E., 1958.—Observations on the Gymnosomatous Pteropod *Clione limacina* (Phipps). *J. mar. biol. Ass. U.K.*, **37**, 287–297.
Mullin, M. M., 1963.—Some factors affecting the feeding of marine copepods of the genus *Calanus*. *Limnol. Oceanogr.*, **8**, 239–250.
Mullin, M. M., 1969.—Production of zooplankton in the ocean: the present status and problems. *Oceanogr. mar. Biol. A. Rev.*, **7**, 293–314.
Mullin, M. M. & Brooks, E. R., 1967.—Laboratory culture, growth rate and feeding behaviour of a planktonic marine copepod. *Limnol. Oceanogr.*, **12**, 657–666.
Mullin, M. M. & Brooks, E. R., 1970.—Growth and metabolism of two planktonic marine copepods as influenced by temperature and type of food. In *Marine Food Chains*, ed. J. H. Steele, Oliver & Boyd, Edinburgh, 74–95.
Mullin, M. M., Sloan, P. R. & Eppley, R. W., 1966.—Relationship between carbon content, cell volume and area in phytoplankton. *Limnol. Oceanogr.*, **11**, 307–311.
Murray, R. G. E. & Watson, S. W., 1963.—An organelle confined within the cell wall of *Nitrosocystis oceanus* (Watson). *Nature*, **197**, 211–212.
Natarajan, K. V., 1970.—Distribution and significance of vitamin B12 and thiamine in the subarctic Pacific ocean. *Limnol. Oceanogr.*, **15**, 655–659.
Natarajan, K. V. & Dugdale, R. C., 1966.—Bioassay and distribution of thiamine in the sea. *Limnol. Oceanogr.*, **11**, 621–629.
Neess, J. C., 1949.—A contribution to aquatic population dynamics. *Ph.D. thesis, Univ. Wis. Madison* (cité par Stavn, 1971).
Nicholls, A. G., 1933.—On the biology of *Calanus finmarchicus*. III. Vertical distribution and diurnal migration in the Clyde Sea area. *J. mar. biol. Ass. U.K.*, **19**, 139–164.
Nicol, J. A. C., 1960.—*The biology of marine animals*. Pitman, Lond., 699 p.
Nival, P., Nival, S. & Leroy, C., 1971.—Essai d'utilisation de la méthode polarographique de dosage de l'oxygène pour l'estimation de la respiration d'*Acartia clausi*. *Rapp. Comm. int. Mer Médit.*, **20**, 301–303.
Nival, P., Nival, S. & Palazzoli, I., 1972.—Données sur la respiration de différents organismes communs dans le plancton de Villefranche-sur-Mer. *Mar. Biol.*, **17**, 63–76.
Nuemann, W., 1957.—Natürliche und künstliche "redwater" mit anschliessen den Fischsterben im Meer. *Arch. Fischereiwiss.*, **8**, 204–209.
Omori, M., 1969.—Weight and chemical composition of some important oceanic zooplankton in the North Pacific Ocean. *Mar. Biol.*, **3**, 4–10.
Paasche, E., 1968.—Biology and physiology of coccolithophorids. *A. Rev. Microbiol.*, **22**, 71–85.
Packard, T. T., Blasco, D., MacIsaac, J. J. & Dugdale, R. C., 1971.—Variations of nitrate reductase activity in marine phytoplankton. *Inv. Pesq.*, **35**, 209–220.

Paffenhöfer, G. A., 1970.—Cultivation of *Calanus helgolandicus* under controlled conditions. *Helgoländer wiss. Meeresunters.*, **20**, 346–359.

Paffenhöfer, G. A., 1971.—Grazing and ingestion rates of nauplii, copepodids and adults of the marine planktonic copepod *Calanus helgolandicus*. *Mar. Biol.*, **11**, 286–298.

Paine, R. T., 1971.—The measurement and application of the calorie to ecological problems. *A. Rev. Ecology Systematic*, **2**, 145–164.

Parke, M., 1949.—Studies on marine flagellates. *J. mar. biol. Ass. U.K.*, **28**, 255–286.

Parke, M. & Adams, I., 1960.—The motile (*Crystallolithus hyalinus* Gaarder & Markali) and non-motile phases in the life history of *Coccolithus pelagicus* (Wallich) Schiller. *J. mar. biol. Ass. U.K.*, **39**, 263–274.

Parke, M. & Dixon, P. S., 1968.—Check-list of british marine algae. Second revision. *J. mar. biol. Ass. U.K.*, **48**, 783–832.

Parsons, T. R., 1961.—On the pigment composition of eleven species of marine phytoplankters. *J. Fish. Res. Bd. Can.*, **18**, 1017–1025.

Parsons, T. R., 1972.—Plankton as a food source. *Underwater journal*, February, 30–37.

Parsons, T. R. & Anderson, G. C., 1970.—Large scale studies of primary production in the north Pacific ocean. *Deep-Sea Res.*, **17**, 765–776.

Parsons, T. R. & LeBrasseur, R. J., 1968.—A discussion of some critical indices of primary and secondary production for large-scale ocean surveys. *Calcofi Rep.*, **12**, 54–63.

Parsons, T. R. & LeBrasseur, R. J., 1970.—The availability of food to different trophic levels in the marine food chains. In *Marine Food Chains*, ed. J. H. Steele, Oliver & Boyd, Edinburgh, 325–343.

Parsons, T. R., LeBrasseur, R. J. & Barraclough, W. E., 1970.—Levels of production in the pelagic environment of the strait of Georgia, British Columbia: a review. *J. Fish. Res. Bd. Can.*, **27**, 1251–1264.

Parsons, T. R., LeBrasseur, R. J. & Fulton, J. D., 1967.—Some observations on the dependence of zooplankton grazing on cell size and concentration of phytoplankton blooms. *J. oceanogr. Soc. Japan.*, **23**, 10–17.

Parsons, T. R., LeBrasseur, R. J., Fulton, J. D. & Kennedy, O. D., 1969.—Production studies in the Strait of Georgia. Part II. Secondary production under the Fraser river plume, february to may 1967. *J. exp. mar. Biol. Ecol.*, **3**, 39–50.

Parsons, T. R., Stephens, K. & LeBrasseur, R. J., 1969.—Production studies in the strait of Georgia. Part I. Primary production under the Fraser river plume, february to may, 1967. *J. exp. mar. Biol. Ecol.*, **3**, 27–38.

Parsons, T. R., Stephens, K. & Strickland, J. D. R., 1961.—On the chemical composition of eleven species of marine phytoplankters. *J. Fish. Res. Bd. Can.*, **18**, 1001–1016.

Parsons, T. R., Stephens, K. & Takahashi, M., 1972.—The fertilization of Great Central Lake. I. Effect of primary production. *Fish. Bull.*, **70**, 13–23.

Parsons, T. R., Stephens, K. & Takahashi, M., 1972.—The fertilization of Great Central Lake. I. Effect of primary production. *Fish. Bull.*, **70**, 13–23.

Parsons, T. R. & Strickland, J. D. H., 1962a.—Oceanic detritus. *Science*, **136**, 313–314.

Parsons, T. R. & Strickland, J. D. H., 1962b.—On the production of particulate organic carbon by heterotrophic processes in sea water. *Deep-Sea Res.*, **8**, 211–222.

Patten, B., 1961.—Plankton energetics of Raritan Bay. *Limnol. Oceanogr.*, **6**, 369–387.

Patten, B. C., 1968.—Mathematical models of plankton production. *Int. Rev. Hydrobiol.*, **53**, 357–408.

Pennak, R. W., 1969.—Colorado semidrainage mountain lakes. *Limnol. Oceanogr.*, **14**, 720–725.

Peres, J. M. & Deveze, L., 1963.—*Océanographie biologique et biologie marine. II. La vie pélagique*. Press. Univ. Fr., Paris, 514 p.

Petipa, T. S., 1959.—Nutrition of *Acartia clausi*. Giesbr. and *A. latisetosa* Kritcz, in the Black Sea. *Trud. Sevastopol biol. Sta.*, **12**, 130–152.

Petipa, T. S., 1967.—On the efficiency of utilization of energy in pelagic ecosystems of the Black Sea. *Fish. Res. Bd. Can. Translation*, 973.

Phifer, L. D. & Thompson, T. G., 1937.—Seasonal variations in the surface waters of San Juan

Channel during the five years period, January 1931 to December 30 1935. *J. mar. Res.*, **1**, 34–59.
Platt, T., 1969.—The concept of energy efficiency in primary production. *Limnol. Oceanogr.*, **14**, 653–659.
Platt, T., 1971.—The annual production by phytoplankton in St. Margaret's bay, Nova Scotia. *J. Cons.*, **33**, 324–334.
Platt, T., 1972.—Local phytoplankton abundance und turbulence. *Deep-Sea Res.*, **19**, 183–187.
Platt, T. T., Dickie, L. M. & Trites, R. W., 1970.—Spatial heterogeneity of phytoplankton in a near-shore environment. *J. Fish. Res. Bd. Can.*, **27**, 1453–1473.
Platt, T. & Subba Rao, D. V., 1970.—Primary production measurements on a natural plankton bloom. *J. Fish. Res. Bd. Can.*, **27**, 887–899.
Pomeroy, L. R. & Johannes, R. E., 1968.—Occurrence and respiration of ultraplankton in the upper 500 meters of the ocean. *Deep-Sea Res.*, **15**, 381–391.
Pomeroy, L. R., Mathews, H. M. & Hong Saik Min, 1963.—Excretion of phosphate and soluble organic phosphoric compounds by zooplankton. *Limnol. Oceanogr.*, **8**, 50–55.
Posta, A., 1963.—Relation entre l'évolution de quelques tintinnides de la rade de Villefranche-sur-Mer et la température de l'eau. *Cah. Biol. mar.*, **4**, 201–210.
Pratt, D. M., 1966.—Competition between *Skeletonema costatum* and *Olisthodiscus luteus* in Narragansett bay and in culture. *Limnol. Oceanogr.*, **11**, 447–455.
Pratt, D. M. & Berkson, H., 1959.—Two sources of error in the oxygen light and dark bottle method. *Limnol. Oceanogr.*, **4**, 328–334.
Prochazkova, L., 1964.—Spectrophotometric determination of ammonia as rubazoic acid and with bispyrazolone reagent. *Anal. Chem.*, **36**, 865–871.
Prochazkova, L., Blazka, P. & Kralova, M., 1970.—Chemical changes involving nitrogen metabolism in water and particulate matter during primary production experiments. *Limnol. Oceanogr.*, **15**, 797–807.
Provasoli, L. & D'Agostino, A., 1962.—Vitamin requirements of *Artemia salina* in aseptic culture. *Am. Zool.*, **2**, 439.
Provasoli, I., McLaughlin, J. J. A. & Droop, M. R., 1957.—The development of artificial media for marine algae. *Arch. Mikrobiol.*, **25**, 392–428.
Provasoli, L. & Shiraishi, K., 1959.—Axenic cultivation of the brine shrimp *Artemia salina*. *Biol. Bull.*, **117**, 347–355.
Puetter, A., 1925.—Die Ernährung der Copepoden. *Arch. Hydrobiol.*, **15**, 70–117.
Pytkowicz, R. M., 1971.—On the apparent oxygen utilization and the preformed phosphate in the oceans. *Limnol. Oceanogr.*, **16**, 39–42.
Rabinowitch, E. I., 1951 & 1956.—*Photosynthesis and related processes*. Vol. II, Part I (1951). Vol. II, Part 2 (1956). Interscience, New York, London.
Raymont, J. E. G., 1963.—*Plankton and productivity in the oceans*. Pergamon Press, Oxford, 660 p.
Raymont, J. E. G., 1970.—Problems of the feeding of zooplankton in the deep-sea. *Proc. int. Symp. biol. sound scattering in the ocean*, ed. Brooke, Rep. 005, Maury Center, Washington, 134–146.
Raymont, J. E. G., Srinivagasam, R. T. & Raymont, J. K. B., 1971.—Biochemical studies on marine zooplankton. VIII. Further investigations on *Meganyctiphanes norvegica* (M. Sars). *Deep-sea Res.*, **18**, 1167–1178.
Razouls, Cl., 1972.—Estimation de la production secondaire (copépodes pélagiques) dans une province néritique méditerranéenne (golfe du Lion). *Thèse d'État, Univ. Paris VI*, 301 p. manusc.
Redfield, A. C., Ketchum, B. H. & Richards, F. A., 1963.—The influence of organisms on the composition of sea water. In *The sea*, ed. Hill, Interscience, New York, Lond., **2**, 26–77.
Redfield, A. C. & Keys, A. B., 1938.—The distribution of ammonia in the waters of the Gulf of Maine. *Biol. Bull.*, **74**, 83–92.
Reid, G. K., 1961.—*Ecology of inland waters and estuaries*. Reinhold, New York, 375 p.
Renn, C. E., 1937.—Bacteria and the phosphorus cycle in the sea. *Biol. Bull.*, **72**, 190–195.
Reyssac, J., 1970.—Phytoplancton et production primaire au large de la Côte d'Ivoire. *Bull. I. F. A. N.*, **32**, Ser. A, 869–981.

Rice, A. L., 1962.—Responses of *Calanus finmarchicus* (Gunnerus) to changes in hydrostatic pressure. *Nature*, **194**, 1189–1190.

Richards, F. A., Anderson, J. J. & Cline, J. D., 1971.—Chemical and physical observations in Golfo Dulce, an anoxic basin on the Pacific coast of Costa Rica. *Limnol. Oceanogr.*, **16**, 43–50.

Richards, F. A. with Thompson, T. G., 1952.—The estimation and characterization of plankton populations by pigment analyses. II. A spectrophotometric method for the estimation of plankton pigments. *J. mar. Res.*, **11**, 156–172.

Richman, S., 1958—The transformation of energy by *Daphnia pulex*. *Ecol. Monogr.*, **28**, 273–291.

Richman, S. & Rogers, J. N., 1969.—The feeding of *Calanus helgolandicus* on synchronously growing populations of the marine diatom *Ditylum brightwellii*. *Limnol. Oceanogr.*, **14**, 701–709.

Rigler, F. H., 1971.—Feeding rates. Zooplankton. In *A Manual on methods for the assessment of secondary productivity in fresh waters*, ed. Edmondson & Winberg, I. B. P., Handbook n° 17, Blackwell, Oxford & Edinburgh, 228–256.

Riley, G. A., 1937.—Significance of the Mississippi river drainage for biological conditions in the northern Gulf of Mexico. *J. mar. Res.*, **1**, 60–74.

Riley, G. A., 1946.—Factors controlling phytoplankton populations on Georges Bank. *J. mar. Res.*, **6**, 54–73.

Riley, G. A., 1947.—A theoretical analysis of the zooplankton population of Georges Bank. *J. mar. Res.*, **6**, 104–113.

Riley, G. A., 1951.—Oxygen, phosphate and nitrate in the Atlantic ocean. *Bull. Bingham oceanogr. Coll.*, **13**, art. 1, 1–126.

Riley, G. A., 1956.—Oceanography of Long Island Sound, 1952–1954. IX. Production and utilization of organic matter. *Bull. Bingham oceanogr. Coll.*, **15**, 325–344.

Riley, G. A., 1957.—Phytoplankton of the north central Sargasso sea. *Limnol. Oceanogr.*, **2**, 252–270.

Riley, G. A., 1963.—Organic aggregates in sea water and the dynamics of their formation and utilization. *Limnol. Oceanogr.*, **8**, 372–381.

Riley, G. A., 1966.—Seasonal cycles and succession of societies. *In Marine Biology*, II, ed. Oppenheimer, Acad. Sci., New York, 145–176.

Riley, G. A., 1970.—Particulate and organic matter in sea water. *Adv. mar. Biol.*, **8**, 1–118.

Riley, G. A., 1971.—Book reviews. *Limnol. Oceanogr.*, **16**, 152–153.

Riley, G. A., Stommel, H. & Bumpus, D. F., 1949.—Quantitative ecology of the plankton of the western North-Atlantic. *Bull. Bingham Oceanogr. Coll.*, **12**, art. 3, 1–169.

Riley, J. P., 1965.—Analytical chemistry of sea water. In *Chemical Oceanography*, ed. Riley & Skirrow, Academic Press, Lond. & New York, **1**, 295–424.

Roger, C., 1971.—Les Euphausiacés du Pacifique équatorial et sud-tropical. Zoogéographie, Écologie, Biologie et situation trophique. *Thèse d'État, Univ. Provence, Marseille*, 331 p. manusc.

Rose, M., 1925.—Contributions à l'étude de la biologie du plancton. Le problème des migrations verticales journalières. *Arch. Zool. exp. gén.*, **64**, 387–542.

Rothschild, Lord, 1961.—*A classification of living animals*. Longmans, Lond., 106 p.

Rotschi, H., 1961.—Oxygène, phosphate et gaz carbonique en Mer de Corail. *Deep-Sea Res.*, **8**, 181–195.

Rudjakov, J. A., 1970.—The possible causes of diel vertical migrations of planktonic animals. *Mar. Biol.*, **6**, 98–105.

Rusness, D. & Burris, R. H., 1970.—Acetylène reduction (nitrogen fixation) in Wisconsin lakes. *Limnol. Oceanogr.*, **15**, 808–813.

Russell, F. S., 1927.—The vertical distribution of plankton in the sea. *Biol. Rev.*, **2**, 213–256.

Russell, F. S., 1931a.—The vertical distribution of marine macroplankton. X. Notes on the behaviour of *Sagitta* in the Plymouth area. *J. mar. biol. Ass. U.K.*, **17**, 391–414.

Russell, F. S., 1931b.—The vertical distribution of marine macroplankton. XI. Further observations on diurnal changes. *J. mar. biol. Ass. U.K.*, **17**, 767–784.

Russell, F. S., 1934.—The vertical distribution of marine macroplankton. XII. Some observations on the vertical distribution of *Calanus finmarchicus* in relation to light intensity. *J. mar. biol. Ass. U.K.*, **19**, 569–584.

Russell, F. S., 1935.—A review of some aspects of zooplankton research. *Rapp. Cons. Explor. Mer*, **95**, 5–30.

Russell, F. S., 1939.—Hydrographical and biological conditions in the North Sea as indicated by plankton organisms. *J. Cons.*, **14**, 171–192.

Ryther, J. H., 1955.—The ratio of photosynthesis to respiration in marine plankton algae and its effects upon the measurement of productivity. *Deep-Sea Res.*, **2**, 134–139.

Ryther, J. H., 1956.—Photosynthesis in the ocean as a function of light intensity. *Limnol. Oceanogr.*, **1**, 61–70.

Ryther, J. H., 1959.—Potential productivity of the sea. *Science*, **130**, 602–608.

Ryther, J. H. & Guillard, R. R. L., 1962.—Studies of marine planktonic diatoms. II. Use of *Cyclotella nana* Hustedt for assays of vitamin B12 in sea water. *Can. J. Microbiol*, **8**, 437–445.

Ryther, J. H. & Hulburt, E. M., 1960.—On winter mixing and the vertical distribution of phytoplankton. *Limnol. Oceanogr.*, **5**, 337–338.

Ryther, J. H. & Kramer, D. D., 1961.—Relative iron requirement of some coastal and offshore plankton algae. *Ecology*, **42**, 444–446.

Ryther, J. H. & Menzel, D. W., 1965.—On the production, composition and distribution of organic matter in the western Arabian sea. *Deep-Sea Res.*, **12**, 199–209.

Ryther, J. H., Menzel, D. W., Hulburt, E. M., Lorenzen, C. J. & Corwin, N., 1971.—The production and utilization of organic matter in the Peru coastal current. *Inv. pesq.*, **35**, 43–60.

Ryther, J. H. & Yentsch, C. S., 1957.—The estimation of phytoplankton production in the ocean from chlorophyll and light data. *Limnol. Oceanogr.*, **2**, 281–286.

Ryther, J. H. & Yentsch, C. S., 1958.—Primary production of continental shelf waters of New York. *Limnol. Oceanogr.*, **3**, 327–335.

Rzepishevsky, I. K., 1962.—Conditions of the mass liberation of the nauplii of the common barnacle *Balanus balanoides* (L), in the eastern Murman, *Int. Rev. Hydrobiol.*, **47**, 471–479.

Saijo, Y., Ijzuka, S. & Asaoka, O., 1969.—Chlorophyll maxima in Kuroshio and adjacent area. *Mar. Biol.*, **4**, 190–196.

Saville, A., 1958.—Mesh selection in plankton nets. *J. Cons.*, **23**, 192–201.

Scholander, P. F., Flagg, W., Walters, V. & Irving, L., 1953.—Climatic adaptation in arctic and tropical poikilotherms. *Physiol. Zool.*, **26**, 67–92.

Sentz, E., 1962.—Étude écologique sur les relations entre les larves planctoniques et les jeunes stades fixés dans la rade de Villefranche-sur-Mer. *Thèse de spécialité, Fac. Sci. Univ. Paris*, 151 p. manusc.

Sentz, E., 1963.—Étude comparative de la richesse en larves planctoniques de différents points de la rade de Villefranche. *Rapp. Comm. int. Mer Médit.*, **17**, 581–584.

Sentz-Braconnot, E., 1965.—Sur la capture des proies par le ptéropode gymnosome *Pneumodermopsis paucidens* (Boas). *Cah. Biol. mar.*, **6**, 191–194.

Shelbourne, J. E., 1956.—The effect of water conservation on the structure of marine fish embryos and larvae. *J. mar. biol. Ass. U.K.*, **35**, 275–286.

Sheldon, R. W., 1972.—Size separation of marine seston by membrane and glass-fiber filters. *Limnol. Oceanogr.*, **17**, 494–498.

Sheldon, R. W. & Parsons, T. R., 1967.—*A practical manual on the use of the Coulter Counter in marine research*. Coulter Electronics Sales Co., Canada, 66 p.

Sheldon, R. W. & Sutcliffe, W. H., 1969.—Retention of marine particles by screen and filters. *Limnol. Oceanogr.*, **14**, 441–444.

Shushkina, E. A., 1968.—Calculation of copepod production based on metabolic features and the coefficient of the utilization of assimilated food for growth. *Oceanology*, **8**, 98–109.

Sieburth, J., McN., 1968.—The influence of algal antibiosis on the ecology of marine microorganisms. *Adv. Microbiol. Sea*, **1**, 63–94.

Sieburth, J., Mc N. & Pratt, D. M., 1962.—Anticoliform activity of sea water associated with the termination of *Skeletonema costatum* blooms. *Trans. N.Y. Acad. Sci.*, Ser. II, **24**, 498–501.

Silliman, R. P., 1946.—A study of variability in plankton townet catches of Pacific pilchard (*Sardinops caerulea*) eggs. *J. mar. Res.*, **6**, 74–83.

Skopintsev, B. A. & Bruck, E. S., 1940.—Contribution to the study of regeneration of nitrogen and phosphorus compounds in the course of decomposition of dead plankton. *C.R. (Doklady) Acad. Sci. U.R.S.S.*, **25**, 807–810.

Smayda, T. J., 1965.—A quantitative analysis of the phytoplankton of the Gulf of Panama. II. On the relationship between 14C assimilation and the diatom standing crop. *Bull. interam. trop. Tuna Comm.*, **9**, 465–531.

Smayda, T. J., 1970.—The suspension and sinking of phytoplankton in the sea. *Oceanogr. mar. Biol. A. Rev.*, **8**, 353–414.

Smith, E. L., 1936.—Photosynthesis in relation to light and carbon dioxide. *Proc. nat. Acad. Sci. Wash.*, **22**, 504–511.

Smith, G. M., 1955.—*Cryptogamic botany. I. Algae and Fungi.* McGraw-Hill, New York, 546 p.

Solorzano, L. & Strickland, J. D. H., 1969.—Polyphosphates in sea water. *Limnol. Oceanogr.*, **13**, 515–518.

Sournia, A., 1968a.—Recherches sur le phytoplancton et la production primaire dans le Canal de Mozambique. *Thèse d'État, Fac. Sci. Univ. Paris*, 96 p. manusc.

Sournia, A., 1968b.—La Cyanophycée *Oscillatoria* (= *Trichodesmium*) dans le plancton marin: taxinomie et observations dans le Canal de Mozambique. *Nova Hedwigia*, **15**, 1–12.

Sournia, A., 1969.—Cycle annuel du phytoplancton et de la production primaire dans les mers tropicales. *Mar. Biol.*, **3**, 287–303.

Sournia, A., 1973.—La production primaire planctonique en Méditerranée. Essai de mise à jour. *Bull. Étude en commun Méditerranée*, **5**, 1–128.

Sournia, A. & Citeau, J., 1972.—Sur la distribution du molybdène en mer et ses relations avec la production primaire. *C.R. Acad. Sc., Paris*, **275**, 1299–1302.

Stanbury, F. A., 1931.—The effects of light of different intensities, reduced selectively or non selectively, upon the rate of growth of *Nitzschia closterium*. *J. mar. biol. Ass. U.K.*, **17**, 633–653.

Stavn, R. H., 1971.—The horizontal-vertical distribution hypothesis: Langmuir circulations and *Daphnia* distributions. *Limnol. Oceanogr.*, **16**, 453–466.

Steele, J. H., 1958.—Plant production in the northern North sea. *Mar. Res.*, **7**, 1–36.

Steele, J. H., 1962.—Environmental control of photosynthesis in the sea. *Limnol. Oceanogr.*, **7**, 137–150.

Steele, J. H. & Menzel, D. W., 1962.—Conditions for maximum primary production in the mixed layer. *Deep-Sea Res.*, **9**, 39–49.

Steemann-Nielsen, E., 1952.—The use of radioactive carbon (14C) for measuring organic production in the sea. *J. Cons.*, **18**, 117–140.

Steemann-Nielsen, E., 1962.—On the maximum quantity of plankton chlorophyll per surface unit of a lake or the sea. *Int. Rev. Hydrobiol.*, **47**, 333–338.

Steemann-Nielsen, E., 1965.—On the determination of the activity in 14C ampoules for measuring primary production. *Limnol. Oceanogr.*, **10** (Suppl.), 247–252.

Steemann-Nielsen, E. & Alkholy, A. A., 1956.—Use of ^{14}C technique in measuring photosynthesis of phosphorus or nitrogen deficient algae. *Physiol. Plant.*, **9**, 144–153.

Steemann-Nielsen, E. & Hansen, V. K., 1959.—Measurement with the carbon 14 technique of the respiration rates in natural population of phytoplankton. *Deep-Sea Res.*, **5**, 222–223.

Steemann-Nielsen, E. & Jensen, E. A., 1957.—Primary oceanic production. The autotrophic production of organic matter in the oceans. *Galathea Rep.* **1**, 49–136.

Steemann-Nielsen, E. & Wium-Andersen, S., 1970.—Copper ions as poison in the sea and in freshwater. *Mar. Biol.*, **6**, 93–97.

Stephens, K., 1970.—Automated measurement of dissolved nutrients. *Deep-Sea Res.*, **17**, 393–396.

Stephens, K., Sheldon, R. W. & Parsons, T. R., 1967.—Seasonal variations in the availability of food for benthos in a coastal environment. *Ecology*, **48**, 852–855.
Steuer, A., 1910.—*Planktonkunde*. Teubner, Leipzig, 723 p.
Stommel, H., 1949.—Trajectories of small bodies sinking slowly through convection cells. *J. mar. Res.*, **8**, 24–29.
Strathman, R. R., 1967.—Estimating the organic carbon content of phytoplankton from cell volume or plasma volume. *Limnol. Oceanogr.*, **12**, 411–418.
Strickland, J. D. H., 1958.—Solar radiation penetrating the ocean. A review of requirements, data and methods of measurement, with particular reference to photosynthetic productivity. *J. Fish. Res. Bd. Can.*, **15**, 453–493.
Strickland, J. D. H., 1960.—Measuring the production of marine phytoplankton. *Bull. Fish. Res. Bd. Can.*, **122**, 172 p.
Strickland, J. D. H., 1965.—Production of organic matter in the primary stages of the marine food chain. In *Chemical Oceanography*, ed. Riley & Skirrow, Academic Press, Lond. & New York, 478–610.
Strickland, J. D. H. & Austin, K. H., 1960.—On the forms, balance and cycle of phosphorus observed in the coastal and oceanic waters of the northeastern Pacific. *J. Fish. Res. Bd. Can.*, **17**, 337–351.
Strickland, J. D. H., Holm-Hansen, O., Eppley, R. W. & Linn, R. J., 1969.—The use of a deep tank in plankton ecology. I. Studies of the growth and composition of phytoplankton crops at low nutrient levels. *Limnol. Oceanogr.*, **14**, 23–34.
Strickland, J. D. H. & Parsons, T. R., 1968.—A practical handbook of sea water analysis. *Bull. Fish. Res. Bd. Can.*, **167**, 1–311.
Strickland, J. D. H. & Solorzano, L., 1966.—Determination of monoesterase hydrolysable phosphate and phosphomonoesterase activity in sea water. In *Some contemporary studies in marine science*, ed. H. Barnes, Allen & Unwin, Lond., 665–674.
Stross, R. G., Neess, J. C. & Hasler, A. D., 1961.—Turnover time and production of planktonic crustacea in limed and reference portion of a bog lake. *Ecology*, **42**, 237–245.
Subba Rao, D. V., 1969.—*Asterionella japonica* bloom and discoloration off Waltair, bay of Bengal. *Limnol. Oceanogr.*, **14**, 632–633.
Sushchenya, L. M., 1970.—Food rations, metabolism and growth of crustaceans. In *Marine Food Chains*, ed. Steele, Oliver & Boyd, Edinburgh, 127–141.
Sutcliffe, W. H., Sheldon, R. W. & Prakash, A., 1970.—Certain aspects of production and standing stock of particulate matter in the surface waters of the northwest Atlantic ocean. *J. Fish. Res. Bd. Can.*, **27**, 1917–1926.
Sverdrup, H. U., 1953.—On conditions for the vernal blooming of phytoplankton. *J. Cons.*, **18**, 287–295.
Sverdrup, H. U., Johnson, M. W., Fleming, R. H., 1946.—*The oceans. Their physics, chemistry and general biology*. 2nd ed., Prentice-Hall, New York, 1087 p.
Syrett, P. J., 1962.—Nitrogen assimilation. In *Physiology and Biochemistry of Algae*, ed. Lewin, Academic Press, Lond. & New York, 171–188.
Talling, J. F., 1957.—Photosynthetic characteristics of some freshwater plankton diatoms in relation to underwater radiation. *New Phytol.*, **56**, 29–50.
Taylor, L. R., 1961.—Aggregation, variance and the mean. *Nature*, **189**, 732–735.
Taylor, F. J. R., Blackbourn, D. J. & Blackbourn, J., 1971.—The red water ciliate *Mesodinium rubrum* and its "incomplete symbionts": a review including new ultrastructural observations. *J. Fish. Res. Bd. Can.*, **28**, 391–407.
Teal, J. M. & Kanwisher, J., 1966.—The use of pCO_2 for the calculation of biological production, with examples from waters of Massachusetts. *J. mar. Res.*, **24**, 4–14.
Thomas, W. H., 1970.—On nitrogen deficiency in tropical pacific oceanic phytoplankton: photosynthetic parameters in poor and rich waters. *Limnol. Oceanogr.*, **15**, 380–385.
Thomas, W. H. & Dodson, A. N., 1968.—Effects of phosphate concentration on cell division rates and yield of a tropical oceanic diatom. *Biol. Bull.*, **134**, 199–208.
Timmis, G. M. & Epstein, S. S., 1959.—New antimetabolites of vitamin B12. *Nature*. **184**, 1383–1384.

Totton, A. K., 1965.—*A synopsis of the Siphonophora*. British Museum, Lond., 230 p.
Tranter, D. J. & Smith, P. E., 1968.—Filtration performance. In *Zooplankton sampling*. Monogr. oceanogr. Methodol., Unesco, **2**, 27–56.
Travers, M., 1971.—Diversité du microplancton du Golfe de Marseille en 1964. *Mar. Biol.*, **8**, 308–343.
Tregouboff, G. & Rose, M., 1957.—*Manuel de Planctonologie méditerranéenne*. Centre National de la Recherche Scientifique, Paris, 587 p.
Tyler, J. E. & Smith, R. C., 1970.—*Measurements of spectral irradiance underwater*. Gordon & Breach, New York, 103 p.
Ullyott, P., 1939.—Die täglichen Wanderungen des planktonischen Susswassercrustaceen. *Int. Rev. Hydrobiol.*, **38**, 262–284.
Vaccaro, R. F., 1962.—The oxidation of ammonia in sea water. *J. Cons.*, **27**, 3–14.
Vaccaro, R. F., 1965.—Inorganic nitrogen in sea water. In *Chemical Oceanography*, ed. Riley & Skirrow, Academic Press, Lond. & New York, 365–408.
Vaccaro, R. F. & Ryther, J. H., 1960.—Marine phytoplankton and the distribution of nitrite in the sea. *J. Cons.*, **25**, 260–271.
Vannucci, M., 1968.—Loss of organisms through the meshes. In *Zooplankton sampling*. Monogr. oceanogr. Methodol., Unesco, **2**, 77–86.
Venrick, E. L., 1971.—The statistics of subsampling. *Limnol. Oceanogr.*, **16**, 811–818.
Vinogradov, A. P., 1953.—The elementary chemical composition of marine organisms. *Mem. Sears Found. Mar. Res.*, **2**, 1–647.
Vinogradov, M. E., 1970.—*Vertical distribution of the oceanic zooplankton*. Israel program for Scientific translations, Jerusalem, 339 p.
Vlymen, W. J., 1970.—Energy expenditure of swimming copepods. *Limnol. Oceanogr.*, **15**, 348–356.
Vollenweider, R. A., 1965.—Calculation model of photosynthesis-depth curves and some implications regarding day rate estimates in primary production estimates. *Mem. Ist. ital. Idrobiol.*, **18** (Suppl.), 425–457.
Vollenweider, R. A., 1969.—*A manual on methods for measuring primary production in aquatic environments*, I. B. P., Handbook, n° 12, Blackwell, Oxford & Edinburgh, 213 p.
Walsh, J. J. & Dugdale, R. C., 1971.—A simulation model of the nitrogen flow in the peruvian upwelling system. *Inv. pesq.*, **35**, 309–330.
Walsh, J. J., Kelley, J. C., Dugdale, R. C. & Frost, B. W., 1971.—Gross features of the peruvian upwelling system with special reference to possible diurnal variation. *Inv. pesq.*, **35**, 25–42.
Wangersky, P. J., 1969.—Distribution of suspended carbonate with depth in the ocean. *Limnol. Oceanogr.*, **14**, 929–933.
Wangersky, P. J. & Gordon, D. C., 1965.—Particulate carbonate, organic carbon and Mn^{++} in the open ocean. *Limnol. Oceanogr.*, **10**, 544–550.
Warren, C. E., 1971.—*Biology and water pollution control*. Saunders, Philadelphia..., 434 p.
Whitledge, T. E. & Packard, T. T., 1971.—Nutrient excretion by anchovies and zooplankton in pacific upwelling regions. *Inv. pesq.*, **35**, 243–250.
Wickstead, J. H., 1959.—A predatory copepod. *J. anim. Ecol.*, **28**, 69–72.
Wiebe, P. H., 1970.—Small-scale spatial distribution in oceanic zooplankton. *Limnol. Oceanogr.*, **15**, 205–217.
Wiebe, P. H. & Holland, W. R., 1968.—Plankton patchiness: effects on repeated net tows. *Limnol. Oceanogr.*, **13**, 315–321.
Williams, P. M., 1969.—The determination of dissolved organic carbon in sea water: a comparison of two methods. *Limnol. Oceanogr.*, **14**, 297–298.
Wilson, D. P., 1951.—A biological difference between natural waters. *J. mar. biol. Ass. U.K.*, **30**, 1–20.
Wimpenny, R. S., 1966.—*The plankton of the sea*. Faber, London, 426 p.
Winberg, G. G., 1965.—The interdependence of metabolic rate and growth rate in animals. In *Trud. Sevastopol Confer.*, Izd. Naukova dumka.

Winberg, G. G., 1966.—Growth rate and metabolic rate in animals. *Uspekhi savrem. biol.*, **61** (2).

Winsor, C. P. & Clarke, G. L., 1940.—A statistical study of variation in the catch of plankton nets. *J. mar. Res.*, **3**, 1–34.

Womersley, H. B. S. & Norris, R. E., 1959.—A free-floating marine red alga. *Nature*, **184** (Suppl.), 828–829.

Wood, E. J. F., 1967.—*Microbiology of oceans and estuaries*. Elsevier, Amsterdam, 319 p.

Yentsch, C. S., 1960.—The influence of phytoplankton pigments on the colours of sea water. *Deep-Sea Res.*, **7**, 1–9.

Yentsch, C. S., 1962.—Marine plankton. In *Physiology and Biochemistry of Algae*, ed. R. Lewin, Academic Press, Lond. & New York, 771–797.

Yentsch, C. S., 1963.—Primary production. *Oceanogr. mar. Biol. A. Rev.*, **1**, 157–175.

Yentsch, C. S. & Scagel, R. F., 1958.—Diurnal study of phytoplankton pigments, an *in situ* study in East Sound, Washington. *J. mar. Res.*, **17**, 567–583.

Zalkina, A. V., 1970.—Vertical distribution and diurnal migration of some Cyclopoida (Copepoda) in the tropical region of the Pacific Ocean. *Mar. Biol.*, **5**, 275–282.

Zaika, V. E., 1968.—Age-structure dependence of the "specific production" in zooplankton populations. *Mar. Biol.*, **1**, 311–315.

Zelickman, E. A., Gelfand, V. I. & Shifrin, M. A., 1969.—Growth, reproduction and nutrition of some Barents Sea hydromedusae in natural aggregations. *Mar. Biol.*, **4**, 167–173.

INDEX

Abidjan, 97, 98
Abylopsis, 170, 311
Acantharia, 158, 309
Acanthochiasma, 158, 311
Acanthometra, 311
Acartia, 170, 192, 202-210, 231, 237-269, 280-290, 313
Actinotrocha, 177, 313
Aegina, 311
Aequorea, 294, 311
Aglantha, 201, 311
Aglaura, 311
Alaurina, 312
Alciope, 165, 313
Aleutian Islands, 84, 147
Allen curves, 275
Amallothrix, 297
Amino acids, 37-48, 264, 296
Ammonium, 36, 40-44, 55, 87, 107, 264
– dosage, 48
Amphidinium, 27, 31, 60, 78, 153, 309
Amphilonche, 311
Amphipoda, 167, 314
Amphisolenia, 12, 309
Angola, 119
Anomalocera, 215, 313
Annelida, 165, 176, 313
Anthomedusa, 160, 311
Antibiotics, 87
Antilles, 139
Antithamnion, 8
AUO, 72
Appendicularia, 174, 198, 203, 314
Arabacia, 197
Arabian Sea, 39, 148
Arcachon Basin, 10
Artemia, 61, 240, 267, 271, 313
Ash, 153
Asterionella, 9, 35, 57, 85, 95, 118, 120, 310
Atlanta, 171, 312
Atlantic Ocean, 47, 67, 71, 75-80, 139, 151, 206, 215, 261
Atolla, 311
ATP, 57, 93, 156
Aulacantha, 159, 311
Aurelia, 161, 242, 311
Auricularia, 175, 314
Australia, 92, 119, 259

Auxosporulation, 8
Azotobacter, 45
Azov Sea, 290

Bacteria, 41, 62, 88, 123, 238, 273, 303
Bacteriastrum, 10, 108, 310
Balanus, 197, 231, 241
Baltic Sea, 76
Barents Sea, 90, 214, 291
Barosensibility, 229
Bathyplankton, 2
Batorine Lake, 286
Bengal Bay, 120
Bering Sea, 259
Bermuda, 20, 82, 151, 292-295
Beroe, 164, 294, 312
Biddulphia, 10, 105, 109, 113, 237, 310
Bimini, 21
Bioluminescence, 182
Biotin, 84, 271
Bipinnaria, 175, 314
Black Sea, 69, 280, 291
Bolina, 164, 312
Bosmina, 281, 313
Bothnia Golf, 76
British Columbia, 133, 145-150
British Islands, 202, 210, 213
Bryozoa (= Endoprocta + Ectoprocta), 177

Cachonina, 35, 58, 156
Calanipeda, 292, 313
Calanus, 4, 61, 305, 313
– assimilation, 252-257
– chemical composition, 292-297
– excretion, 264
– feeding regime, 237-240
– food ration, 243-249
– growth, 266-269
– morphology, 168
– production, 276-291
– respiration, 256-264
– sampling, 183-193
– spatio-temporal variations, 198-212
– vertical migrations, 219-236
California, 21, 40, 68, 95, 107, 119, 219, 225, 299, 323

Callinectes, 260
Calytopis, 171, 314
Canar Islands, 218, 231-235
Candacia, 204, 242, 294, 313
Cap-Breton Island, 268
Cape Cod, 150, 286
Cape Martin, 142
Carbohydrates, 153-156, 297
Carbon – 14, 126-129
Carbon dioxide, 15
Carinaria, 171, 312
Carotenoids, 91, 156
Carteria, 31, 310
Carybdea, 311
Castellon, 97
Celtic Sea, 71
Centropages, 167, 187-193, 203-210, 241, 283, 294, 313
Ceratium, 10, 108, 309, 238
Cestus, 164, 312
Chaetoceros, 4, 9, 14, 57, 78, 95, 105, 156, 310, 239-248, 305
Chaetognatha, 165, 314
Chelophyes, 162, 311
Chesapeake Bay, 291
Chlamydomonas, 27, 37, 127, 310, 245, 310
Chlorella, 35, 58, 80, 310
Chlorophyceae, 14, 31
Chlorophyll, 8, 11, 14, 23, 91, 113, 143, 156
Chromatophore, 8-13
Chrysaora, 161, 311
Chrysophyceae, 13
Ciliata, 159, 304, 311
Cirripedia, 176, 313
Cladocera, 166, 203, 313
Cladonema, 311
Clarke-Bumpus net, 179
Clausocalanus, 313
Clione, 173, 204, 210, 294, 312
Clostridium, 45
Clyde, 97, 119, 220, 257, 295
Cnidaria, 160, 311
Coccoliths, 12, 153
Coccolithophorids, 12, 94, 109
Coccolithus, 12, 59, 78, 95, 107, 118, 309
Cochlodinium, 119, 309
Cohorts, method of, 274
Coilodesme, 24, 27
Collozoum, 159, 311
Conchoecia, 167, 313
Copepoda, 167, 176, 203, 313
Copepodites, 176, 193, 203
Copper, 81
Corsica, 70, 142
Corycaeus, 191, 203, 210, 313
Coscinodiscus, 10, 23, 28, 80, 106, 153, 237, 240, 252

Costa Rica, 44
Cotylorhiza, 311
Coulter counter, 93, 129, 145, 179
Crassostrea, 61
Creatin, 42
Creseis, 171, 214, 312
Cromwell Current, 81
Crustacea, 166, 175, 260, 264, 313
Cryptococcus, 83
Cryptomonas, 14, 309
Cryptophyceae, 14
Crystallolithus, 13, 309
Ctenophora, 162, 312
Cumacea, 220, 314
Cunina, 160, 311
Cyanea, 4, 161, 267, 294, 311
Cyanophyceae, 14, 39, 88
Cyclops, 226, 285, 313
Cyclotella, 27, 60, 82, 118, 240, 252, 310
Cymbulia, 312
Cyphonautus, 177, 313
Cypridina, 313
Cypris, 176, 313
Cystine, 86
Cythere, 313
Cytherella, 313

Dactyliosolen, 10, 310
Daphnia, 228, 267, 313
Decapoda, 176, 314
Denitrification, 43-45
Diaptomus, 259, 263, 313
Diastylis, 314
Diatomea, 7, 31, 75, 105
Dicrateria, 238
Dictyocha, 14, 310
Dictyocysta, 159, 311
Dinocaryon, 11
Dinoflagellata = see Peridinia,
Dinophysis, 11, 309
Distephanus, 305
Ditylum, 10, 36, 58, 93, 105, 155, 237-244, 252, 269, 310
DNA, 156
Doliolum, 173, 314
Doublings/day, 316
Dunaliella, 14, 31, 80, 92, 107, 127, 134, 153, 251, 310

Echinoderms, 174, 314
Echinopluteus, 175, 314
Elaphocaris, 176, 314
Encounter feeding, 239
Endospore, 8
Energy flux, 151

English Channel, 12, 47-78, 90, 97, 132, 139, 203, 212, 282, 291
Engraulis, 40, 68
Enteropneusta, 178, 314
Enzymes, 48, 253
Epineuston, 2
Epiplankton, 2
Eucalanus, 204
Euchaeta, 167, 183, 193, 202, 207, 313
Euchirella, 253, 313
Euclio, 294, 312
Eucopia, 168, 314
Eudoxia, 162
Euglena, 14, 37, 82, 310
Eukrohnia, 208, 314
Euphausia, 40, 169, 191, 221, 225, 232, 238, 248, 263, 267, 270, 290-294, 304, 314
Euphausiacea, 169, 214, 233, 236, 306, 314
Euplotes, 61
Eurylume, 226
Euterpina, 167, 203, 313
Euthemisto, 168, 314
Eutintinnus, 159, 311
Euthrophic, 38, 127
Evadne, 167, 313
Excretion, 61, 264
Extinction coefficient, 20, 136, 144, 152, 317
Exuviella, 11, 31, 118, 153, 252, 309

Faecal pellets, 79, 237, 253
Favella, 159, 311
Feeding, superfluous, 46, 64, 101, 253
Fiji Islands, 385
Fish eggs, 178, 184
Fixation, 179
Florida, 94, 119
Food ration, 243
Food web, 303
Foraminifera, 158, 311
Formosa, 120
Forskalia, 161, 311
Fraser River, 143, 145, 276
Friday Harbor 75
Fritillaria, 174, 198, 314
Frustule, 7, 153
Fucoxanthin, 8, 23, 25, 91
Fucus, 88
Furcilia, 176, 248, 305, 314
Fyne Loch, 220, 223

Gaetanus, 260, 297, 313
Gammarus, 260
Gascognian Gulf, 78
Gaussia, 167, 222, 263, 297, 313
Georges Bank, 94, 133, 141, 150, 222, 228, 286, 288, 291

Georgia, Strait of, 143, 145, 150, 276, 291
Geotactism, 230
Geryonia, 311
Gigantocypris, 167, 313
Globigerina, 159, 311
Globigerinella, 158, 311
Globigerinidae, 158
Glucose, 154
Glycerophosphate, 59
Gnathophausia, 205, 314
Golfo Dulce, 44
Gonionemus, 311
Gonyaulax, 12, 36, 58, 94, 119, 156, 309
Grazing, 55, 101, 234, 250
Growth, 266, 285
Guadalupe, Island of, 189, 297
Guinardia, 106, 310
Guinea Current, 300
Guinea, Gulf of, 97
Gulf Stream, 21, 94
Gullmar Fjord, 318
Gymnodinium, 12, 31, 119, 238, 245, 309
Gyrodinium, 31, 309

Halifax, 112, 117, 139, 151
Haloptilus, 285, 313
Halospharea, 14, 310
Haptonema, 12
Haptophyceae, 13
Hemiselmis, 14, 84, 309
Heteronereis, 165
Heteropoda, 169, 312
Hexacontium, 159, 311
Holoplankton, 2
Hydrocarbons, 155, 297
Hydrogen sulphide, 44
Hydromedusae, 160, 193
Hyperidae, 167, 170, 314
Hyperia, 168, 170, 314
Hyponeuston, 2, 215

Index
 – of accuracy, 188
 – of aggregation, 188, 195
 – of dispersion, 189
 – of diversity, 109
 – of goodness, 190
 – of similarity, 233
Indian Arm, 133
Indian Ocean, 14, 51, 68, 76
Infraplankton, 2
Inhibition, 29, 32
Ionian Sea, 218
Iron, 77
Irradiance, 17, 31, 107, 134, 146, 318, 325

Isias, 204, 313
Isochrysis, 13, 35, 57, 78, 309
Isoleucine, 84
Isopoda, 260

Janthina, 2, 172, 312
Japan, 95, 214, 259, 306
Jaranyshnaya Fjord, 214

K_1, 266
K_2, 266, 285
Kamtchatka Trench, 216
Kourile Trench, 216
Krivoe Lake, 281
Krogh curves, 278, 326
Kyu-Shiu Island, 306

Labidocera, 204, 209, 264, 294, 313
Lactobacillus, 82
La Jolla, 94, 119, 244
Land's End Cape, 71
Langmuir cells, 190
Laodicea, 311
Lauderia, 95, 155, 239, 245, 310
Lembos, 61
Lepidurus, 260
Leptocylindrus, 95, 108, 118, 310
Leptodora, 288, 313
Leptomedusa, 160, 313
Limacina, 171, 191, 203, 294, 312
Limnomedusa, 160, 311
Lipids, 153, 155, 296
Liriope, 311
Lithoptera, 159, 311
Littorina, 61
Loanda Bay, 120
Long Island Sound, 40, 141, 151, 202, 291
Lophius, 40
Los Angeles, 82
Low Island, 119

Macroplankton, 1, 179
Maine, Gulf of, 41, 44, 50, 105, 224, 228, 261
Man, Island of, 95
Manganese, 80
Margarita Island, 119
Marseille, 111, 216, 282
Martin Cape, 142
Mediterranean Sea, 14, 51, 68, 94, 97, 113, 134, 142, 217, 259, 291, 303
Megacalanus, 263, 313
Megalopa, 176, 313
Meganyctiphanus, 169, 215, 231, 253, 293-297, 314

Megaplankton, 1
Melosira, 10, 106, 310
Menai Straits, 300, 302
Meroplankton, 2, 10, 109, 196
Mesh
 – selection, 184
 – velocity, 185, 322
 – width, 184, 319
Mesochaetopterus, 177, 224
Mesodinium, 122
Mesoplankton, 2
Mesozooplankton, 2
Mesoporos (Peridinian), 119
Metanauplius, 176, 313
Metazoa, 176, 314
Metridia, 193, 207, 243, 262-265, 293, 313
Michigan Lake, 229
Microcalanus, 192, 313
Microcystis, 39
Micronecton, 1
Microplankton, 1, 89
Millport, 97, 238
Mischococcus, 31
Mississippi, 46
Mitraria, 176, 196, 313
Mixotrophe, 11
Mnemiopsis, 294, 312
Modiolus, 61
Mollusca, 169, 177, 312
Molybdenum, 87
Monaco, 142
Monochrysis, 83-86, 153, 310
Monodus, 35
Motoda bottle, 180
Mourmansk Sea, 196
Muggiaea, 163, 311
Munnopsis, 314
Mysidacea, 168, 220, 306, 314
Mysis, 228, 314
Mysis (larval stage), 176, 314

Nanaiamo, 143, 145
Nannochloris, 31, 92, 238, 310
Nanomia, 161, 311
Nanoplankton, 1, 90, 157
Nantucket Sound, 133, 141
Narcomedusa, 160, 311
Narragansett Bay, 88, 108
Nauplius, 175, 196, 203, 313
Nausithoe, 311
Navicula, 31, 84
Nectochaeta, 176, 313
Necton, 1
Nematoscelis, 225, 232, 314
Nemertea (= Nemertina), 176, 312
Neritic distribution, 209

Neuston, 2
Newport, 19, 34
New York, 142, 151
Niacin, 84
Nitrate, 36-53, 87, 107, 146-149
Nitrification, 42
Nitrite, 36-48
Nitrobacter, 43
Nitrosocystis, 43
Nitrogen, 35
 – balance of, 51
 – cycle of, 45
 – dissolved organic, 46, 264
 – fixation of, 39, 45
 – ^{15}N, 38
Nitrosomonas, 43
Nitzschia, 9, 31, 81-85, 106-118, 310
Noctiluca, 12, 154, 203, 309, 311
North Sea, 10, 46-54, 82, 112, 204-213, 288, 291
Norwegian Coast, 14
Norwegian Sea, 71, 101
Nostoc, 14, 309
Nosy-Be, 224
N/P ratio, 58
Nucleic acids, 48, 57
Nyctiphanes, 192, 314

Obelia, 160, 311
Oceanic water, 21
Oikopleura, 174, 193, 198
Oithona, 167, 184, 192, 199, 203, 268, 313
Oligoelements, 39
Olindias, 160, 311
Olisthodiscus, 88, 108, 310
Oncaea, 233, 313
Oncorhynchus, 305
O/N ratio, 265
Ophiopluteus, 175, 314
Oregon, 51
Ornithocercus, 12, 309
Oscillatoria, 14, 39, 118, 309
Oslo Fjord, 108
Osmotrophy, 157
Ostracoda, 167, 313
Oweniidea, 196, 197
Oxygen, 44, 72, 98, 121-133, 149
Oxyrrhis, 157, 311

Pacific Ocean, 33, 44, 51-83, 103, 143, 147, 218-233, 291
Panama, Gulf of, 151
Pandea, 160, 311
Pantothenic acid, 84, 271
Para-amino benzoic acid, 84
Paracalanus, 192-210, 241, 313

Paracentrotus, 175
Parathemisto, 294, 314
Pareuchaeta, 204, 210, 237-242, 263, 296, 313
Patches, 112, 133, 189
Patchiness, 188, 190
Patuxent, Estuary of, 284
P/B ratio, 291, 292
Pelagia, 160, 294, 311
Pelagonemertes, 312
Pelagosphaera, 313
Pelagothuria, 314
Penaeus, 61
Penilia, 167, 267, 313
Peridinia, 11, 31, 114, 154
Peridinium, 12, 35, 57, 58, 83, 118, 238, 241, 309
Periphylla, 311
Persa, 215, 311
Peru, 23, 40, 43, 54, 68-79, 93, 147
Phaeocystis, 13, 309
Phaeodactylum, 10, 25, 28, 57, 59, 83, 153, 310
Phaeophytin, 92, 94
Phagotrophy, 157
Phalacroma, 12, 309
Phialidium, 160, 170, 201, 311
Philippins, 259
Phoronis, 177, 313
Phosphatase, 59-62
Phosphate, 57-66, 132, 146
Phosphorus, 57-66
Photorespiration, 32
Photosynthesis, gross, 23-30, 124-137
Phototaxis, 225
Phronima, 168, 170, 265, 314
Phylliroe, 172, 313
Phyllosoma, 176, 314
Physalia, 161, 311
Pigments, various, 11, 90, 153
Pilidium, 176, 312
Plankton calendar, 198
Plankton nets, 1, 179, 180, 319
Plankton recorder, 112, 181, 185, 195, 199, 201, 204-210
Planktosphaera, 314
Plasma volume, 89
Platymonas, 31, 310
Pleurobrachia, 164, 201, 206, 209, 312
Pleurocondes, 220, 306, 314
Pleuromamma, 202-210, 294, 313
Pleuronectes, 177
Pleuston, 2, 215
Plume, 54, 143, 145
Pluteus, 174, 196, 314
Plymouth, 30, 48, 67-76, 90, 119, 202-226, 296
Pneumodermopsis, 172, 312
Podon, 167, 313
Polychaeta, larve of, 176, 203, 313

Polypeptids, 48
Polyphosphate, 8, 60, 63, 68
Pontellina, 294, 313
Pontosphaera, 12, 309
Porosira, 105, 310
Porosity, 319
Port Erin, 95, 105
Port Hacking, 92
Portumnus, 176
Prasinophyceae, 14, 32
Proplectella, 197, 199, 311
Prorocentrum, 11, 35, 57, 85, 108, 119, 238, 241, 309
Proteins, 48, 59, 153-156, 295
Prymnesium, 13, 83, 309
Pseudocalanus, 192, 199-210, 241, 248, 268, 304, 313
Pseudolibrotus, 260
Pseudomonas, 42, 43
Pteridin, 88
Pteropada, 236
  – Gymnosoma, 169, 173, 312
  – Thecosoma, 169, 172, 312
Pterotrachea, 171
Pyramimonas, 238
Pyrosoma, 173, 294, 314

Q_{10}, 258

Radiolaria, 75, 159, 311
Rate of,
  – assimilation, 251, 287
  – filtration, 243
  – grazing, 250, 287
  – relative growth, 39, 143, 145, 250, 315, 316, 317
  – relative mortality, 250
  – respiration, 287
  – relative respiration, 260
  – specific growth, 316
Rathkea, 214, 311
Red Sea, 14
Respiration of phytoplankton, 124
  – of zooplankton, 256
Rhabdosphaera, 13, 309
Rhinocalanus, 204, 210, 241, 266-270, 294, 297, 313
Rhizosolenia, 10, 14, 105-118, 238, 252, 310
Rhizostoma, 161, 311
Rhodosorus, 25
Rhône, 53
Rhopalonema, 311
Richelia, 14, 309
Roscoff, 14, 105, 134, 139
Rotifera, 203, 312

Saanich Inlet, 305
Sagitta, 165, 191-219, 267, 294, 314
Saint Esprit Island, 21
Saint Margaret's Bay, 112-117, 139-151
Salinity, 85
Salpa, 2, 61, 173, 197, 212, 294, 314
Sampling, 179, 194
San Diego, 68, 300, 302, 323
Sapphirina, 4, 313
Sardinops, 192
Sargasso Sea, 6, 22, 39, 82, 94, 139, 141, 201, 291-295, 303
Sargassum, 6
Sarsia, 160, 205, 311
Saturation, 31, 32
Scenedesmus, 87
Schroederella, 108-110, 310
Scyphomedusa (= Scyphozoa), 160, 311
Seattle, 75, 147
Sergestes, 169, 176
Seston, 3, 93, 146, 296-301
Shoal, 214
Silicium, 75, 87, 144, 156
Silicoflagellates, 13, 75
Simulation, 54, 55
Sinking, 55, 94, 98
Siphonophora, 161, 229, 235, 311
Siriella, 294, 314
Skeletonema, 10, 32, 57, 77-99, 105-111, 118, 153, 155, 237-252, 296, 310
Solen, 274
Söllerod Sö, 149
Solmaris, 160, 311
Solmissus, 311
Southampton, 79
Spiratella, 202-210, 312
Squalus, 71
Stenolume, 226
Stichococcus, 31
Streptotheca, 106
Striven Loch, 97-100, 119, 200
Strontium sulphate, 159
Stylocheiron, 232, 314
Submergency, 205, 215
Sulculeolaria, 162, 163, 311
Sulphur, 85
Sverdrup's theory, 100, 317
Swarm, 214
Syracosphaera, 12, 153, 238, 309
Syracuse, 94

Tabellaria, 59
Temora, 167, 192-210, 238, 241, 257, 268, 291, 313
Terebellid larvae, 193
Tetraselmis, 153, 310

Thalassicolla, 311
Thalassionema, 10, 118, 310
Thalassiosira, 10, 105-118, 240-252, 269, 310
Thalassiothrix, 10, 106, 310
Thysanoessa, 171, 192, 215, 232, 257, 314
Thysanopoda, 205, 215, 232, 314
Thermocline, 229
Tiaropsis, 214, 311
Tintinnidae, 159, 197, 311
Tintinnopsis, 159, 198, 199, 311
Tomopteris, 165, 193, 220, 313
Tornaria, 178, 314
Tortanus, 193, 313
Trachymedusa, 160, 311
Trichodesmium = see *Oscillatoria*
Trichromatic method, 92
Tridacna, 61
Trimethylamine, 42, 296
Tripton, 3, 93, 269, 297, 300, 301
Trochophora, 176, 313
Tunicata, 172, 178, 314
Turbulence, 117
Turnover, time method, 283

Uca, 61, 260
Ultramicroplankton, 1
Ultraplankton, 303
Upwelling, 54, 68-81, 93, 97, 107, 120, 147-149

Uracil, 84
Urea, 40, 87, 264
Uronema, 61

Vancouver Island, 18, 143, 145, 307
Velella, 2, 162, 164, 172, 311
Veliger, 177, 312
Venezuela, 118
Venice, 119
Vibilia, 168, 314
Vicariant species, 208
Vigo, Ria de, 108-112
Villefranche, 70, 142, 196, 198, 211, 215
Vineyard Sound, 132, 141, 187
Vitamin, 39, 107, 271
 – A, 297
 – B_1, 81
 – B_{12}, 83

Walfish Bay, 139, 149
Windermere Lake, 227
Woods Hole, 31, 132, 134

Zanclea, 311
Zoea, 176, 231, 314
Zoothamnium, 311